ISNM
International Series of Numerical Mathematics
Vol. 131

Applications and Computation of Orthogonal Polynomials

Conference at the Mathematical Research Institute Oberwolfach, Germany, March 22–28, 1998

Edited by

W. Gautschi
G.H. Golub
G. Opfer

Birkhäuser Verlag
Basel · Boston · Berlin

Editors:

Walter Gautschi
Department of Computer Sciences
Purdue University
West Lafayette, IN 47907-1398
USA
e-mail: wxg@cs.purdue.edu

Gene H. Golub
Computer Science Department
Stanford University
Stanford, CA 94305
USA
e-mail: golub@sccm.stanford.edu

Gerhard Opfer
Institute for Applied Mathematics
University of Hamburg
Bundesstr. 55
D-20146 Hamburg
Germany
e-mail: opfer@math.uni-hamburg.de

1991 Mathematics Subject Classification 65-06

A CIP catalogue record for this book is available from the Library of Congress, Washington D.C., USA

Deutsche Bibliothek Cataloging-in-Publication Data

Applications and computation of orthogonal polynomials : conference at the Mathematical Research Institute Oberwolfach, Germany, March 22 - 28, 1998 / ed. by W. Gautschi ... - Basel ; Boston ; Berlin : Birkhäuser, 1999
(International series of numerical mathematics ; Vol. 131)
ISBN 3-7643-6137-9 (Basel ...)
ISBN 0-8176-6137-9 (Boston)

Printed on acid-free paper produced of chlorine-free pulp. TCF ∞
Cover design: Heinz Hiltbrunner, Basel
Printed in Germany
ISBN 3-7643-6137-9
ISBN 0-8176-6137-9

Dedicated to the memory of GÜNTHER HÄMMERLIN
(1928–1997)

Contents

Preface

The workshop on Applications and Computation of Orthogonal Polynomials took place March 22–28, 1998 at the Oberwolfach Mathematical Research Institute. It was the first workshop on this topic ever held at Oberwolfach. There were 46 participants from 13 countries, more than half coming from Germany and the United States, and a substantial number from Italy. A total of 23 plenary lectures were presented and 4 short informal talks. Open problems were discussed during an evening session. This volume contains refereed versions of 18 papers presented at, or submitted to, the conference.

The theory of orthogonal polynomials, as a branch of classical analysis, is well established. But orthogonal polynomials play also an important role in many areas of scientific computing, such as least squares fitting, numerical integration, and solving linear algebraic systems. Though the basic tenets have their roots in 19th-century mathematics, the use of modern computers has required the development and study of new algorithms that are accurate and robust.

The computational methods and applications represented in this volume, of necessity, are incomplete, yet sufficiently varied to convey an impression of current activities in this area.

There are papers on least squares approximation, where a prominent role is played by discrete orthogonal polynomials. They are generated here using updating and downdating techniques. The sensitivity to small perturbations in the data is an important aspect of the problem, the study of which leads to Vandermonde-like matrices and their numerical condition. This in turn invokes techniques from complex potential theory. The numerical solution of Vandermonde-like systems is the topic of another paper, in which numerator polynomials and fast algorithms for the arithmetic of orthogonal polynomials find useful applications.

Linear algebra is the focal point of two other papers. One deals with Lanczos's method for solving systems of linear algebraic equations. Three-term and coupled two-term recurrence relations satisfied by formal orthogonal polynomials provide the underpinning of this work. Occurrences of breakdown (division by zero) are given special attention. The other paper has as objective the computation of lower and upper bounds for matrix functionals. Here quadrature techniques, in particular Gaussian quadrature and "anti-Gaussian" rules, recently developed by D.P. Laurie, are the key ingredients of the approach taken. The relative merits of three-term vs. two-term recurrences is the topic of yet another paper.

The area of numerical quadrature, historically closely allied with orthogonal polynomials, is represented by three papers. Two investigate Gauss-Kronrod quadrature, an extension of Gauss quadrature of interest, for example, in automatic and adaptive integration. The first is concerned with error bounds and requires a detailed analytic study of the underlying Stieltjes polynomials. The other addresses questions of positivity in the case of ultraspherical measures of integration. The third paper employs multiple-nodes quadrature rules and associated s-orthogonal polynomials to compute Cauchy principal value and Hadamard finite part integrals, a task of some importance in boundary element methods.

Recurrence relations are at the heart of a paper on Szegö polynomials, i.e., polynomials orthogonal on the unit circle. Even though these recurrence relations are more complicated than those in the real case, they still allow, as is shown, the development of Christoffel-Darboux-type identities. There are contributions on other novel types of orthogonal polynomial systems. One is generated by a given orthogonal polynomial of fixed degree, which appears as a factor raised to an even power in the underlying weight function. The numerical computation of these "induced" polynomials is effected by modern techniques of computational linear algebra, notably the QR algorithm. Another system involves Müntz polynomials appropriately orthogonalized. Here methods from complex analysis are used for their computation.

Three contributions are more in the direction of analysis and approximation theory, dealing with the moment problem, piecewise analytic functions, and inequalities. In the first, indeterminate moment problems are at the center of attention. Associated with them are certain entire functions, introduced by R. Nevanlinna, whose growth properties are analyzed. The second defines a sequence of fast decreasing algebraic resp. trigonometric polynomials for the purpose of detecting the singularities of functions known only to be piecewise analytic on a compact interval resp. on the circle. The third presents a set of variations on the theme of Markov's inequality.

The remaining papers discuss applications to differential equations and the physical sciences. One of them uses nonstandard orthogonal polynomials in spectral collocation methods for solving boundary and eigenvalue problems. Another tackles Schrödinger's equation and requires Jacobi matrices associated with singular measures. Finally, multidimensional Hermite polynomials make their appearance in the resolution of certain balance equations in the kinetic theory of gases and liquids.

We fondly dedicate this volume to the memory of Günther Hämmerlin, who in the 1980s and 1990s initiated, directed, and codirected an eminently successful series of Oberwolfach conferences on the closely related topic of numerical integration.

WALTER GAUTSCHI
GENE H. GOLUB
GERHARD OPFER

Participants

GREGORY S. AMMAR, Northern Illinois University, DeKalb, Illinois, USA

WALTER VAN ASSCHE, Katholieke Universiteit Leuven, Heverlee, Belgium

MARC VAN BAREL, Katholieke Universiteit Leuven, Heverlee, Belgium

BERNHARD BECKERMANN, Université des Sciences et Technologies de Lille, Villeneuve d'Ascq, France

CLAUDE BREZINSKI, Université des Sciences et Technologies de Lille, Villeneuve d'Ascq, France

MARTIN BUHMANN, Justus-Liebig Universität, Gießen, Germany

DANIELA CALVETTI, Case Westen Reserve University, Cleveland, Ohio, USA

SVEN EHRICH, GSF-National Research Center for Environment and Health, Neuherberg, Germany

BERND R.W. FISCHER, Medizinische Universität zu Lübeck, Lübeck, Germany

HANS-JÜRGEN FISCHER, Technische Universität Chemnitz, Chemnitz, Germany

KLAUS-JÜRGEN FÖRSTER, Universität Hildesheim, Hildesheim, Germany

MAMA FOUPOUAGNIGNI, Université du Benin, Benin, Africa

ROLAND W. FREUND, Bell Laboratories, Murray Hill, New Jersey, USA

ERICH FUCHS, Universität Passau, Passau, Germany

WALTER GAUTSCHI, Purdue University, West Lafayette, Indiana, USA

GENE H. GOLUB, Stanford University, Stanford, California, USA

LAURA GORI, Università di Roma "La Sapienza", Roma, Italy

WILLIAM B. GRAGG, Naval Postgraduate School, Monterey, California, USA

MARTIN H. GUTKNECHT, Eidgenössische Technische Hochschule, Zürich, Switzerland

MARTIN HANKE, Johannes-Gutenberg-Universität, Mainz, Germany

DAVID B. HUNTER, University of Bradford, Bradford, England

ARIEH ISERLES, University of Cambridge, Cambridge, England

DIRK P. LAURIE, Potchefstroom University for Christian Higher Education, Vanderbijlpark, South Africa

SHIKANG LI, Southeastern Louisiana University, Hammond, Louisiana, USA

GIORGIO MANTICA, Università degli Studi di Milano, Como, Italy

GIUSEPPE MASTROIANNI, Università degli Studi della Basilicata, Potenza, Italy

HRUSHIKESH N. MHASKAR, California State University, Los Angeles, USA

GRADIMIR V. MILOVANOVIĆ, University of Niš, Niš, Serbia, Yugoslavia
GIOVANNI MONEGATO, Politecnico di Torino, Torino, Italy
SOTIRIOS E. NOTARIS, University of Athens, Athens, Greece
GERHARD OPFER, Universität Hamburg, Hamburg, Germany
FRANZ PEHERSTORFER, Universität Linz, Linz, Austria
KNUT PETRAS, Ludwig-Maximilians-Universität München, München, Germany
MICHELA REDIVO-ZAGLIA, Università degli Studi di Padova, Padova, Italy
LOTHAR REICHEL, Kent State University, Kent, Ohio, USA
WOLFGANG RIPKEN, Zesenstrasse 10, Hamburg, Germany
HANS-JOACHIM RUNCKEL, Universität Ulm, Ulm, Germany
STEPHAN RUSCHEWEYH, Universität Würzburg, Würzburg, Germany
EDWARD B. SAFF, University of South Florida, Tampa, Florida, USA
ELISABETTA SANTI, Università di L'Aquila, L'Aquila, Italy
MICHAEL-RALF SKRZIPEK, Fernuniversität Gesamthochschule in Hagen, Hagen, Germany
HERBERT STAHL, Technische Fachhochschule Berlin, Berlin, Germany
KURT SUCHY, Universität Düsseldorf, Düsseldorf, Germany
VILMOS TOTIK, Jozsef Attila University, Szeged, Hungary
GALLIANO VALENT, Université Paris 7, Paris, France
J. ANDRÉ C. WEIDEMAN, University of Stellenbosch, Matieland, South Africa

Oberwolfach March 25, 1998, near Landeck, Meeting on Applications and computation of orthogonal polynomials. Photo: Michael Skrzipek.

From left (nose counts): Klaus-Jürgen Förster, Hans-Jürgen Fischer, Bernd R. W. Fischer, Kut Petras, Hrushikesh Mhaskar, Franz Peherstorfer, Kurt Suchy, (?), Dirk Laurie, (?), Gerhard Opfer, Andre Weidemann, Marc Van Barel, Sven Ehrich, Martin Gutknecht, Wolfgang Ripken, Mama Foupouagnigni, Shikang Li, (?), Vilmos Totik, Laura Gori, Walter van Assche, David B. Hunter, Sotirios Notaris, Giovanni Monegato, Walter Gautschi

International Series of Numerical Mathematics
Vol. 131, © 1999 Birkhäuser Verlag Basel/Switzerland

The Sensitivity of Least Squares Polynomial Approximation

Bernhard Beckermann and Edward B. Saff[1]

Abstract. Given integers $N \geq n \geq 0$, we consider the least squares problem of finding the vector of coefficients $\vec{P}$ with respect to a polynomial basis $\{p_0, \dots, p_n\}$, $\deg p_j = j$, of a polynomial P, $\deg P \leq n$, which is of best approximation to a given function f with respect to some weighted discrete norm, i.e., which minimizes $\sum_{j=0}^{N} w_n(z_j)^2 |f(z_j) - P(z_j)|^2$. Here a perturbation of the values $f(z_j)$ leads to some perturbation of the coefficient vector $\vec{P}$. We denote by κ_n the maximal magnification of relative errors, i.e., the Euclidean condition number of the underlying weighted Vandermonde-like matrix.

For the basis of monomials ($p_j(z) = z^j$), the quantity κ_n equals one when the abscissas are the roots of unity; however, it is known that κ_n increases exponentially in the case of real abscissas. Here we investigate the nth-root behavior of κ_n for some fixed basis and a fixed distribution of (complex) abscissas. An estimate for the nth-root limit of κ_n is given in terms of the solution to a weighted constrained energy problem in complex potential theory.

1. Introduction

Given a sequence of polynomials $(p_j)_{j\geq 0}$, $\deg p_j = j$ for all j, some integers $N \geq n \geq 0$, $E_n := \{z_0, \dots, z_N\} \subset \mathbb{C}$, and a weight function w_n taking only positive values on E_n, the corresponding *weighted Vandermonde-like matrix* $V_n(w_n, E_n)$ of size $(N+1) \times (n+1)$ is defined by

$$\begin{pmatrix} w_n(z_0)p_0(z_0) & w_n(z_0)p_1(z_0) & w_n(z_0)p_2(z_0) & \cdots & w_n(z_0)p_n(z_0) \\ w_n(z_1)p_0(z_1) & w_n(z_1)p_1(z_1) & w_n(z_1)p_2(z_1) & \cdots & w_n(z_1)p_n(z_1) \\ \vdots & \vdots & \vdots & & \vdots \\ w_n(z_N)p_0(z_N) & w_n(z_N)p_1(z_N) & w_n(z_N)p_2(z_N) & \cdots & w_n(z_N)p_n(z_N) \end{pmatrix}.$$

1991 *Mathematics Subject Classification.* 15A12, 31A15, 65F35. *CR.* G1.3, G1.6.
Key words and phrases. Least squares polynomial approximation, condition number, Vandermonde matrices, complex potential theory.
[1]The research of E.B. Saff was supported, in part, by the National Science Foundation research grant DMS 980 1677.

In this paper we will be interested in determining the size of the Euclidean condition number $\kappa(V_n(w_n, E_n))$ of $V_n(w_n, E_n)$, which equals the square root of the ratio of the largest and the smallest eigenvalue of the Hermitian positive definite matrix $V_n(w_n, E_n)^H \cdot V_n(w_n, E_n)$, where $V_n(w_n, E_n)^H$ denotes the Hermitian counterpart of $V_n(w_n, E_n)$. It is well known that the Euclidean condition number is a measure for the relative distance to matrices not having full rank [14, p.80]. Moreover, writing $||\cdot||$ for the Euclidean vector and matrix norm, and $V = V_n(w_n, E_n)$, we have [14, §2.7.2]

$$\kappa(V) = \max_{y\neq 0} \frac{||y||}{||Vy||} \cdot \max_{x\neq 0} \frac{||Vx||}{||x||} = ||V|| \cdot ||V^+||, \tag{1}$$

where $V^+ = (V^H V)^{-1} V^H$ denotes the pseudoinverse of V. The study of the condition number of weighted Vandermonde-like matrices is very much related to a study of the condition number of more general classes of structured matrices such as (modified) Gram matrices, positive definite Hankel matrices, or (modified) Krylov matrices (see, e.g., [23, 24, 2, 3, 5]).

The quantity $\kappa(V_n(w_n, E_n))$ may serve to measure the sensitivity of least squares polynomial approximation: Given some function f defined on E_n, consider the problem of finding a polynomial P of degree at most n with minimal deviation from f with respect to some discrete L_2 norm. Writing $\vec{P} = (a_0, \dots, a_n)^T$ for the polynomial $P = \sum_{j=0}^n a_j p_j$, we are left with the problem of determining $a_0, \dots, a_n \in \mathbb{C}$ minimizing the expression

$$\sum_{k=0}^{N} w_n(z_k)^2 |f(z_k) - \sum_{j=0}^{n} a_j p_j(z_k)|^2,$$

with its unique solution given by

$$\vec{P} = V_n(w_n, E_n)^+ (w_n(z_0)f(z_0), \dots, w_n(z_N)f(z_N))^T.$$

Suppose now that the vector $b = (w_n(z_0)f(z_0), \dots, w_n(z_N)f(z_N))^T$ is perturbed slightly; what happens to the vector of coefficients of the corresponding best approximant? The factor of magnification of the corresponding relative errors is given by

$$\frac{||V_n(w_n, E_n)^+(b + \Delta b) - V_n(w_n, E_n)^+ b||}{||V_n(w_n, E_n)^+ b||} \cdot \left[\frac{||\Delta b||}{||b||}\right]^{-1}$$

for some vectors of weighted data values $b, \Delta b$. Suppose now that b is the vector of weighted data values resulting from some polynomial of degree at most n, i.e., $b = V_n(w_n, E_n) \cdot \vec{P}$ with $\deg P \leq n$, which is perturbed by an arbitrary $\Delta b \in \mathbb{C}^{N+1}$. From (1) we may conclude that the maximal factor of magnification of relative errors is just given by $\kappa(V_n(w_n, E_n))$.

For given bases of polynomials and a given sequence of weights and abscissas, one observes quite often that the quantity $\kappa(V_n(w_n, E_n))$ grows exponentially in n. In the present paper we provide a lower bound of the nth-root limit in terms of complex potential theory, and describe necessary and sufficient conditions for the data in order to insure subexponential growth. In the case of square Vandermonde-like matrices, such a study may be based on properties of an underlying weighted Lebesgue function (see, e.g., [2, Appendix B]). Here we will be interested in the more involved case where $\#E_n/n$ tends to some constant larger than one, a rather typical situation for least squares approximation.

We conclude this section by describing our assumptions on the input data (or for some subsequence) which are appropriate for describing the nth-root behavior of $\kappa(V_n(w_n, E_n))$:

(i) E is some compact subset of the complex plane $\mathbb{C}$, and $E_n \subset E$ for all $n \geq 0$. Furthermore, there exists some finite positive Borel measure σ with finite logarithmic energy, $\operatorname{supp}(\sigma) = E$, such that

$$\lim_{n\to\infty} \frac{1}{n} \sum_{z\in E_n} f(z) = \int f(z)\, d\sigma(z)$$

for all functions f continuous on E.

(ii) $w_n : E \to (0, \infty)$, and $(w_n^{1/n})_{n\geq 0}$ converges to some positive continuous function w uniformly in E.

(iii) Let $\rho_n(z) := \sqrt{|p_0(z)|^2 + |p_1(z)|^2 + \cdots + |p_n(z)|^2}$; then $(\rho_n^{1/n})_{n\geq 0}$ tends to some function ρ uniformly on compact subsets of $\mathbb{C}$.

Examples of polynomials satisfying assumption (iii) include monomials, Chebyshev polynomials or other suitable sequences of orthonormal polynomials; see also the more detailed discussion in §4.

The rest of this paper is organized as follows: We recall in §2 some related recent estimates for special structured matrices. Subsequently, we state our main results: on the one hand we obtain in Theorem 2.1 exponentially increasing condition numbers for most of the configurations described in (i)–(iii). In contrast, we give in Theorem 2.2 necessary and sufficient conditions on the data for insuring subexponential growth. The proofs of these properties are given in §3. Section 4 is devoted to studying an illustrating example.

2. Statement of our Main Results and Related Estimates

The numerical condition of (weighted) Vandermonde(-like) matrices has received much attention in the past 25 years. In a number of papers [10, 11, 12, 13], Gautschi investigated the condition number of square Vandermonde matrices (i.e., $p_j(z) = z^j$, $w_n = 1$) with real abscissas, showing that $\kappa_n(V_n(1, E_n))$ is bounded from below by some function increasing exponentially in n. Further results in this area have been given in [23, 24, 2]; it is shown in [3, Theorem 4.1] that

$$\kappa(V_n(1, E_n)) \geq \sqrt{\frac{2}{n+1}} \cdot (1 + \sqrt{2})^{n-1} \tag{2}$$

for all $n \geq 2$ and for all $E_n \subset \mathbb{R}$, $\#E_n = n+1$; it is also proved that bound (2) may be improved at most by a factor $(n+1)^{3/2}$. Similar results may be stated for nonnegative abscissas [3, Theorem 4.1]. It is important to notice that the choice of real abscissas is not appropriate for the basis of monomials: If E_n is the set of $(N+1)$st roots of unity, then obviously $\kappa(V_n(1, E_n)) = 1$. Similarly, if E_n results from a Van der Corput enumeration of particular roots of unity, then $\kappa(V_n(1, E_n)) \leq \sqrt{2(n+1)}$ [6, Corollary 3].

The condition number of (unweighted) Vandermonde-like matrices has also been investigated for other bases of polynomials, e.g., the basis of Newton polynomials [9, 19], the basis of Faber polynomials of some ellipses [20], or the basis of a family of orthogonal polynomials [12, 20]. Here, in general, subexponential growth of $(\kappa(V_n(1, E_n)))_{n\geq 0}$ is established; however, each time the choice of the abscissas was motivated by asymptotic properties of the corresponding basis.

One might expect to be able to decrease the condition number by allowing for an additional weight function — such a (in general unknown) weight occurs naturally in the context of Krylov or Gram matrices (see [2, 3, 5]). However, for the basis of monomials it is shown in [3, Theorem 3.6] that

$$\kappa(V_n(w_n, E_n)) \geq \frac{\gamma^n}{4\sqrt{n+1}}, \quad \gamma := \exp\left(\frac{2\text{Catalan}}{\pi}\right) \approx 1.792 \tag{3}$$

for all $n \geq 2$, for all $E_n \subset \mathbb{R}$, $\#E_n \geq n+1$, and for all $w_n : E_n \to \mathbb{R}$; it is also proved that bound (3) may be improved at most by a factor $\gamma \cdot (8n+8)^{1/2}$. Similar results are given in [3, Corollary 3.2, Remark 3.4 and Remark 3.5] for the case of abscissas located in some real interval.

The numerical condition of weighted Vandermonde-like matrices for arbitrary "admissible" bases $(p_j)_{j\geq 0}$ (see assumption (iii) above) is discussed in [5, Theorem 1.2 and Theorem 1.3]. In order to describe their findings and the findings of the present paper, we will need some facts from complex potential theory. Here we follow [21]; however, for ease of presentation, we will impose some quite strong regularity assumptions which simplify some of the characterization statements.

For an arbitrary finite Borel measure μ with compact support $\text{supp}(\mu)$, the logarithmic potential of μ is defined by

$$U^\mu(z) = \int \log \frac{1}{|z-t|} \, d\mu(t).$$

Let E and w satisfy assumptions (i), (ii), and suppose in addition that E is regular, i.e., the connected components of $\mathbb{C} \setminus E$ are regular with respect to the Dirichlet problem. We denote by $\mathcal{M}(E)$ the set of all positive unit Borel measures, and define for $\mu \in \mathcal{M}(E)$ the weighted energy integral

$$I_w(\mu) := \int\int \log \frac{1}{|z-t| w(z) w(t)} \, d\mu(t) \, d\mu(z).$$

Then there exists a unique *extremal measure* $\lambda_w \in \mathcal{M}(E)$ with $I_w(\lambda_w) = \inf\{I_w(\mu) : \mu \in \mathcal{M}(E)\}$ (see [21, Theorem I.1.3]). According to our regularity assumptions, we know from [21, Theorem I.4.8 and Theorem I.5.1] that the potential U^{λ_w} is continuous in $\mathbb{C}$. Moreover, by [21, Theorem I.1.3] there exists a constant $F =: F_w$ such that for $\mu = \lambda_w$

$$U^\mu(z) - \log w(z) \begin{cases} \geq F & \text{for all } z \in E, \\ = F & \text{for all } z \in \operatorname{supp}(\mu). \end{cases} \tag{4}$$

In addition [21, Theorem I.3.1], if $\mu \in \mathcal{M}(E)$ satisfies (4), then necessarily (μ, F) coincides with (λ_w, F_w). We refer the reader to [21] for various applications of the weighted energy problem. For regular compact sets E and for admissible bases, it is shown in [5, Theorem 1.2 and Theorem 1.3] that

$$\begin{aligned} &\liminf_{n\to\infty} \log \inf\{\kappa(V_n(w_n, E_n)) : E_n \subset E,\ w_n \text{ positive on } E_n\} \\ &\geq \sup_{z\in\operatorname{supp}(\lambda_{1/\rho})} \big[U^{\lambda_{1/\rho}}(z) + \log \rho(z)\big] - \inf_{z\in\mathbb{C}} \big[U^{\lambda_{1/\rho}}(z) + \log \rho(z)\big], \end{aligned} \tag{5}$$

with ρ as in (iii). Also, various bases are discussed where actually equality holds (e.g., the basis of monomials, compare with (3)). Notice that the term on the right-hand side is necessarily nonnegative by (4), and that only a careful choice of a basis in terms of E (as well as of w_n and E_n) will enable us to obtain subexponential growth of $(\kappa(V_n(w_n, E_n))_{n\geq 0}$.

In the present paper we derive similar estimates for the case where the basis as well as a configuration of abscissas (and possibly the weights) are given. We will show that the constrained weighted energy problem plays an important role [18, 8]: here one tries to minimize $I_w(\mu)$ with respect to all $\mu \in \mathcal{M}(E)$ satisfying the additional constraint $\mu \leq \sigma$. Recently, this energy problem has been introduced by Rakhmanov and further studied by several other authors [7, 15, 16] for describing the asymptotic behavior of so-called ray sequences of orthonormal polynomials with respect to some discrete measure (such as discrete Chebyshev or Krawtchouk polynomials). Let E, w, and σ be as in assumptions (i), (ii), with $\sigma(\mathbb{C}) > 1$ (the case $\sigma(\mathbb{C}) = 1$ is also allowed but is trivial). We denote by $\mathcal{M}^\sigma$ the set of all positive unit Borel measures satisfying the additional constraint $\mu \leq \sigma$, that is, $\sigma - \mu$ is a positive Borel measure. Then there exists again a unique *constrained extremal measure* $\lambda_w^\sigma \in \mathcal{M}^\sigma$ with $I_w(\lambda_w^\sigma) = \inf\{I_w(\mu) : \mu \in \mathcal{M}^\sigma\}$ (see [8, Theorem 2.1]).

We shall prove the following:

Theorem 2.1. *Let $E_n, E, \sigma, p_n, \rho$ be as in* (i),(iii). *Then*

$$\begin{aligned} &\liminf_{n\to\infty} \log \kappa(V_n(w_n, E_n))^{1/n} \geq \\ &\sup_{z\in\operatorname{supp}(\lambda)} \big[U^\lambda(z) + \log \rho(z)\big] - \inf_{z\in\mathbb{C}} \big[U^\lambda(z) + \log \rho(z)\big] \end{aligned} \tag{6}$$

with $\lambda = \lambda_{1/\rho}^\sigma$. If in addition w_n, w are as in (ii), *then*

$$\begin{aligned} \liminf_{n\to\infty} \log \kappa(V_n(w_n, E_n))^{1/n} &\geq \sup_{z\in\operatorname{supp}(\lambda)} \big[U^\lambda(z) - \log w(z)\big] \\ &- \inf_{z\in\mathbb{C}} \big[U^\lambda(z) + \log \rho(z)\big] + \max_{z\in E} \big[\log w(z) + \log \rho(z)\big] \end{aligned} \tag{7}$$

with $\lambda = \lambda_w^\sigma$.

Conversely, provided that the additional assumptions (iv),(v) *described below hold, and* $\text{supp}(\lambda_w^\sigma) \cap \text{supp}(\sigma - \lambda_w^\sigma)$ *is nonempty, the limit of the sequence on the left-hand side of* (7) *exists and coincides with the right-hand side of* (7).

The proof of Theorem 2.1 is based on several observations: first we notice that for a polynomial $P = a_0 p_0 + \cdots + a_n p_n$ we have $\vec{P} = (a_0, \ldots, a_n)^T$, and from the Cauchy-Schwarz inequality we get $|P(z)| \leq ||\vec{P}|| \cdot \rho_n(z)$ for all $z \in \mathbb{C}$. This inequality enables us to relate the norm of the pseudoinverse of $V_n(w_n, E_n)$ to some polynomial extremal problem as described in Lemma 3.1 below. However, we may only expect to have equality in (7) provided that

$$\textbf{(iv)} \qquad \lim_{n\to\infty} \epsilon_n(H)^{1/n} = 1 \quad \text{for some } H \subset \mathbb{C} \text{ being compact},$$

$$\epsilon_n(H) := \max_{\deg P \leq n} \frac{||\vec{P}||}{||P/\rho_n||_H} \geq 1,$$

where $|| \cdot ||_H$ denotes the usual supremum norm on H.

A second key observation is that $||V_n(w_n, E_n)^+||$ approximately equals the norm of the inverse of the square submatrix of order $n+1$ which has maximal determinant. This submatrix is given by $V_n(w_n, E_n^*)$, where E_n^* is the set of the $n+1$ weighted *Fekete points* out of E_n. Asymptotic properties of weighted Fekete points formed from discrete sets have been described in [4]. Here one necessarily requires an additional separation property for the sets E_n: we denote the *scaled counting measure* of some finite $A \subset \mathbb{C}$ by

$$\nu_n(A) := \frac{1}{n} \sum_{a \in A} \delta_a,$$

where δ_a denotes the usual Dirac measure at a. Notice that $(\nu_n(E_n))_{n\geq 0}$ has the weak* limit σ by assumption (i). Here, following [8], we will have to add the property

(v) for any sequence $(\zeta_n)_{n\geq 0}$, $\zeta_n \in E_n$, with limit ζ, there holds
$$\lim_{n\to\infty} U^{\nu_n(E_n \setminus \{\zeta_n\})}(\zeta_n) = U^\sigma(\zeta) < \infty.$$

Sequences of sets E_n as described in (i) satisfying condition (v) are described in [8, Lemma 3.2]; examples are equidistant nodes, or sets of zeros of polynomials orthogonal with respect to some measure in the class **Reg** on an interval.[2] One may easily construct[3] sets E_n satisfying (i) even with continuous U^σ, where (v) is violated (and the last part of Theorem 2.1 does not remain valid). We should however mention that we may relax (v) by allowing for an exceptional set of capacity zero (see [7, 16]). Also, by following [4] it is possible to replace condition (v) by

[2] Abscissas satisfying both (i), (v) are for instance given by $E_n = \{f(j/N_n) : j = 0, \ldots, N_n := \#E_n - 1\}$, where $f : [0,1] \to \mathbb{C}$ is continuous and injective, and $f^{-1} \in \text{Lip}_\alpha$ for some $0 < \alpha \leq 1$.
[3] Take, e.g., $E_n = \{j/n, j/n - \delta_n : j = 1, 2, \ldots, n\}$, where δ_n tends rapidly to zero.

the regularity assumption

(v′) U^σ is continuous, and

$$\lim_{n\to\infty} \frac{1}{n^2} \sum_{x,y\in E_n, x\neq y} \log \frac{1}{|x-y|} = \int\int \log \frac{1}{|x-y|}\, d\sigma(x)\, d\sigma(y),$$

as conjectured by Rakhmanov [15] to be sufficient for establishing the asymptotic behavior of related discrete orthogonal polynomials. Note that condition (v') has an interpretation in terms of the asymptotic behavior of the determinants of square Vandermonde matrices of order $\#E_n$.

Theorem 2.1 is of some theoretical interest for identifying configurations of data where the condition number grows exponentially in n. Of more practical interest, however, is the case where we observe subexponential growth. Here we have the following result.

Theorem 2.2. *Let $E_n, E, \sigma, w_n, n, p_n, \rho$ be as in* (i), (ii), (iii), (v), *with* $\mathrm{supp}(\lambda_w^\sigma) \cap \mathrm{supp}(\sigma - \lambda_w^\sigma)$ *being nonempty. We have subexponential growth of* $(\kappa(V_n(w_n, E_n)))_{n\geq 0}$ *if and only if*

(a) $\log[w(z)\cdot\rho(z)] = \max_{t\in E} \log[w(t)\cdot\rho(t)] =: F'$ *for all* $z \in \mathrm{supp}(\lambda_w^\sigma)$;

(b) $U^{\lambda_w^\sigma} + \log\rho$ *equals some constant* F^* *in* $\mathbb{C}$;

(c) *condition* (iv) *holds.*

In this case, we have the implications

(d) $\lambda_w^\sigma = \lambda_w$ *and* $F^* - F' = F_w$ *(and thus* $\lambda_w \leq \sigma$*)*;

(e) *for the leading coefficient* b_j *of* p_j, *there holds* $\lim_{n\to\infty} |b_n|^{1/n} = \exp(F^*)$;

(f) $\lim_{n\to\infty} \left[\max_{z\in E} |w(z)^n \cdot P_n(z)|\right]^{1/n} = \exp(-F_w), \quad P_n := p_n/b_n.$

For numerical reasons, one also wants in general that $\|V_n(w_n, E_n)\|^{1/n} \to 1$, and hence the constant F' in Theorem 2.2(a) should be equal to 0.

Following [21, Chapter III.4], we may conclude from Theorem 2.2(f) that P_n are asymptotically extremal monic polynomials. For such polynomials, many results about zero distributions in terms of $\lambda_w^\sigma = \lambda_w$ are known.

To conclude this section, we mention that many of the above results remain valid in the case of an unbounded set $E = \mathrm{supp}(\sigma)$. Here one requires a particular decay rate of w at infinity, and a suitable reformulation of assumptions (i), (ii), (v′) to insure that the corresponding weighted Fekete points remain uniformly bounded. For further details we refer the reader to [16, 15, 7, 4].

3. Proofs

In all of this section we will assume that $E_n, E, \sigma, p_n, \rho$ are as in (i), (iii), and $w_n : E \to (0,\infty)$. Further assumptions will be mentioned explicitly. A basic observation in our proofs of Theorems 2.1 and 2.2 is that condition numbers of weighted Vandermonde-like matrices are closely related to some weighted extremal problems for polynomials, as shown in the following result.

Lemma 3.1. *Let $H \subset \mathbb{C}$ be compact, and define*

$$\delta_n(w_n, E_n, H) := \max_{\deg P \leq n} \frac{||P/\rho_n||_H}{||w_n P||_{E_n}}.$$

Then, with $N+1 := \#E_n$ and $\epsilon_n(H)$ as defined in (iv), *there holds*

$$\kappa(V_n(w_n, E_n))\sqrt{N+1} \geq ||w_n \rho_n||_{E_n} \delta_n(w_n, E_n, H) \geq \frac{\kappa(V_n(w_n, E_n))}{\epsilon_n(H)\sqrt{N+1}}, \quad (8)$$

$$||w_n \rho_n||_{E_n} \delta_n(w_n, E_n, H) \geq \delta_n(1/\rho_n, E_n, H). \quad (9)$$

Furthermore, if we denote by $E_n^* \subset E_n$ a set of weighted Fekete nodes, i.e., a set where the maximum is attained in $\max\{|\det V_n(w_n, E_n')| : E_n' \subset E_n, \#E_n' = n+1\}$, *then*

$$(n+1) \cdot \delta_n(w_n, E_n, H) \geq \delta_n(w_n, E_n^*, H) \geq \delta_n(w_n, E_n, H). \quad (10)$$

Proof. Writing $E_n = \{z_0, \ldots, z_N\}$, we first notice that $w_n(z_j) \cdot \rho_n(z_j)$ equals the Euclidean norm of the jth row of $V_n(w_n, E_n)$. Consequently,

$$||w_n \rho_n||_{E_n} \leq ||V_n(w_n, E_n)|| \leq \Big[\sum_{j=0}^{N} w_n(z_j)^2 \rho_n(z_j)^2\Big]^{1/2} \leq \sqrt{N+1} \cdot ||w_n \rho_n||_{E_n}. \quad (11)$$

Also, for any polynomial P of degree at most n and for any $z \in \mathbb{C}$ there holds

$$|P(z)| \leq ||\vec{P}|| \cdot \rho_n(z), \quad ||w_n P||_{E_n} \leq ||V_n(w_n, E_n)\vec{P}|| \leq \sqrt{N+1} \cdot ||w_n P||_{E_n}.$$

Thus we get, using (1) and (11), that

$$\kappa(V_n(w_n, E_n)) \geq ||w_n \rho_n||_{E_n} \cdot \frac{||\vec{P}||}{||V_n(w_n, E_n)\vec{P}||} \geq \frac{||w_n \rho_n||_{E_n}}{\sqrt{N+1}} \cdot \frac{|P(z)/\rho_n(z)|}{||w_n P||_{E_n}},$$

as required for the first part of (8). The other part follows by observing that

$$\epsilon_n(H) \cdot \delta_n(w_n, E_n, H) \geq \max_{\deg P \leq n} \frac{||\vec{P}||}{||w_n P||_{E_n}} \geq \frac{\kappa(V_n(w_n, E_n))}{\sqrt{N+1} \cdot ||w_n \rho_n||_{E_n}},$$

where for the second step we have applied (1) and (11). The inequality (9) is trivial. It remains to show (10). Here the inequality $\delta_n(w_n, E_n^*, H) \geq \delta_n(w_n, E_n, H)$ is a trivial consequence of the fact that $E_n^* \subset E_n$. In order to obtain the other inequality of (10), let $E^* := \{x_0, \ldots, x_n\}$, and consider the corresponding Lagrange polynomials

$$\ell_j(z) = \prod_{k \neq j} \frac{z - x_k}{x_j - x_k}, \quad j = 0, \ldots, n.$$

By the triangle inequality we have

$$\max_{\deg P \leq n} \frac{||w_n P||_{E_n}}{||w_n P||_{E_n^*}} = \max_{z \in E_n} \sum_{j=0}^{n} \frac{w_n(z)}{w_n(x_j)} \cdot |\ell_j(z)|,$$

and it is sufficient to show that the right-hand side is $\leq n+1$. This, however, is an immediate consequence of the construction of E_n^* since for any $z \in E_n$ and for any $0 \leq j \leq n$ there holds

$$\left|\frac{w_n(z)}{w_n(x_j)} \cdot \ell_j(z)\right| = \frac{|\det V_n(w_n, \{x_0, \dots, x_{j-1}, z, x_{j+1}, \dots, x_n\})|}{|\det V_n(w_n, \{x_0, \dots, x_{j-1}, x_j, x_{j+1}, \dots, x_n\})|} \leq 1. \tag{12}$$

□

We see from Lemma 3.1 that — at least for bases satisfying (iv) — the asymptotic behavior of $(\kappa(V_n(w_n, E_n))^{1/n})_{n\in\Lambda}$ is completely determined by that of $(\delta_n(w_n, E_n, H)^{1/n})_{n\in\Lambda}$ (or of $(\delta_n(w_n, E_n^*, H)^{1/n})_{n\in\Lambda}$) since the asymptotic behavior of $(||w_n\rho_n||_{E_n}^{1/n})_{n\in\Lambda}$ is known according to assumptions (i)–(iii). Also, though the determination of a weight w_n minimizing $\kappa(V_n(w_n, E_n))$ in general is a nontrivial task (see, e.g., [1]), the simpler expression $||w_n\rho_n||_{E_n} \cdot \delta_n(w_n, E_n, H)$ is clearly minimized for the choice $w_n = 1/\rho_n$ by (9). Finally, the occurrence of weighted Fekete points is quite natural since, as in the proof of Lemma 3.1, one shows that

$$||V_n(w_n, E_n)^+|| \leq ||V_n(w_n, E_n^*)^{-1}|| \leq (N+1) \cdot ||V_n(w_n, E_n)^+||.$$

In other words, the sensitivity of polynomial least squares approximation is closely related to the sensitivity of polynomial interpolation at a suitable subset of abscissas.

With Lemma 3.1 taken into account, it remains to discuss the asymptotic behavior of $(\delta_n(w_n, E_n^*, H)^{1/n})_{n\geq 0}$. Here the constrained energy problem with external field plays an important role.

Lemma 3.2. *Let $R > 0$ and $\Delta_R := \{z \in \mathbb{C} : |z| \leq R\}$. Furthermore, suppose that assumption* (ii) *holds, and denote by $\lambda = \lambda_w^\sigma$ the extremal measure of the constrained weighted energy problem as introduced before Theorem* 2.1. *Then*

$$\liminf_{n\to\infty} \log \delta_n(w_n, E_n^*, \Delta_R)^{1/n} \geq \sup_{z\in\text{supp}(\lambda)} \left[U^\lambda(z) - \log w(z)\right] - \min_{z\in\Delta_R}\left[U^\lambda(z) + \log\rho(z)\right].$$

Proof. We write more explicitly $E_n^* = \{x_{0,n}, \dots, x_{n,n}\}$ for the set of weighted Fekete points as introduced in Lemma 3.1, denote by $\ell_{0,n}, \dots, \ell_{n,n}$ the corresponding Lagrange polynomials, and consider the measures $\mu_{j,n} := \nu_n(E_n^* \setminus \{x_{j,n}\})$, $\mu_n := \nu_n(E_n^*)$, $0 \leq j \leq n$. First, as in the proof of Lemma 3.1, one shows that

$$\delta_n(w_n, E_n^*, \Delta_R) = \max_{z\in\Delta_R} \sum_{j=0}^{n} \frac{|\ell_{j,n}(z)|/\rho_n(z)}{w_n(z_j)}.$$

Taking into account that $\log |\ell_{j,n}(z)|^{1/n} = U^{\mu_{j,n}}(x_{j,n}) - U^{\mu_{j,n}}(z)$ and writing $a_{j,n} := U^{\mu_{j,n}}(x_{j,n}) - \log w_n(x_{j,n})^{1/n}$, $0 \leq j \leq n$, $a_n := \max_{0 \leq j \leq n} a_{j,n}$, we obtain

$$\begin{aligned}
&\frac{1}{n} \log \delta_n(w_n, E_n^*, \Delta_R) \\
&= \frac{1}{n} \cdot \log \max_{z \in \Delta_R} \sum_{j=0}^{n} \exp\Big(n \cdot \big(a_{j,n} - U^{\mu_{j,n}}(z) - \log \rho_n(z)^{1/n}\big)\Big) \\
&= \frac{\log \eta_n}{n} + \max_{0 \leq j \leq n} \max_{z \in \Delta_R} \big(a_{j,n} - U^{\mu_{j,n}}(z) - \log \rho_n(z)^{1/n}\big) \qquad (13)
\end{aligned}$$

for some $\eta_n \in [1, (n+1)]$.

Since $\mathrm{supp}(\mu_n) \subset E$, Helly's Theorem asserts that, given some infinite set Λ_0 of integers, we may find $\Lambda_1 \subset \Lambda_0$ such that $(\mu_n)_{n \in \Lambda_1}$ converges weak* to a probability measure μ. One easily verifies, using assumption (i), that $\mu \in \mathcal{M}^\sigma$. Moreover, for any $0 \leq j_n \leq n$, the sequence $(\mu_{j_n,n})_{n \in \Lambda_1}$ also has the weak* limit μ. By construction, for any $z \in \mathrm{supp}(\mu)$ we may find a sequence $(x_{j_n,n})_{n \in \Lambda_1}$ tending to z, and therefore

$$U^\mu(z) - \log w(z) \leq \liminf_{n \to \infty, n \in \Lambda_1} a_{j_n,n} \leq \liminf_{n \to \infty, n \in \Lambda_1} a_n \qquad (14)$$

for all $z \in \mathrm{supp}(\mu)$ by the principle of descent (see, e.g., [21, Theorem I.6.8]) and assumption (ii). Also, the closed disk Δ_R has the K-property, and thus

$$\lim_{n \to \infty, n \in \Lambda_1} \min_{z \in \Delta_R} \big[U^{\mu_{j_n,n}}(z) + \log \rho_n(z)^{1/n}\big] = \min_{z \in \Delta_R} \big[U^\mu(z) + \log \rho(z)\big] \qquad (15)$$

for any $0 \leq j_n \leq n$ according to [17, Theorem V.4.3, p.182] and assumption (iii). Combining (13), (14), and (15), we obtain

$$\begin{aligned}
&\liminf_{n \to \infty, n \in \Lambda_1} \frac{1}{n} \log \delta_n(w_n, E_n^*, \Delta_R) \\
&\geq \sup_{z \in \mathrm{supp}(\mu)} \big[U^\mu(z) - \log w(z)\big] - \min_{z \in \Delta_R} \big[U^\mu(z) + \log \rho(z)\big].
\end{aligned}$$

Since the choice of Λ_0 was arbitrary, the assertion of Lemma 3.2 now follows by showing that for all $z \in \mathbb{C}$ we have

$$U^\mu(z) \leq \sup_{t \in \mathrm{supp}(\mu)} \big[U^\mu(t) - \log w(t)\big] + U^\lambda(z) - \sup_{t \in \mathrm{supp}(\lambda)} \big[U^\lambda(t) - \log w(t)\big]. \qquad (16)$$

In fact, from [8, Theorem 2.1.(c)] we know that there exists a set $K \subset E$ with $(\sigma - \lambda)(K) = 0$ such that

$$U^\lambda(z) - \log w(z) \geq \sup_{t \in \mathrm{supp}(\lambda)} \big[U^\lambda(t) - \log w(t)\big], \quad z \in \mathrm{supp}(\sigma - \lambda) \setminus K. \qquad (17)$$

Denote by μ' the restriction of μ to $\mathrm{supp}(\sigma - \lambda)$. If $\mu' = 0$, then necessarily $\lambda = \mu$ (cf. [8, Lemma 5.1]), and thus (16) trivially holds. Also, (16) trivially is true if $\sup\{[U^\mu(t) - \log w(t)] : t \in \operatorname{supp}(\mu)\} = +\infty$, the latter being equivalent to $\sup\{U^\mu(t) : t \in \mathbb{C}\} = +\infty$ by the maximum principle for potentials [21, Corollary II.3.3]. It therefore remains to discuss the case where $\mu \neq 0$ has a finite potential, and thus $\mu' \neq 0$ has both a finite potential and finite logarithmic energy. Then one verifies, using Fubini's Theorem, that also $\mu'(K) = 0$. Furthermore, from (17) we obtain for $z \in \operatorname{supp}(\mu') \setminus K \subset \operatorname{supp}(\mu)$

$$\begin{aligned} U^{\mu'}(z) &\leq \sup_{t\in\operatorname{supp}(\mu)} \big[U^\mu(t) - \log w(t)\big] + \log w(z) - U^{\mu-\mu'}(z) \\ &\leq \sup_{t\in\operatorname{supp}(\mu)} \big[U^\mu(t) - \log w(t)\big] - \sup_{t\in\operatorname{supp}(\lambda)} \big[U^\lambda(t) - \log w(t)\big] \\ &\qquad + U^\lambda(z) - U^{\mu-\mu'}(z). \end{aligned} \tag{18}$$

Also, with $S := E\setminus\operatorname{supp}(\sigma-\lambda)$ we have by construction $\mu-\mu' = \mu|_S \leq \sigma|_S = \lambda|_S$, showing that $\lambda-\mu+\mu'$ is a nonnegative finite Borel measure with compact support, with its total mass not exceeding that of μ'. From the principle of domination [21, Theorem II.3.2] we may conclude that (18) holds for all $z \in \mathbb{C}$, as claimed in (16). □

Observe that, for $z \to \infty$, inequality (16) provides a new characterization of the extremal measure λ_w^σ in the case of compact $\operatorname{supp}(\sigma)$ and continuous w which is complementary to [8, Theorem 2.1.(e)]. Here the uniqueness result follows from the uniqueness theorem [21, Theorem 2.1] and the maximum principle for subharmonic functions.

Corollary 3.3. *We have for* $\lambda = \lambda_w^\sigma$

$$\sup_{t\in\operatorname{supp}(\lambda)} \big[U^\lambda(t) - \log w(t)\big] = \min_{\mu\in\mathcal{M}^\sigma} \sup_{t\in\operatorname{supp}(\mu)} \big[U^\mu(t) - \log w(t)\big]. \tag{19}$$

If, in addition, the polynomial convex hull of $\operatorname{supp}(\lambda_w^\sigma)$ *is of two-dimensional Lebesgue measure zero, then any measure* $\lambda \in \mathcal{M}^\sigma$ *satisfying* (19) *necessarily coincides with* λ_w^σ.

For the second part of Theorem 2.1 we need a sharper version of Lemma 3.2 which is attainable if we add some separation property such as assumption (v) or (v′). In fact, it follows from [4, Theorem 1.5(a),(c)] that equality holds in Lemma 3.2 provided that $E = \operatorname{supp}(\sigma)$ is connected and (v′) holds, and that this result may even be generalized for measures σ with unbounded support. Here we will restrict ourselves to the simpler condition (v).

Lemma 3.4. *Under the assumptions of Lemma* 3.2, *suppose in addition that* (v) *holds, and that* $\operatorname{supp}(\lambda) \cap \operatorname{supp}(\sigma - \lambda)$ *is nonempty, where* $\lambda = \lambda_w^\sigma$. *Then*[4]

$$\lim_{n\to\infty} \log \delta_n(w_n, E_n^*, \Delta_R)^{1/n} = \sup_{z\in\operatorname{supp}(\lambda)} \big[U^\lambda(z) - \log w(z)\big] - \min_{z\in\Delta_R} \big[U^\lambda(z) + \log\rho(z)\big].$$

[4]The set Δ_R in Lemma 3.2 and Lemma 3.4 may be replaced, e.g., by any compact set having an empty intersection with E.

Proof. Let $\mu_n, \mu_{j,n}, a_n, a_{j,n}, \Lambda_0, \Lambda_1, \mu$ be as in the proof of Lemma 3.2. We choose $k_n \in \{0, \dots, n\}$, $\zeta \in \operatorname{supp}(\sigma - \mu)$, and $\Lambda_2 \subset \Lambda_1$ with

$$\limsup_{n\to\infty, n\in\Lambda_1} a_n = \lim_{n\to\infty, n\in\Lambda_2} a_{k_n,n}$$
$$\text{and} \quad \min_{t\in\operatorname{supp}(\sigma-\mu)} \big[U^\mu(t) - \log w(t)\big] = U^\mu(\zeta) - \log w(\zeta).$$

By assumption (i), we may find $\zeta_n \in E_n \setminus E_n^*$ with $(\zeta_n)_{n\in\Lambda_2}$ tending to ζ. Set $\sigma_n' := \mu_{k_n,n}$, $\sigma_n'' := \nu_n(E_n \setminus \{\zeta_n\}) - \sigma_n'$. Then the sequence $(\sigma_n'')_{n\in\Lambda_1}$ has the weak* limit $\sigma - \mu \geq 0$. Applying twice the principle of descent and assumption (v), we obtain

$$\begin{aligned} U^\mu(\zeta) &\leq \liminf_{n\to\infty, n\in\Lambda_2} U^{\sigma_n'}(\zeta_n) \\ &\leq \limsup_{n\to\infty, n\in\Lambda_2} U^{\sigma_n'}(\zeta_n) = U^\sigma(\zeta) - \liminf_{n\to\infty, n\in\Lambda_2} U^{\sigma_n''}(\zeta_n) \leq U^\mu(\zeta). \end{aligned}$$

With (12) and assumption (ii) taken into account, it follows that

$$U^\mu(\zeta) - \log w(\zeta) = \lim_{n\to\infty, n\in\Lambda_2} \big[U^{\sigma_n'}(\zeta_n) - \log w_n(\zeta_n)^{1/n}\big] \geq \lim_{n\to\infty, n\in\Lambda_2} a_{k_n,n},$$

and a combination with (14) leads to

$$\begin{aligned} \min_{t\in\operatorname{supp}(\sigma-\mu)} \big[U^\mu(t) - \log w(t)\big] &\geq \limsup_{n\to\infty, n\in\Lambda_1} a_n \geq \liminf_{n\to\infty, n\in\Lambda_1} a_n \\ &\geq \max_{t\in\operatorname{supp}(\mu)} \big[U^\mu(t) - \log w(t)\big]. \end{aligned} \tag{20}$$

In particular, the equilibrium condition (17) holds for the measure $\mu \in \mathcal{M}^\sigma$ with K being empty. From the uniqueness result [8, Theorem 2.1(d)] we may conclude that $\mu = \lambda$. Recalling that the set Λ_0 was arbitrary, we may conclude that the sequence of normalized counting measures of Fekete points $(\mu_n)_{n\geq 0}$ has the weak* limit λ (see also [4, Theorem 1.5(a)]). Also, since $\operatorname{supp}(\lambda) \cap \operatorname{supp}(\sigma - \lambda)$ is nonempty, we obtain from (20) the convergence[5] of $(a_n)_{n\geq 0}$, with limit described in (20). Finally, the assertion of Lemma 3.4 now follows from (13) together with (15). □

We are now prepared to establish our main theorems.

Proof of Theorem 2.1. In order to establish (6), recall from (8), (9), and (10) that

$$\kappa(V_n(w_n, E_n)) \geq \frac{1}{\sqrt{\#E_n}} \cdot \delta_n(1/\rho_n, E_n, \Delta_R) \geq \frac{1}{(n+1)\cdot\sqrt{\#E_n}} \cdot \delta_n(1/\rho_n, E_n^{**}, \Delta_R)$$

for every $R > 0$, where E_n^{**} is an $(n+1)$-point Fekete set for the weight $1/\rho_n$. Thus, it just remains to apply Lemma 3.2 with $w_n = 1/\rho_n$. Similarly, for a proof of (7) we apply (8), (10), and Lemma 3.2, and observe that

$$\lim_{n\to\infty} \frac{1}{n} \log ||w_n \rho_n||_{E_n} = \max_{z\in E} \big[\log w(z) + \log \rho(z)\big]$$

by assumptions (i), (ii), and (iii).

[5]Using [8, Example 2.4], one may construct examples where $\mu = \lambda$, but $(a_n)_{n\geq 0}$ does not necessarily converge.

Now let (iv) and (v) hold, and let $\mathrm{supp}(\lambda_w^\sigma) \cap \mathrm{supp}(\sigma - \lambda_w^\sigma)$ be nonempty. We choose a sufficiently large $R > 0$ such that the set H of assumption (iv) is contained in the disk Δ_R. Since $\epsilon_n(H) \geq \epsilon_n(\Delta_R) \geq 1$ by construction, we may conclude that (iv) is also true for $H = \Delta_R$. Thus we obtain from (8) and (10)

$$\limsup_{n\to\infty} \log \kappa(V_n(w_n, E_n))^{1/n}$$
$$\leq \max_{z\in E}\big[\log w(z) + \log \rho(z)\big] + \limsup_{n\to\infty} \log \delta_n(w_n, E_n^*, H)^{1/n},$$

with the right-hand side being computed in Lemma 3.4. Letting $R \to \infty$ and combining with (7) yields the final claim of Theorem 2.1. □

Proof of Theorem 2.2. In the first part of the proof we assume that there is subexponential growth, and write $\lambda = \lambda_w^\sigma$. First, recalling the inequalities of Lemma 3.1, we have

$$1 \leq \epsilon_n(E) \leq \epsilon_n(E_n) = \max_{\deg P \leq n} \frac{||\vec{P}|| \cdot ||w_n \rho_n||_{E_n}}{||P/\rho_n||_{E_n} \cdot ||w_n\rho_n||_{E_n}}$$
$$\leq \max_{\deg P\leq n} \frac{||\vec{P}|| \cdot ||w_n\rho_n||_{E_n}}{||w_n P||_{E_n}} \leq \sqrt{\#E_n} \cdot \kappa(V_n(w_n, E_n)),$$

and thus (iv) holds with $H = E$, as claimed in part (c). Moreover, because of the subexponential growth, the right-hand side of (6) has to be ≤ 0. In particular, the function $f(z) := U^\lambda(z) + \log\rho(z)$ has to be equal to some constant F^* on $\mathrm{supp}(\lambda)$, and $f(z) \geq F^*$ for $z \in \mathbb{C} \setminus \mathrm{supp}(\lambda)$. One verifies (see, e.g., [5, Lemma 2.1]), using assumption (iii), that $\log\rho$ is continuous and subharmonic in $\mathbb{C}$, and $\log\rho(z) - \log|z|$ is bounded above around infinity. It follows from the principle of continuity [21, Theorem II.3.5] that U^λ is continuous in $\mathbb{C}$. Consequently, f is subharmonic in $\mathbb{C}\setminus\mathrm{supp}(\lambda)$, continuous in $\mathbb{C}$, bounded above by F^* on $\mathrm{supp}(\lambda)$, and bounded above around infinity. From the maximum principle for subharmonic functions it follows that $f(z) \leq F^*$, and thus $f(z) = F^*$ for all $z \in \mathbb{C}$, which yields property (b). In addition, since the right-hand side of (7) has to be ≤ 0, we get

$$\sup_{z\in\mathrm{supp}(\lambda)} \big[U^\lambda(z) - \log w(z)\big] + \max_{z\in E}\big[\log w(z) - U^\lambda(z)\big] \leq 0,$$

that is, $U^\lambda(z) - \log w(z) - F^* = -\log w(z) - \log\rho(z)$ is equal to some constant $-F'$ in $\mathrm{supp}(\lambda)$, and $\geq -F'$ in $E \setminus \mathrm{supp}(\lambda)$, as claimed in part (a).

Conversely, if (a), (b) and (c) hold, then subexponential growth follows from the second part of Theorem 2.1. Part (d) now is an immediate consequence of (a), (b), and [21, Theorem I.3.1] (see equation (4)), with $F^* - F' = F_w$. In order to show part (e), define $\tilde{\rho}_n(z) := \max_{0\leq j\leq n} |p_j(z)|$. Then $\log \tilde{\rho}_n^{1/n}$ tends to $\log\rho$ uniformly on compact subsets of $\mathbb{C}$ by assumption (iii). Furthermore, $\log\tilde{\rho}_n^{1/n} + U^\lambda$ is subharmonic in $(\mathbb{C} \cup \{\infty\}) \cap \{|z| \geq R\}$, for R large enough. Consequently, we get from the maximum principle for subharmonic functions

$$\log|b_n|^{1/n} = \lim_{|z|\to\infty}\left[\log\tilde{\rho}_n(z)^{1/n} + U^{\lambda}(z)\right] \le \max_{|z|=R}\left[\log\tilde{\rho}_n(z)^{1/n} + U^{\lambda}(z)\right],$$

with the right-hand side tending to F^* for $n \to \infty$. Thus, $\limsup_{n\to\infty}\log|b_n|^{1/n} \le F^*$. On the other hand, we may conclude from assumptions (i)–(iii) and (a) that $\log||V_n(w_n, E_n)||^{1/n}$ tends to F' (see (11)), and thus $\log||V_n(w_n, E_n)^+||^{1/n}$ tends to $-F'$. Furthermore, for any monic polynomial P of degree n there holds $|b_n| \cdot ||\vec{P}|| \ge 1$, and by [21, Theorem III.3.1]

$$\lim_{n\to\infty}\left[\min_{P(z)=z^n+\cdots}||w^n P||_E\right]^{1/n} = \exp(-F_w). \tag{21}$$

Therefore, using again assumption (ii), we obtain

$$\begin{aligned} F_w &\le \liminf_{n\to\infty}\frac{1}{n}\log\max_{\deg P=n}\frac{|b_n|\cdot||\vec{P}||}{||w^n P||_E} \le \liminf_{n\to\infty}\frac{1}{n}\log\max_{\deg P=n}\frac{|b_n|\cdot||\vec{P}||}{||w_n P||_{E_n}} \\ &\le \liminf_{n\to\infty}\left[\log|b_n|^{1/n} + \log||V_n(w_n, E_n)^+||^{1/n}\right] = \liminf_{n\to\infty}\log|b_n|^{1/n} - F', \end{aligned}$$

showing that $\liminf_{n\to\infty}\log|b_n|^{1/n} \ge F_w + F' = F^*$, as required for part (e).

In order to establish part (f), notice first that $|P_n(z)| \le \rho_n(z)/|b_n|$, and thus by (21), assumption (ii), part (a), and part (e),

$$e^{-F_w} \le \liminf_{n\to\infty}||w^n P_n||_E^{1/n} \le \liminf_{n\to\infty}|b_n|^{-1/n}\cdot||w_n\rho_n||_E^{1/n} = e^{-F^*+F'} = e^{-F_w}.$$

□

We complete this section by discussing two special cases of Theorem 2.2 which are of major interest for applications: let $w = 1$ and denote by H the polynomial convex hull of E. From Theorem 2.2(d) we may conclude that $\lambda_w^\sigma = \lambda_1 =: \omega_E$, the equilibrium measure of E, and Theorem 2.2(b) implies that $\log\rho$ is the Green's function for the unbounded component of $\mathbb{C}\setminus E$ (or of $\mathbb{C}\setminus H$), with pole at infinity, denoted by $g_E(z,\infty)$. Conversely, these data satisfy parts (a), (b) of Theorem 2.2 (provided of course that $\omega_E \le \sigma$). As a second case, suppose that $\log\rho(z) = g_H(z,\infty)$ for some compact set H of positive logarithmic capacity having a connected complement. Then Theorem 2.2(b) may equivalently be written as $\lambda_w^\sigma = \omega_H$. We summarize our findings in

Corollary 3.5. *Let $E_n, E, \sigma, w_n, n, p_n, \rho$ be as in* (i), (ii), (iii), (v), *with* $\mathrm{supp}(\lambda_w^\sigma)\cap \mathrm{supp}(\sigma-\lambda_w^\sigma)$ *being nonempty.*

(a) *In the case $w = 1$ (e.g., $w_n = 1$ for all $n \ge 0$), we have subexponential growth of $(\kappa(V_n(w_n, E_n)))_{n\ge 0}$ if and only if $\omega_E \le \sigma$, $\log\rho = g_E(\cdot,\infty)$, and condition* (iv) *holds.*

(b) *Let $\log\rho(z) = g_H(z,\infty)$ with some compact set H of positive logarithmic capacity having a connected complement. Then we have subexponential growth of $(\kappa(V_n(w_n, E_n)))_{n\ge 0}$ if and only if $\omega_H \le \sigma$, condition* (iv) *holds, w equals some constant* $\exp(F')$ *on the boundary ∂H of H, and $w\cdot\rho$ is less than or equal to* $\exp(F')$ *in $E\setminus\partial H$.* □

4. Examples

In order to illustrate our main results, we will restrict ourselves to the case of an asymptotically trivial weight $w = 1$, and consider only real abscissas with $E = \text{supp}(\sigma) = [-1, 1]$. Furthermore, in order to be able to give some integral representation for the constant occurring in Theorem 2.1, we wish to restrict ourselves to measures σ having an even potential which is concave on $(-1, 1)$.

Let us first describe how to obtain the corresponding equilibrium measure $\lambda = \lambda_1^\sigma$ of the constrained energy problem as described before Theorem 2.1. According to [8, Corollary 2.15], the probability measure $\tau := (\sigma - \lambda)/(\sigma(E) - 1)$ is the solution to the (unconstrained) weighted energy problem on E with weight $v(z) := \exp(U^\sigma(z)/(\sigma(E) - 1))$. Since $\log(1/v)$ is convex and even, by assumption, we may conclude from [21, Theorem IV.1.10] that $\text{supp}(\tau)$ is an interval[6] of the form $[-r, r]$. Having determined the shape of the support, we may find the corresponding parameter r by maximizing the Mhaskar-Saff functional [21, Theorem IV.1.5]

$$F(r) = -\log(\text{cap}([-r, r]) - \int g_{[-r,r]}(t, \infty)\, d\sigma(t).$$

The equation $F'(r) = 0$ allows us to determine r as the unique solution of

$$1 = 2 \int_r^1 \frac{t}{\sqrt{t^2 - r^2}}\, d\sigma(t). \tag{22}$$

According to [8, Corollary 2.15], we have $\sigma - \lambda = \hat{\sigma} - \omega_{[-r,r]}$, where $\hat{\sigma}$ is the balayage measure of σ onto $[-r, r]$. By [21, §II.4, Eq. (4.47)], $\hat{\sigma}$ may be rewritten as $\hat{\sigma} = \sigma|_{[-r,r]} + \mu$, with

$$\frac{d\mu}{dx}(x) = \frac{2}{\pi} \int_r^1 \frac{t\sqrt{t^2 - r^2}}{(t^2 - x^2) \cdot \sqrt{r^2 - x^2}}\, d\sigma(t), \quad x \in [-r, r],$$

and thus

$$\frac{d\lambda}{dx}(x) = \frac{d(\omega_{[-r,r]} - \mu)}{dx}(x) = \frac{1}{\pi} \cdot \int_r^1 \frac{2t}{t^2 - x^2} \cdot \frac{\sqrt{r^2 - x^2}}{\sqrt{t^2 - r^2}}\, d\sigma(t), \quad x \in [-r, r], \tag{23}$$

whereas $d\lambda(x) = d\sigma(x)$ for $x \in E \setminus [-r, r]$.

We now turn our attention to suitable bases of polynomials. Let H be some compact set with connected complement and positive logarithmic capacity. It is shown in [19] that the basis $(p_j)_{n \geq 0}$ of Newton polynomials at Leja points of H satisfies both assumptions (iii) and (iv) provided H has capacity 1, with $\rho = g_H(\cdot, \infty)$ (this result remains valid in the case of capacity different from one if one divides the jth Newton polynomial by its maximum norm on H). More generally [2, Theorem 2.11], provided H is regular with respect to the Dirichlet problem, we may replace Leja points by any other sequence of points if the weak* limit of the corresponding sequence of normalized zero counting measures coincides with the

[6] This property can also be derived under weaker assumptions on U^σ; see, e.g., [21, Theorem IV.1.10], [21, Corollary IV.1.10] or [18, Theorem 4].

equilibrium measure of H, e.g., we may take a Van der Corput enumeration of Fejér points [9]. Another family of polynomials satisfying (iii), (iv) with $\log \rho = g_H(\cdot, \infty)$ for some domain is given by the corresponding sequence of Faber polynomials [2, §2.4.3].

As a final class of polynomials let μ be some positive Borel measure with compact support, and denote by p_n the corresponding nth orthonormal polynomial. Then condition (iv) holds with H being the polynomial convex hull of $\text{supp}(\mu)$. To see this, notice that for a polynomial P of degree at most n there holds

$$||\vec{P}||^2 = \int |P(z)|^2 \, d\mu(z) \leq ||P/\rho_n||_H^2 \cdot \int \rho_n(z)^2 \, d\mu(z) \leq (n+1) \cdot ||P/\rho_n||_H^2,$$

and thus $\epsilon_n(H) \leq \sqrt{n+1}$. Now if H is regular, then assumption (iii) holds with $\log \rho = g_H(\cdot, \infty)$ if and only if $\mu \in \mathbf{Reg}$ (see [22, Theorem 3.2.3] where further equivalent descriptions are given). Examples include the sequence of monomials ($H = \mathbb{D} := \{z \in \mathbb{C} : |z| \leq 1\}$) and the sequence of Chebyshev polynomials ($H = [-1, 1]$).

Let us now determine the constants occurring in Theorem 2.1 for the special case $w = 1$ and $\log \rho = g_H$ for some compact set $H \supset E = [-1, 1]$. First notice that $\log w + \log \rho$ equals zero on E. Also, $U^\lambda + g_H$ is superharmonic in $(\mathbb{C} \cup \{\infty\}) \setminus \partial H$, and thus

$$\begin{aligned} \Gamma(\sigma, H) &:= \sup_{z \in \text{supp}(\lambda)} \left[U^\lambda(z) - \log w(z)\right] \\ &\quad - \inf_{z \in \mathbb{C}} \left[U^\lambda(z) + \log \rho(z)\right] + \max_{z \in E} \left[\log w(z) + \log \rho(z)\right] \\ &= \sup_{z \in \text{supp}(\lambda)} U^\lambda(z) - \inf_{z \in \partial H} U^\lambda(z), \end{aligned}$$

in accordance with the observations of Corollary 3.5. Recall from (23) that we have at our disposal an integral representation for the potential U^λ. From (4) we know that $U^{\sigma-\lambda}$ is constant on $[-r, r]$. Furthermore, $U^{\sigma-\lambda}$ is convex outside the support of $\sigma - \lambda$. Using the representation $U^\lambda = U^\sigma - U^{\sigma-\lambda}$, we may conclude that U^λ is concave on $[-1, 1]$, decreasing on $[1, +\infty)$, and even. Consequently, $\Gamma(\sigma, H) = U^\lambda(0) - U^\lambda(c)$, with $c = \max(|a|, |b|)$ in the case of a real interval $H = [a, b] \supset E$, and $c = i$ in the case $H = \mathbb{D}$.

To be more concrete, consider the case $\sigma = \alpha \cdot \omega_E + \beta \cdot \tau$ with α, β some nonnegative real constants, and $d\tau(x) = dx$ on $[-1, 1]$, with $\sigma(E) = \alpha + 2\beta > 1$. One easily verifies that U^τ is concave on E, and so is U^σ. In the case $\alpha \geq 1$ we have $\omega_E \leq \sigma$, and thus $\Gamma(\sigma, H) = g_E(c, \infty)$, which is obviously equal to 0 in the case $H = [-1, 1]$ and equal to $\log(1 + \sqrt{2})$ in the case $H = \mathbb{D}$. Otherwise, we may determine r via (22), and obtain the equation

$$1 = 2 \int_r^1 \left[\frac{\alpha t}{\pi \sqrt{t^2 - r^2} \sqrt{1 - t^2}} + \frac{\beta t}{\sqrt{t^2 - r^2}} \right] dt = \alpha + 2\beta \cdot \sqrt{1 - r^2},$$

that is, $r = \sqrt{1 - ((1 - \alpha)/(2\beta))^2}$. The integral occurring in (23) may be calculated explicitly; for the sake of simplicity we will restrict ourselves to the case $\alpha = 0$.

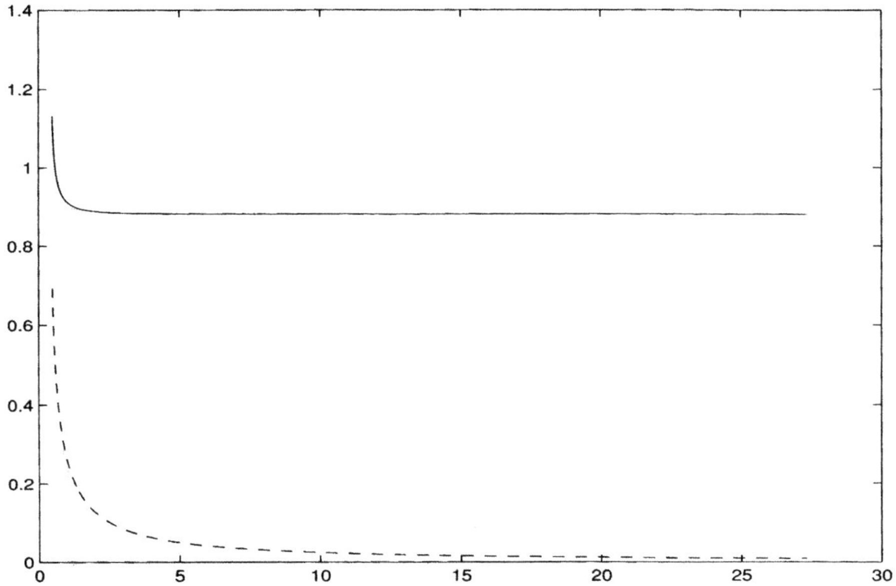

FIGURE 1. The quantities $\Gamma(\beta \cdot \tau, \mathbb{D})$ (solid) and $\Gamma(\beta \cdot \tau, [-1,1])$ (dashed) for the measure $d\tau(x) = dx$ on $[-1,1]$ of equidistant nodes in $[-1,1]$ for $\beta \in [0.5, 27]$

Then for $x \in [-r,r]$ (compare with [8, Example 4.1])

$$\frac{d\lambda}{dx}(x) = \begin{cases} \beta, & x \in E \setminus [-r,r], \\ \frac{2\beta}{\pi} \arctan(\frac{\sqrt{1-r^2}}{\sqrt{r^2-x^2}}), & x \in [-r,r], \end{cases}$$

leading to

$$\Gamma(\sigma, H) = U^\lambda(0) - U^\lambda(c) = \beta \cdot \int_r^1 \log\left|1 - \frac{c^2}{x^2}\right| dx$$

$$+\frac{2\beta}{\pi} \cdot \int_0^r \log\left|1 - \frac{c^2}{x^2}\right| \cdot \arctan\left(\frac{\sqrt{1-r^2}}{\sqrt{r^2-x^2}}\right) dx.$$

In the case $H = [-1,1]$ (i.e., $c = 1$) of, e.g., Chebyshev polynomials, we obtain a function decreasing in β (see Fig. 1), which for $\beta \to \infty$ tends to zero (then we obtain Fekete nodes which approximately are distributed like the arcsin measure), and for $\beta \to 1/2$ (the case of square Vandermonde-Chebyshev matrices with equidistant nodes) tends to $\log 2 \approx 0.693$, describing the classical behavior of the Lebesgue constant for equidistant nodes on $[-1,1]$.

In contrast, in the case $H = \mathbb{D}$ (i.e., $c = i$) of, e.g., monomials we obtain a function decreasing in β (see Fig. 1), which for $\beta \to \infty$ tends to $\log(1+\sqrt{2}) \approx 0.881$ (Vandermonde matrices with optimal choice of abscissas in $[-1,1]$ being distributed like the arcsin measure [2, 3]), and for $\beta \to 1/2$ (the case of square Vandermonde matrices with equidistant nodes) tends to $\log(\sqrt{2}) + \pi/4 \approx 1.132$, confirming a result of Gautschi [12, Example 3.3].

References

[1] F.L. Bauer, *Optimally scaled matrices*, Numer. Math., **5** (1963), 73–87.

[2] B. Beckermann, *On the numerical condition of polynomial bases: estimates for the condition number of Vandermonde, Krylov and Hankel matrices*, Habilitationsschrift, Universität Hannover, April 1996.

[3] B. Beckermann, *The condition number of real Vandermonde, Krylov and positive definite Hankel matrices*, Publication ANO 380, Université de Lille (1997), submitted for publication.

[4] B. Beckermann, *On a conjecture of E.A. Rakhmanov*. Publication ANO 385, Université de Lille (1998), submitted for publication.

[5] B. Beckermann and H. Stahl, *Some n-th root estimates for the condition number of Krylov, Krylov-like and Gram matrices*, in preparation.

[6] A. Córdova, W. Gautschi and S. Ruscheweyh, *Vandermonde matrices on the circle: spectral properties and conditioning*, Numer. Math., **57** (1990), 577–591.

[7] S.B. Damelin and E.B. Saff, *Asymptotics of weighted polynomials on varying sets*, manuscript (1998).

[8] P.D. Dragnev and E.B. Saff, *Constrained energy problems with applications to orthogonal polynomials of a discrete variable*, J. Anal. Math., **72** (1997), 223–259.

[9] B. Fischer and L. Reichel, *Newton interpolation in Fejér and Chebyshev points*, Math. Comp., **53** (1989), 265–278.

[10] W. Gautschi, *Norm estimates for inverses of Vandermonde matrices*, Numer. Math., **23** (1975), 337–347.

[11] W. Gautschi, *Optimally conditioned Vandermonde matrices*, Numer. Math., **24** (1975), 1–12.

[12] W. Gautschi, *How (un)stable are Vandermonde systems?*, in: R. Wong, Ed., *Asymptotic and computational analysis*, Lecture Notes in Pure and Appl. Math. **124**, Dekker, New York, 1990, 193–210.

[13] W. Gautschi and G. Inglese, *Lower bounds for the condition number of Vandermonde matrices*, Numer. Math., **52** (1988), 241–250.

[14] G.H. Golub and C.F. Van Loan, *Matrix computations*, 2d ed., Johns Hopkins University Press, Baltimore, 1993.

[15] A.B.J. Kuijlaars and E.A. Rakhmanov, *Zero distributions for discrete orthogonal polynomials*, Proceedings of the VIII SPOA, Sevilla 1997, to appear in: J. Comput. Appl. Math..

[16] A.B.J. Kuijlaars and W. Van Assche, *Extremal polynomials on discrete sets*, manuscript (1998).

[17] E.M. Nikishin and V.N. Sorokin, *Rational approximations and orthogonality*, Translations of Mathematical Monographs **92**, Amer. Math. Soc., Providence, RI, 1991.

[18] E.A. Rakhmanov, *Equilibrium measure and the distribution of zeros of the extremal polynomials of a discrete variable*, Math. Sbornik, **187** (1996), 1213–1228.

[19] L. Reichel, *Newton interpolation at Leja points*, BIT, **30** (1990), 332–346.

[20] L. Reichel and G. Opfer, *Chebyshev-Vandermonde systems*, Math. Comp., **57** (1991), 703–721.

[21] E.B. Saff and V. Totik, *Logarithmic potentials with external fields*, Grundlehren der Mathematischen Wissenschaften **316**, Springer, Berlin, 1997.

[22] H. Stahl and V. Totik, *General orthogonal polynomials*, Encyclopedia of mathematics and its applications **43**, Cambridge Univ. Press, New York, 1992.

[23] J.M. Taylor, *The condition of Gram matrices and related problems*, Proc. Roy. Soc. Edinburgh, Sect. A, **80** (1978), 45–56.

[24] E.E. Tyrtyshnikov, *How bad are Hankel matrices?*, Numer. Math., **67** (1994), 261–269.

Bernhard Beckermann
Laboratoire d'Analyse Numérique et d'Optimisation
UFR IEEA - M3
Université des Sciences et Technologies de Lille
F-59655 Villeneuve d'Ascq Cedex, France
E-mail address: bbecker@ano.univ-lille1.fr

Edward B. Saff
Institute for Constructive Mathematics
Department of Mathematics
University of South Florida
Tampa, FL 33620, USA
E-mail address: esaff@math.usf.edu

International Series of Numerical Mathematics
Vol. 131,

Transpose-Free Look-Ahead Algorithms for Lanczos' Method

Claude Brezinski
Michela Redivo-Zaglia

Abstract. The method of Lanczos for solving systems of linear equations is implemented by various recurrence relationships. Sometimes, a division by a quantity close to zero occurs, thus producing numerical instability in the algorithm. It is shown how to treat such a near-breakdown by a look-ahead technique which, at the same time, avoids the use of the transpose of the matrix. The main idea of the algorithms given in this paper consists of computing the coefficients of these recurrences by a look-ahead algorithm for the CGS or the BiCGSTAB and, then, to use them also in Lanczos' method. Thus, transpose-free look-ahead algorithms for treating simultaneously breakdowns and near-breakdowns in Lanczos' method and in the CGS or the BiCGSTAB will be obtained.

1. Introduction

Lanczos' method [38] for solving a system of linear equations $Ax = b$ is implemented via three-term or coupled two-term recurrence relationships. In these algorithms, an important problem is division by zero (called *breakdown*) or by a quantity close to zero (called *near-breakdown*). In the case of breakdown, the algorithm has to be stopped while, in the case of near-breakdown, it can become numerically unstable. A second problem is due to the necessity of computing products of the form $A^T v$. When the system is large, indirect addressing makes such a computation difficult and costly, and even impossible in the case where A is not known (such as in the solution of systems of nonlinear equations). These two problems were solved separately in several papers by Brezinski, Redivo-Zaglia and/or Sadok [9, 10, 11, 12, 13, 14, 15]; see [32] for a review of the various techniques which can be found in the literature. In this paper we will solve these problems simultaneously by giving transpose-free algorithms for avoiding breakdowns and near-breakdowns in the implementation of Lanczos' method. The idea is quite simple and was first proposed in [20]: since the coefficients used in the recurrences for Lanczos' method are the same as those constructed in a Lanczos-type product method (LTPM), such as CGS or BiCGSTAB, without the use of the transpose, it consists of computing these coefficients only once for the two methods. So, for the price of only one additional matrix-vector product, Lanczos' method and a LTPM can be implemented simultaneously without using A^T. Obviously, only algorithms for LTPM where breakdowns and near-breakdowns are cured can be used for that purpose. Moreover, a quite important advantage of this strategy is that, at each step, the iterates can be combined together by a hybrid procedure [7] for obtaining a better result. When several methods are implemented simultaneously, the iterations are stopped as soon as one of them reaches the desired accuracy.

Our approach is based on the fact that the recurrence relationships for implementing Lanczos' method derive directly from the theory of formal orthogonal polynomials [4]. We mention that some approaches, such as those of [28, 29, 31], make use of the connection between Lanczos' method and Padé approximants. However, as shown in [6], this introduces an unnecessary complication since the numerators of the Padé approximants are not involved and, moreover, the block structure of the table of formal orthogonal polynomials (which are strongly connected to the denominators of the Padé approximants) mimics the block structure of the Padé table as explained in [21].

The method of Lanczos will be recalled in §2 together with its connections to formal orthogonal polynomials. In §3, we will present our transpose-free look-ahead algorithm for avoiding breakdowns and near-breakdowns in the implementation of Lanczos' method. Numerical experiments will be reported in the last section.

2. Lanczos' Method and Orthogonal Polynomials

In this section, we will review Lanczos' method in order to show how formal orthogonal polynomials come into play. We will derive some algorithms from the theory of formal orthogonal polynomials for its implementation. We will see the reason for using the transpose of the matrix and show how to avoid it via a Lanczos-type product method. Finally, we will discuss the problems of breakdown and near-breakdown and explain how to cure them using formal orthogonal polynomials.

2.1. Lanczos' method

Consider the $n \times n$ system of nonsingular linear equations $Ax = b$. The method of Lanczos [38] consists in constructing the sequence of iterates $x_k \in \mathbb{R}^n$ defined by

$$x_k - x_0 \in K_k(A, r_0) = \mathrm{span}(r_0, Ar_0, \ldots, A^{k-1}r_0),$$
$$r_k = b - Ax_k \perp K_k(A^T, y) = \mathrm{span}(y, A^T y, \ldots, A^{T^{k-1}} y),$$

where x_0 is an arbitrary vector, y an almost arbitrary one, and $r_0 = b - Ax_0$.

These two conditions define entirely the vector x_k, and we have

$$r_k = r_0 + a_1 Ar_0 + \cdots + a_k A^k r_0 = P_k(A)r_0.$$

From the orthogonality conditions for r_k, we see that the polynomial P_k satisfies

$$c(\xi^i P_k) = 0 \quad \text{for} \quad i = 0, \ldots, k-1,$$

where c is the linear functional on the vector space of polynomials defined by $c(\xi^i) = (y, A^i r_0)$. These conditions show that $\{P_k\}$ is the family of formal orthogonal polynomials (FOP) with respect to c, normalized by the condition $P_k(0) = 1$. If these polynomials can be recursively computed, then the same is true for the vectors r_k and x_k. There exist various recurrences for computing the polynomials P_k. They were all given in [17]. We will now review those relevant for our purpose.

Let $c^{(1)}$ be the linear functional on the vector space of polynomials defined by

$$c^{(1)}(\xi^i) = c(\xi^{i+1}),$$

and let $\{P_k^{(1)}\}$ denote the family of monic orthogonal polynomials with respect to $c^{(1)}$.

From the theory of FOP, we recall that both P_k and $P_k^{(1)}$ exist if and only if $H_k^{(1)} \neq 0$, where

$$H_k^{(i)} = \begin{vmatrix} c_i & c_{i+1} & \cdots & c_{i+k-1} \\ c_{i+1} & c_{i+2} & \cdots & c_{i+k} \\ \vdots & \vdots & & \vdots \\ c_{i+k-1} & c_{i+k} & \cdots & c_{i+2k-2} \end{vmatrix}.$$

We will now present some recursive algorithms for the computation of these two families of FOP. They lead to various recursive algorithms for computing the sequence (r_k) of residuals in Lanczos' method and the corresponding sequence of iterates (x_k).

It is known that the usual three-term recurrence relationship for computing orthogonal polynomials can be numerically unstable and that it is preferable to use coupled two-term recurrences [34]. So, we will only describe algorithms based on such coupled relationships.

2.2. Lanczos/Orthomin

Let

$$Q_k(\xi) = (-1)^k \frac{H_k^{(0)}}{H_k^{(1)}} P_k^{(1)}(\xi).$$

As explained, for example, in [17], the coefficients of ξ^k in Q_k and P_k are equal and Q_k is proportional to $P_k^{(1)}$. We have the following relations between these two families of FOP:

$$\begin{aligned} P_{k+1}(\xi) &= P_k(\xi) - \lambda_{k+1}\xi Q_k(\xi), \\ Q_{k+1}(\xi) &= P_{k+1}(\xi) + \alpha_{k+1} Q_k(\xi), \end{aligned} \tag{1}$$

with $P_0(\xi) = Q_0(\xi) = 1$ and the coefficients given by

$$\begin{aligned} \lambda_{k+1} &= c(U_k P_k)/c(\xi U_k Q_k), \\ \alpha_{k+1} &= -c(\xi V_k P_{k+1})/c(\xi V_k Q_k), \end{aligned} \tag{2}$$

where U_k and V_k are arbitrary polynomials of exact degree k.

We now use these two recurrence relationships for implementing the method of Lanczos. We have $r_k = P_k(A)r_0$ and we set $p_k = Q_k(A)r_0$. Replacing the variable ξ by the matrix A in the recurrence relationships (1), and multiplying by r_0, leads to

$$\begin{aligned} r_{k+1} &= r_k - \lambda_{k+1}Ap_k, \\ x_{k+1} &= x_k + \lambda_{k+1}p_k, \\ p_{k+1} &= r_{k+1} + \alpha_{k+1}p_k, \end{aligned}$$

with $p_0 = r_0 = b - Ax_0$ and

$$\begin{aligned} \lambda_{k+1} &= (y, U_k(A)r_k)/(y, AU_k(A)p_k), \\ \alpha_{k+1} &= -(y, AV_k(A)r_{k+1})/(y, AV_k(A)p_k). \end{aligned} \tag{3}$$

This algorithm is due to Vinsome [42]. For the choice $U_k \equiv V_k \equiv P_k$ it is called *Lanczos/Orthomin* [43], and it is equivalent to the *biconjugate gradient* (BCG) of Lanczos [37, 38], which was written in algorithmic form, and made popular, by Fletcher [22].

2.3. Lanczos/Orthosym

A variant of Lanczos/Orthodir was proposed in [17]. It consists of replacing the recurrence relation for the polynomials $P_k^{(1)}$ by a different one. These relations were called A8/B8 and correspond to

$$\begin{aligned} P_{k+1}(\xi) &= P_k(\xi) - \lambda'_{k+1}\xi P_k^{(1)}(\xi), \\ P_{k+1}^{(1)}(\xi) &= (\xi + \beta'_{k+1})P_k^{(1)}(\xi) - \gamma'_{k+1}P_k(\xi), \end{aligned} \tag{4}$$

with $P_0(\xi) = P_0^{(1)}(\xi) = 1$. The coefficients are given by

$$\begin{aligned} \lambda'_{k+1} &= c(U_kP_k)/c(\xi U_kP_k^{(1)}), \\ \gamma'_{k+1} &= c(\xi^2V_{k-1}P_k^{(1)})/c(\xi V_{k-1}P_k), \\ \beta'_{k+1} &= [\gamma'_{k+1}c(\xi V_kP_k) - c(\xi^2V_kP_k^{(1)})]/c(\xi V_kP_k^{(1)}). \end{aligned} \tag{5}$$

For the choice $U_k \equiv V_k \equiv P_k$, this algorithm will be called *Lanczos/Orthosym*, where *sym* stands for *symmetric*, since both $P_{k+1}^{(1)}$ and P_{k+1} are computed from the same polynomials. This algorithm is, in fact, the non-look-ahead version of the BSMRZ [12]. The interest of this algorithm lies in the fact that it can be generalized for treating near-breakdowns while Lanczos/Orthomin cannot. This will be explained in §3.

Setting $z_k = P_k^{(1)}(A)r_0$, we immediately obtain from the relations (4)

$$\begin{aligned} r_{k+1} &= r_k - \lambda'_{k+1}Az_k, \\ x_{k+1} &= x_k + \lambda'_{k+1}z_k, \\ z_{k+1} &= (A + \beta'_{k+1}I)z_k - \gamma'_{k+1}r_k, \end{aligned}$$

with $z_0 = r_0 = b - Ax_0$. Using the definition of the linear functional c, we have

$$\lambda'_{k+1} = (y, U_k(A)r_k)/(y, AU_k(A)z_k),$$
$$\gamma'_{k+1} = (y, A^2V_{k-1}(A)z_k)/(y, AV_{k-1}(A)r_k),$$
$$\beta'_{k+1} = [\gamma'_{k+1}(y, AV_k(A)r_k) - (y, A^2V_k(A)z_k)]/(y, AV_k(A)z_k).$$

Remark 2.1. *By using always the same relation for the polynomials P_k, but the relation B10 given in* [17] *for the polynomials $P_k^{(1)}$, we obtain an algorithm similar to Lanczos/Orthomin. It is based on the recurrence relationships*

$$P_{k+1}(\xi) = P_k(\xi) - \lambda'_{k+1}\xi P_k^{(1)}(\xi),$$
$$P_{k+1}^{(1)}(\xi) = \delta_{k+1}P_{k+1}(\xi) + \beta_{k+1}P_k^{(1)}(\xi),$$

with $P_0(\xi) = P_0^{(1)}(\xi) = 1$. The coefficients are given by

$$\lambda'_{k+1} = c(U_kP_k)/c(\xi U_kP_k^{(1)}),$$
$$\delta_{k+1} = -1/\lambda'_{k+1},$$
$$\beta_{k+1} = -\delta_{k+1}c(\xi V_kP_{k+1})/c(\xi V_kP_k^{(1)}).$$

The corresponding algorithm for implementing Lanczos' method will be called Lanczos/Orthobal, *where* bal *stands for* balancing, *and it is, in fact, a scaled version of Lanczos/Orthomin and a simplified version of Lanczos/Orthosym. Indeed, if we substitute the relation for P_{k+1} in that of $P_{k+1}^{(1)}$, we obtain*

$$P_{k+1}^{(1)}(\xi) = (-\lambda'_{k+1}\delta_{k+1}\xi + \beta_{k+1})P_k^{(1)}(\xi) + \delta_{k+1}P_k(\xi) = (\xi + \beta_{k+1})P_k^{(1)}(\xi) + \delta_{k+1}P_k(\xi).$$

2.4. Numerical examples

We now compare Lanczos/Orthomin and Lanczos/Orthosym to show that their stability properties are quite similar. This comparison is made necessary because, as we will explain later, Lanczos/Orthomin cannot be extended to cure near-breakdowns.

All the examples have dimension 100. The solution was chosen randomly, and then the right-hand side was computed accordingly. The matrices were taken from the Matlab matrix toolbox of [35]. In each case, we compare the iterative residuals of both methods in the first frame of the figures and the actual residuals in the second. The residuals of Lanczos/Orthomin are shown in solid lines and those of Lanczos/Orthosym in dashed lines.

Fig. 1 is for the matrix `riemann`, whose condition number is 480.5; Fig. 2 for the matrix `minij`, whose condition number is 16370; and Fig. 3 for the matrix `moler`, whose condition number is 1.43×10^{17}.

2.5. LTPM and transpose-free algorithms

We now explain why, in order to obtain a cheap algorithm, it is necessary to make use of A^T. For example, to compute the numerator of λ_{k+1} with the choice $U_k(\xi) = \xi^k$ in Lanczos/Orthomin, we need to compute (y, A^kr_k). Since the vector

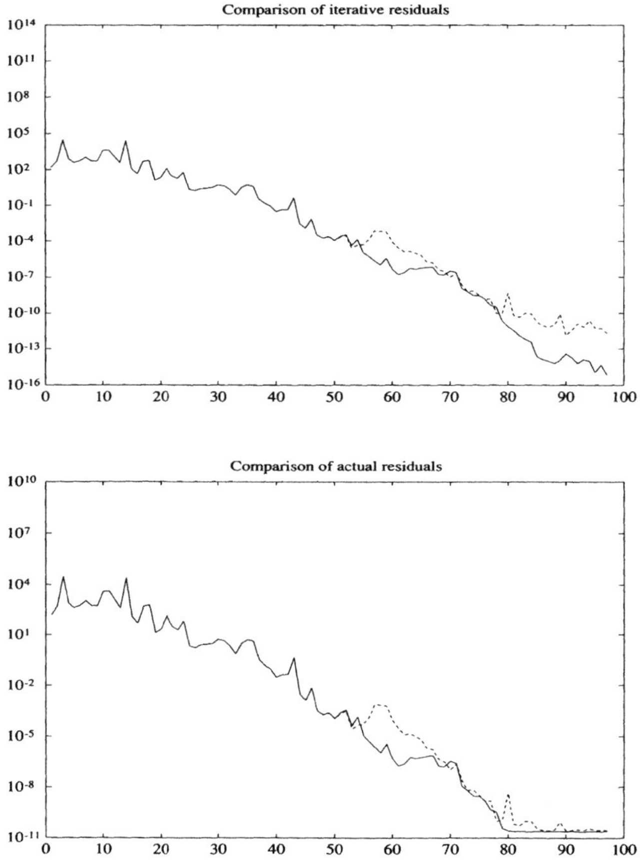

FIGURE 1. Iterative and actual residuals for the matrix `riemann`

r_k depends on k, the computation of $A^k r_k$ requires k matrix-vector products and the algorithm becomes costly. Thus, we will replace the preceding scalar product by $(A^{T^k} y, r_k)$, that is, by (y_k, r_k) with $y_k = A^T y_{k-1}$ and $y_0 = y$. So, the price to pay for having a cheap algorithm is the use of A^T. The same drawback arises in the computation of the coefficients of the other recursive algorithms for implementing the method of Lanczos.

We remark that A^T is only needed in the computation of the coefficients of the recurrence relationships. If these coefficients could be computed without it, then Lanczos' method could also be implemented without using A^T. This idea was introduced in [20] and is fully exploited in [11].

Lanczos-type product methods (LTPM), also called CGM methods [5, 10], are transpose-free variants of Lanczos' method. In these methods the residual has the form $\tilde{r}_k = W_k(A) P_k(A) r_0$, where W_k is an arbitrary polynomial satisfying

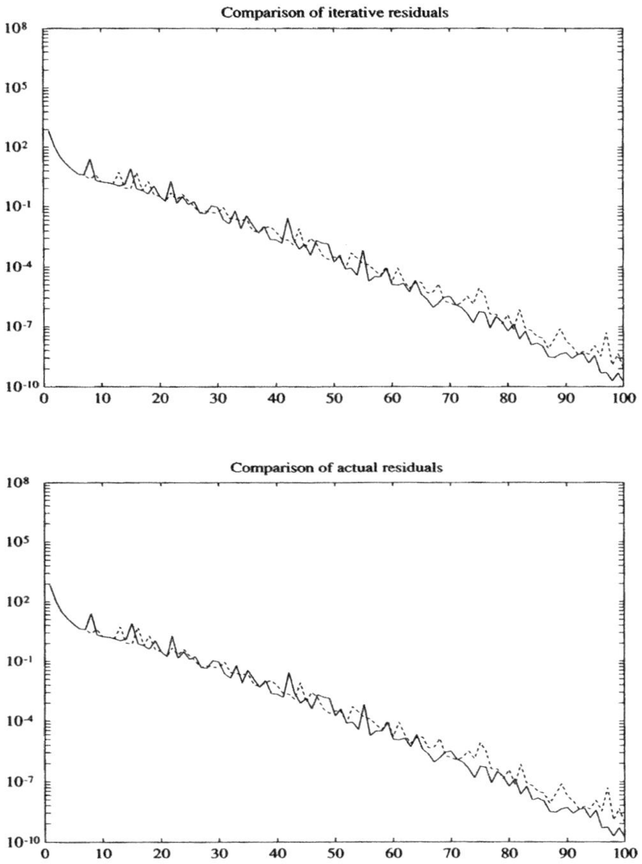

FIGURE 2. Iterative and actual residuals for the matrix `minij`

$W_k(0) = 1$. This condition is mandatory for being able to compute the vector $\tilde{x}_k$ defined by $\tilde{r}_k = b - A\tilde{x}_k$ without using A^{-1}. Thus, the choice $W_k(\xi) = \xi^k$ does not correspond to a LTPM. Among the best-known LTPM are the CGS of Sonneveld [40], and the BiCGSTAB of van der Vorst [41] and its variants [30]. In the algorithms for implementing LTPM, the polynomials W_k and P_k are not computed separately and then multiplied together, but the products W_kP_k are directly computed recursively. The recurrence relation for W_kP_k is obtained by multiplying together the recurrence relationships of W_k and P_k. The products W_kQ_k or $W_kP_k^{(1)}$, and all the other products needed, are also computed recursively in a similar way. Thus, the vectors $\tilde{r}_k$ and the vectors $\tilde{p}_k = W_k(A)Q_k(A)r_0$ (and the vectors $\tilde{z}_k = W_k(A)P_k^{(1)}(A)r_0$) can be computed recursively by a transpose-free algorithm. If the polynomials W_k have exact degree k, we can take $U_k \equiv V_k \equiv W_k$.

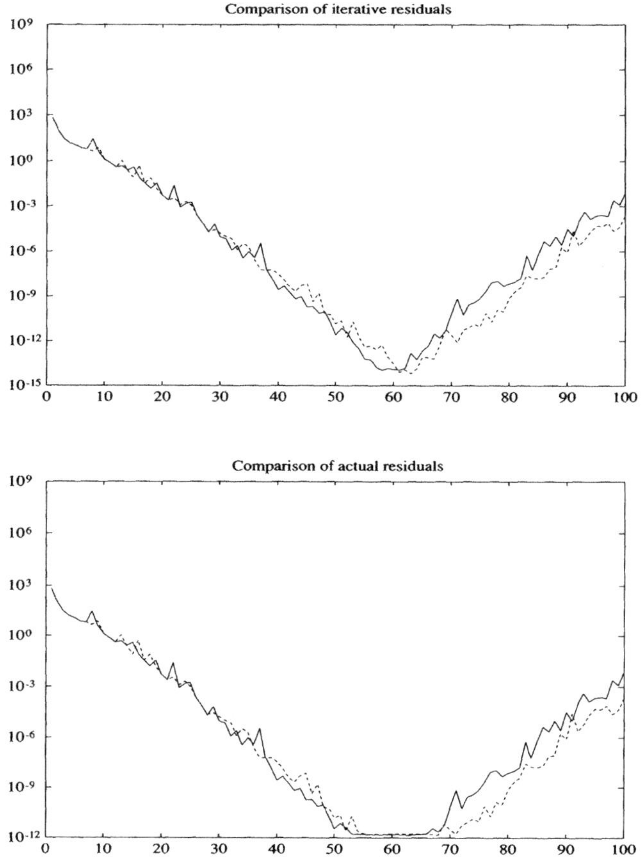

FIGURE 3. Iterative and actual residuals for the matrix `moler`

For example, using this choice in (3), we obtain for Lanczos/Orthomin

$$\begin{aligned}\lambda_{k+1} &= (y, \tilde{r}_k)/(y, A\tilde{p}_k),\\ \alpha_{k+1} &= -(y, A\tilde{q}_k)/(y, A\tilde{p}_k),\end{aligned}$$

where $\tilde{q}_k = W_k(A)P_{k+1}(A)r_0$.

If we take $W_k \equiv P_k$, then $\tilde{r}_k$ is the residual of the CGS and the expressions for λ_{k+1} and α_{k+1} simplify because P_k and Q_k have the same leading coefficient and, as shown in [17, 28, 34], we obtain

$$\begin{aligned}\lambda_{k+1} &= c(P_k^2)/c(\xi Q_k^2) = (y, \tilde{r}_k)/(y, A\tilde{s}_k),\\ \alpha_{k+1} &= c(P_{k+1}^2)/c(P_k^2) = (y, \tilde{r}_{k+1})/(y, \tilde{r}_k),\end{aligned}$$

with $\tilde{s}_k = Q_k^2(A)r_0$.

The coefficients of the recurrences of Lanczos/Orthomin can then be computed via the CGS. Thus, Lanczos' method and the CGS can be implemented simultaneously by a transpose-free algorithm, and similarly for Lanczos' method and the BiCGSTAB, where $W_k(\xi) = (1 - \theta_k \xi) W_{k-1}(\xi)$, with $W_0(\xi) = 1$ and θ_k chosen to minimize $\|\tilde{r}_k\|$. Such algorithms were given in [11], following a strategy already proposed in [20].

2.6. Breakdowns and near-breakdowns

The second important problem in all the algorithms for implementing Lanczos' method concerns breakdowns and near-breakdowns. In these algorithms, for some values of k, the scalar products in the denominators of the coefficients of the recurrence relationships can be zero (a situation called *breakdown*) or close to it (*near-breakdown*). There are two types of breakdowns. The first, called *true* breakdown, occurs when the polynomials P_k and $P_k^{(1)}$ do not exist, which corresponds to $H_k^{(1)} = 0$. In such a situation, the remedy consists in jumping over the polynomials which do not exist and to compute only the existing ones (called *regular*). This is possible by using more complicated recurrence relationships between consecutive regular polynomials. Such a strategy is usually referred to as *look-ahead* [39]. The second type of breakdown, called *ghost* breakdown, can occur even if the polynomial under consideration exists but the recurrence relationship cannot be used for computing the next orthogonal polynomial. In that case, the remedy is the same as before and consists of jumping over the polynomials which cannot be computed (that is over the polynomials which do not exist, but also over existing polynomials for which a ghost breakdown occurs). Again, this is possible by jumping ahead one iteration and using more complicated recurrence relationships between consecutive regular polynomials which do not produce ghost breakdowns. When a scalar product in the denominator of a coefficient is close to zero, the next polynomial will be badly computed and numerical instability will pollute the computations. In that case, the remedy consists in jumping over these polynomials by using the same recurrence relationships as for treating ghost breakdowns. Algorithms for curing breakdowns and near-breakdowns in Lanczos' method were obtained from the theory of FOP in [28, 12, 13, 29, 23, 31, 14]. Similarly, transpose-free algorithms for breakdowns and near-breakdowns were given in [9, 15, 33] for the CGS, and in [10, 33] for the BiCGSTAB. In these transpose-free algorithms, the coefficients in the recurrence relationships are again the same as those appearing in Lanczos' method. Thus, the main idea of the algorithms given in this paper consists of computing these coefficients by a look-ahead algorithm for the CGS or the BiCGSTAB and, then, to use them also in Lanczos' method. Thus, transpose-free look-ahead algorithms for treating simultaneously breakdowns and near-breakdowns in Lanczos' method and in the CGS or the BiCGSTAB will be obtained.

We remark that breakdowns and near-breakdowns can also be treated by another technique, called *look-around*, which consists of going around a block in the Padé table of a function associated with Lanczos' method. This technique, due to Graves-Morris [26], does not make use of the transpose of the matrix. However, since it is equivalent to the computation of $G^i r_0$, where $G = I - A$, it seems to be less stable than our approach. It has been extended to the BiCGSTAB in [27]. Other techniques, where the gaps are filled by polynomials which do not satisfy all the orthogonality conditions or by polynomials satisfying different ones, have been proposed by Freund and Nachtigal [24, 25], Gutknecht [31], and Hochbruck [36]. Another strategy, due to Ayachour [1], is a kind of pivoting, where the orthogonality conditions are considered in a modified order.

3. Transpose-Free Look-Ahead Algorithms

As explained in the previous section, a transpose-free look-ahead algorithm for the implementation of Lanczos' method consists of

1. selecting an algorithm for the implementation of Lanczos' method,
2. extending the algorithm for treating breakdowns and near-breakdowns,
3. building the algorithm for implementing a LTPM.

Among the three main algorithms for the implementation of Lanczos' method, Lanczos/Orthomin seems to be, on average, the most stable, as is shown by the numerical examples reported in [11] and the analysis presented in [34]. Unfortunately, as explained in [8], although the algorithm can be extended for treating breakdowns, near-breakdowns cannot be cured by the technique which consists of jumping over the polynomials which do not exist and also over the existing ones which are badly computed. This is due to the nonexistence of the corresponding recurrence relationship for the orthogonal polynomials. The same is true for Lanczos/Orthobal.

Thus, the first step was to find another algorithm, almost as stable as Lanczos/Orthomin, and where the treatment of near-breakdowns was possible by our technique. In §2.4, we gave some experimental results showing that Lanczos/ Orthosym seems to be almost as stable as Lanczos/Orthomin. Moreover, for this algorithm, we can find generalized recurrence relationships for treating near-breakdowns, and it can also be used for implementing the CGS and the BiCGSTAB, thus leading to a transpose-free algorithm for Lanczos' method.

As explained in §2.6, among the orthogonal polynomials of the two families, some do not exist and others cannot be used because they are badly computed. So, we will change our notations, and the subscript k will from now on only refer to the index of the polynomial in the sequence of those which are used, and it will no longer correspond to the degree. The degree of $P_k^{(1)}$ will be n_k, and P_k will have degree at most n_k. These two polynomials were previously denoted by $P_{n_k}^{(1)}$ and P_{n_k}, respectively. We will also set $n_{k+1} = n_k + m_k$, where m_k is the length of the jump between the degrees of two consecutive polynomials used in the recurrence relationships.

The recurrence relationships on which the look-ahead algorithm for treating breakdowns and near-breakdowns in Lanczos/Orthosym (called the BSMRZ) are based were given in [12]. They are as follows. We first have

$$P_{k+1}(\xi) = P_k(\xi) - \xi w_k(\xi) P_k^{(1)}(\xi) - \xi v_k(\xi) P_k(\xi), \tag{6}$$

where w_k is a polynomial of degree at most $m_k - 1$ and v_k a polynomial of degree at most $m_k - 2$.

Thus, we have $c(\xi^i P_{k+1}) = 0$ for $i = 0, \ldots, n_k - m_k$. Writing this condition for $i = n_k - m_k + 1, \ldots, n_k + m_k - 1$ yields $2m_k - 1$ relations for determining the $2m_k - 1$ unknowns which are the m_k coefficients of w_k and the $m_k - 1$ coefficients of v_k. Of course, we must have $n_k - m_k + 1 \geq 0$. If $n_k - m_k + 1 < 0$, we have $n_k + m_k$ equations, with $n_k + m_k < 2m_k - 1$, for determining the $2m_k - 1$ unknown coefficients. This is the reason why, in this case, v_k must be a polynomial of degree at most $n_k - 1$.

The second recurrence relationship is

$$P_{k+1}^{(1)}(\xi) = q_k(\xi) P_k^{(1)}(\xi) + t_k(\xi) P_k(\xi), \tag{7}$$

where q_k is a monic polynomial of degree m_k and t_k a polynomial of degree at most $m_k - 1$. Thus, we have $2m_k$ unknown coefficients. Writing the orthogonality conditions of $P_{k+1}^{(1)}$ for $i = n_k - m_k, \ldots, n_k + m_k - 1$ gives $2m_k$ relations. Obviously, we must assume that $n_k - m_k \geq 0$. In the case $n_k - m_k < 0$, we must have $n_k + m_k < 2m_k$ equations for computing the $2m_k$ unknowns. Thus, t_k will be a polynomial of degree at most $n_k - 1$.

If the systems giving these coefficients are singular or badly conditioned according to some tests, it means that $P_{k+1}^{(1)}$ (and P_{k+1}) does not exist or cannot be computed by the recurrence relationship under consideration or is badly computed, and the value of m_k has to be increased until a well-computed regular polynomial $P_{k+1}^{(1)}$ is found. At each iteration, it is also necessary to check that $c(\xi^{n_{k+1}} P_{k+1}) \neq 0$. If this condition is not satisfied, then a ghost breakdown will occur at the next iteration because P_{k+2} will have a degree strictly less than n_{k+2} and, therefore, P_{k+2} and $P_{k+2}^{(1)}$ will be linearly dependent. If $c(\xi^{n_{k+1}} P_{k+1}) = 0$, then m_k has to be increased until this quantity becomes different from zero. This strategy is explained in detail in [9, 10]. So, with this additional test, the only possible breakdown in this algorithm is the *incurable hard* one, which occurs when the dimension of the Krylov subspace used is equal to n and a zero denominator is encountered somewhere in the algorithm. However, owing to the arithmetic of the computer, this situation will almost never arise.

Squaring the relations of the BSMRZ leads to a look-ahead algorithm for curing breakdowns and near-breakdowns in the CGS. This algorithm was called the BSMRZS [9].

Since the CGS algorithm consists in taking $\tilde{r}_k = P_k^2(A) r_0$, we shall take the relation (6) of the BSMRZ and square it:

$$P_{k+1}^2 = (1 - \xi v_k)^2 P_k^2 - 2(1 - \xi v_k) \xi w_k P_k P_k^{(1)} + \xi^2 w_k^2 {P_k^{(1)}}^2.$$

We see that it is also necessary to compute recursively the polynomials $P_{k+1}^{(1)^2}$ and $P_{k+1}P_{k+1}^{(1)}$. Thus, by squaring the relation (7) of the BSMRZ, we obtain

$$P_{k+1}^{(1)^2} = q_k^2 P_k^{(1)^2} + 2q_k t_k P_k P_k^{(1)} + t_k^2 P_k^2.$$

Finally, multiplying (6) by (7) leads to

$$P_{k+1}P_{k+1}^{(1)} = (q_k - \xi q_k v_k - \xi t_k w_k) P_k P_k^{(1)} - \xi q_k w_k P_k^{(1)^2} + t_k(1 - \xi v_k) P_k^2.$$

Thus, if we set

$$\begin{aligned} \tilde{r}_k &= P_k^2(A) r_0, \\ \tilde{z}_k &= P_k^{(1)^2}(A)\, r_0, \\ \tilde{s}_k &= P_k(A) P_k^{(1)}(A)\, r_0, \end{aligned}$$

we obtain the BSMRZS

$$\begin{aligned} \tilde{r}_{k+1} &= (I - Av_k(A))^2\, \tilde{r}_k - 2\,(I - Av_k(A))\, Aw_k(A)\tilde{s}_k + A^2 w_k^2(A)\tilde{z}_k, \\ \tilde{x}_{k+1} &= \tilde{x}_k - (Av_k(A) - 2I)\, v_k(A)\tilde{r}_k + 2\,(I - Av_k(A))\, w_k(A)\tilde{s}_k - A\, w_k^2(A)\tilde{z}_k, \\ \tilde{z}_{k+1} &= q_k^2(A)\tilde{z}_k + 2q_k(A)t_k(A)\tilde{s}_k + t_k^2(A)\tilde{r}_k, \qquad (8) \\ \tilde{s}_{k+1} &= (q_k(A) - Aq_k(A)v_k(A) - At_k(A)w_k(A))\, \tilde{s}_k - Aq_k(A)w_k(A)\tilde{z}_k \\ &\quad + t_k(A)\,(I - Av_k(A))\, \tilde{r}_k. \end{aligned}$$

The main interest in the CGS, and in fact the reason for its construction, is that the matrix A^T is not needed in the scalar products giving the coefficients of the recurrence relations. As shown in [9], this feature can be preserved in the BSMRZS, where we have to compute, for $i = 0, \ldots, 2m_k - 1$, the quantities

$$c^{(1)}\left(\xi^i P_k^{(1)^2}\right) = \left(y, A^{i+1} P_k^{(1)^2}(A) r_0\right) = \left(y, A^{i+1}\tilde{z}_k\right)$$

and the following quantities for $i = 0, \ldots, 2m_k - 1$ (in the case $m_k \leq n_k$) or $i = 0, \ldots, n_k + m_k - 1$ (in the case $m_k > n_k$),

$$c\left(\xi^i P_k^{(1)} P_k\right) = \left(y, A^i P_k^{(1)}(A) P_k(A) r_0\right) = \left(y, A^i \tilde{s}_k\right).$$

It must be noted that Sonneveld [40] implemented his method by squaring the recurrence relationships of the Lanczos/Orthomin algorithm. Since, in the BSMRZS, one of the recurrence relationships we are squaring is different, our algorithm does not reduce to Sonneveld's algorithm when $m_k = 1$ for all k. The look-ahead procedure, called CSCGS, proposed by Chan and Szeto [19], is also based on the Lanczos/Orthomin algorithm, and it cures only the true breakdowns by a 2×2 composite step. Since these authors assume that no ghost breakdown occurs, then, as shown by the discussion in [12] (see also [21, 28]), a jump of length 2 is sufficient. The goal of these authors was not to cure all the possible breakdowns, but to obtain only a partial cure at a relatively modest modification of Sonneveld's algorithm. Their procedure reduces to Sonneveld's when no jump occurs. The CSCGS was obtained by squaring the algorithm CSBCG proposed in [3, 2] for treating true breakdowns in Lanczos/Orthomin.

Incorporating in the code of the BSMRZS the recursive computation of the iterates and of the residuals of Lanczos' method, and also the recurrence for the auxiliary vectors, we get a look-ahead transpose-free algorithm for implementing simultaneously Lanczos' method and the CGS. Its pseudocode is given in the next subsection.

3.1. TFLA Lanczos/Orthosym and Look-Ahead CGS

Algorithm TFLACGS $(A, b, x_0, y, n, \varepsilon)$

1. **Initializations**
   $r_0 \leftarrow b - Ax_0$
   $z_0 = r_0$
   $\tilde{x}_0 = x_0$
   $\tilde{r}_0 = \tilde{z}_0 = \tilde{s}_0 = r_0$
   $n_0 \leftarrow 0$
   $m_{-1} \leftarrow 0$
   $\sigma_0^{(0)} \leftarrow (y, \tilde{r}_0)$
2. **For** $k = 0, 1, 2, \ldots$ **until** convergence **do**
   $m_k \leftarrow 1$
   $d_0 \leftarrow \sigma_k^{(m_{k-1})} = (y, \tilde{s}_k)$
   $c_0 \leftarrow (y, A\tilde{z}_k)$
   $d_1 \leftarrow (y, A\tilde{s}_k)$
   $c_1 \leftarrow (y, A^2\tilde{z}_k)$
   compute Az_k
3. **If** $c_0 = (y, A\tilde{z}_k) = 0$ **then** breakdown at $m_k = 1$
   else
   compute β_0 and α_0 (γ_0 and η_0)
   $\tilde{r}_{k+1} = \tilde{r}_k - 2\gamma_0 A\tilde{s}_k + \gamma_0^2 A^2 \tilde{z}_k$
   $\tilde{x}_{k+1} = \tilde{x}_k + 2\gamma_0 \tilde{s}_k - \gamma_0^2 A\tilde{z}_k$
   TF $r_{k+1} = r_k - \gamma_0 A z_k$
   TF $x_{k+1} = x_k + \gamma_0 z_k$
   If $\|\tilde{r}_{k+1}\| \le \varepsilon$ **or** $\|r_{k+1}\| \le \varepsilon$ **then** solution obtained. Stop.
   If $n_k = 0$ **then**
   $\tilde{s}_{k+1} = (\eta_0 + A)\tilde{s}_k - \eta_0\gamma_0 A\tilde{z}_k - \gamma_0 A^2\tilde{z}_k$
   else
   compute α_0' and η_0'
   $\tilde{s}_{k+1} = (\eta_0 + 2A)\tilde{s}_k - \eta_0\gamma_0 A\tilde{z}_k - \gamma_0 A^2 \tilde{z}_k + \eta_0' \tilde{r}_k$
   end if
   $\sigma_{k+1}^{(1)} \leftarrow (y, \tilde{s}_{k+1})$
   end if
4. Execute the tests for breakdown and near-breakdown
5. **While** breakdown **or** near-breakdown **do**
6. **Repeat**
   $m_k \leftarrow m_k + 1$

$c_{2m_k-2} \leftarrow \left(y, A^{2m_k-1}\tilde{z}_k\right)$
$c_{2m_k-1} \leftarrow \left(y, A^{2m_k}\tilde{z}_k\right)$
compute $A^{m_k}z_k$
If $m_k \le n_k + 1$ **then**
  compute $A^{2m_k-3}\tilde{r}_k$, $A^{2m_k-2}\tilde{r}_k$, $A^{m_k-1}r_k$
else
  compute $A^{n_k+m_k}\tilde{s}_k$
end if
If $m_k \le n_k$ **then**
  $d_{2m_k-2} \leftarrow \left(y, A^{2m_k-2}\tilde{s}_k\right)$
  $d_{2m_k-1} \leftarrow \left(y, A^{2m_k-1}\tilde{s}_k\right)$
else
  $d_{n_k+m_k-1} \leftarrow \left(y, A^{n_k+m_k-1}\tilde{s}_k\right)$
end if
compute β_i and the coefficients γ_i of w_k, $i = 0, \ldots, m_k - 1$
compute α_i and the coefficients η_i of q_k, $i = 0, \ldots, m_k - 1$
If $m_k \le n_k$ **then**
  compute α_i' and the coefficients η_i' of t_k, $i = 0, \ldots, m_k - 1$
  compute β_i' and the coefficients γ_i' of v_k, $i = 0, \ldots, m_k - 2$
else if $n_k \ne 0$ **then**
  compute α_i' and the coefficients η_i' of t_k, $i = 0, \ldots, n_k - 1$
  compute β_i' and the coefficients γ_i' of v_k, $i = 0, \ldots, n_k - 1$
end if
until not nearly singular system.
$\tilde{r}_{k+1} = (I - A\,v_k(A))^2\tilde{r}_k - 2(I - A\,v_k(A))\,A\,w_k(A)\,\tilde{s}_k + A^2\,w_k^2(A)\tilde{z}_k$
$\tilde{x}_{k+1} = \tilde{x}_k - (A\,v_k(A) - 2I)v_k(A)\tilde{r}_k + 2(I - A\,v_k(A))w_k(A)\,\tilde{s}_k$
  $-A\,w_k^2(A)\,\tilde{z}_k$
TF $r_{k+1} = (I - A\,v_k(A))r_k - Aw_k(A)z_k$
TF $x_{k+1} = x_x + v_k(A)r_k + w_k(A)z_k$
If $\|\tilde{r}_{k+1}\| \le \varepsilon$ **or** $\|r_{k+1}\| \le \varepsilon$ **then** solution obtained. Stop.
$\tilde{s}_{k+1} = (I - A\,v_k(A))\;q_k(A)\,\tilde{s}_k - A\,t_k(A)\,w_k(A)\,\tilde{s}_k -$
  $Aq_k(A)\,w_k(A)\,\tilde{z}_k + t_k(A)(I - Av_k(A))\,\tilde{r}_k$
$\sigma_{k+1}^{(m_k)} \leftarrow (y, \tilde{s}_{k+1})$
Execute the tests for breakdown and near-breakdown
end while
7. $n_{k+1} \leftarrow n_k + m_k$
$\tilde{z}_{k+1} = q_k^2(A)\,\tilde{z}_k + 2q_k(A)\,t_k(A)\,\tilde{s}_k + t_k^2(A)\,\tilde{r}_k$
TF $z_{k+1} = q_k(A)z_k + t_k(A)r_k$
compute the coefficients of $P_{k+1}^{(1)}$
compute the coefficients of P_{k+1}
end for

The computation of the coefficients $\alpha_i, \beta_i, \alpha_i'$ and β_i' is explained in [9].

4. Numerical Experiments

The following examples have been tested on a SUN Ultra 1 workstation with the FORTRAN version of the TFLACGS algorithm. For testing the near-breakdown we used a threshold ε_1.

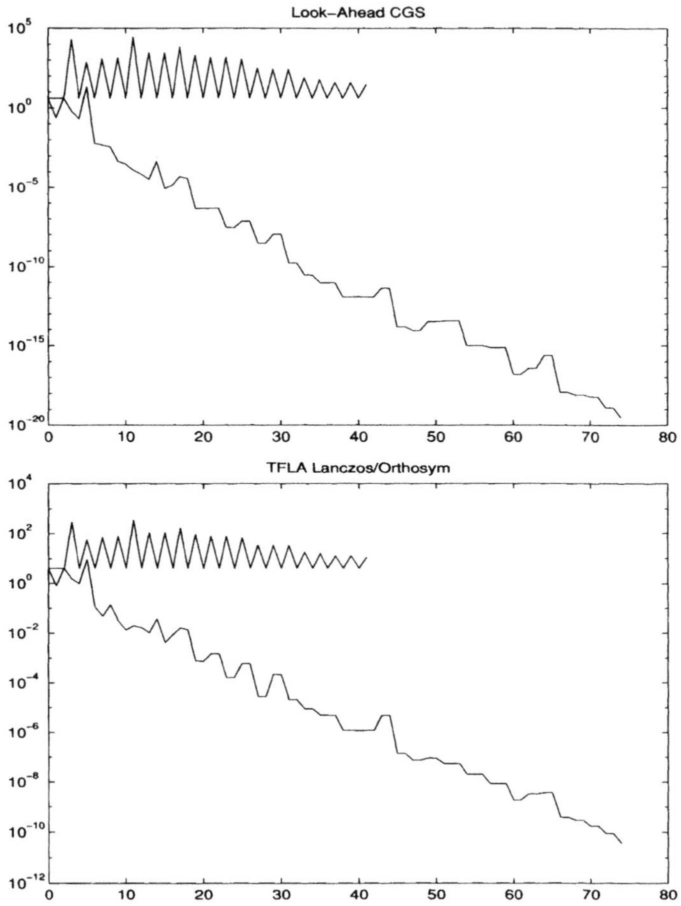

FIGURE 4. Example 1 with $n = 500$

4.1. Example 1

Consider the system

$$\begin{pmatrix} 2 & 1 & & & & & \\ 0 & 2 & 1 & & & & \\ 1 & 0 & 2 & \ddots & & & \\ & 1 & \ddots & \ddots & \ddots & & \\ & & \ddots & \ddots & \ddots & \ddots & \\ & & & 1 & 0 & 2 & 1 \\ & & & & 1 & 0 & 2 \end{pmatrix} \begin{pmatrix} 1 \\ 1 \\ 1 \\ \vdots \\ \vdots \\ 1 \\ 1 \end{pmatrix} = \begin{pmatrix} 3 \\ 3 \\ 4 \\ \vdots \\ \vdots \\ 4 \\ 3 \end{pmatrix}.$$

This system was given by Gutknecht [30] and, for $x_0 = 0$ and $y = (0, 0, 0, -1, 1, \ldots, -(-1)^{n-1}, 0)^T$, a breakdown occurs in the first iteration when n is even, since $(y, r_0) = (y, Ar_0) = 0$. When n is odd, a breakdown also occurs in the first iteration, but for a different reason.

Now replace the last component of the vector y (which was 0) by 10^{-8} in order to produce a near-breakdown.

For $n = 500$, the results can be seen in Fig. 4. For $\varepsilon_1 = 10^{-200}$, there is no jump, the norm of the residual oscillates (between 10^1 and 10^5 for the CGS, and between 10^0 and 10^3 for Lanczos/Orthosym), and an overflow occurs at $n_{41} = 41$ (top curves in Fig. 4).

When $\varepsilon_1 = 10^{-7}$, we have a jump of length 2 in the first iteration. Then, there are several additional jumps. When $n_{42} = 74$, we obtain $\|r_{\mathrm{CGS}}\| = 0.31 \cdot 10^{-19}$ (with an actual residual of $0.97 \cdot 10^{-14}$), and $\|r_{\mathrm{O/Sym}}\| = 0.37 \cdot 10^{-10}$, which coincides with the actual residual (bottom curves in Fig. 4).

For $n = 501$, the results are given in Fig. 5. For $\varepsilon_1 = 10^{-200}$, there is no jump, the norm of the residual oscillates (between 10^0 and 10^4 for the CGS, and between 10^0 and 10^2 for the Lanczos/Orthosym), and there is no convergence (top curves in Fig. 5).

When $\varepsilon_1 = 10^{-7}$, we have a jump of length 2 in the first iteration, and then several jumps. When $n_{51} = 97$, we obtain $\|r_{\mathrm{CGS}}\| = 0.92 \cdot 10^{-19}$ (with an actual residual of $0.35 \cdot 10^{-14}$), and $\|r_{\mathrm{O/Sym}}\| = 0.82 \cdot 10^{-10}$, which coincides with the actual residual (bottom curves in Fig. 5).

4.2. Example 2

Consider the system

$$\begin{pmatrix} 0 & 1 & & & & \\ & 0 & 1 & & & \\ & & \ddots & \ddots & & \\ & & & \ddots & 1 & \\ & & & & 0 & 1 \\ a & & & & & 0 \end{pmatrix} \begin{pmatrix} 1 \\ 1 \\ \vdots \\ \vdots \\ 1 \\ 1 \end{pmatrix} = \begin{pmatrix} 1 \\ 1 \\ \vdots \\ \vdots \\ 1 \\ a \end{pmatrix}.$$

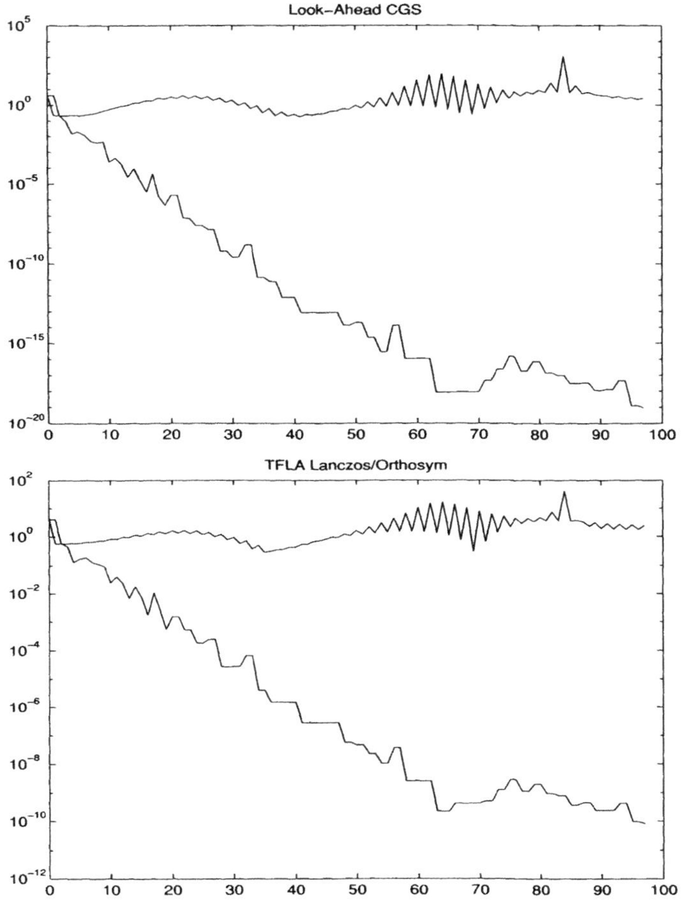

FIGURE 5. Example 1 with $n = 501$

For $n = 200, a = 0.95, x_0 = 0, y = r_0$, and $\varepsilon_1 = 10^{-200}$, we have no jump, the residuals do not converge, and we obtain, at $n_{200} = 200$, $\|r_{\mathrm{CGS}}\| = 0.12 \cdot 10^{95}$ and $\|r_{\mathrm{O/Sym}}\| = 0.86 \cdot 10^{46}$.

When $\varepsilon_1 = 10^{-8}$ we have a jump from $n_1 = 1$ to $n_2 = 198$, and we obtain, at $n_4 = 200$, $\|r_{\mathrm{CGS}}\| = 0.58 \cdot 10^{-7}$ and $\|r_{\mathrm{O/Sym}}\| = 0.13 \cdot 10^{-6}$, which both coincide with the actual residuals.

4.3. Example 3

Now consider the following system:

$$\begin{pmatrix} a & 1 & & & & & \\ -1 & a & 1 & & & & \\ & -1 & a & \ddots & & & \\ & & \ddots & \ddots & \ddots & & \\ & & & \ddots & \ddots & \ddots & \\ & & & & -1 & a & 1 \\ & & & & & -1 & a \end{pmatrix} \begin{pmatrix} 1 \\ 1 \\ 1 \\ \vdots \\ \vdots \\ 1 \\ 1 \end{pmatrix} = \begin{pmatrix} a+1 \\ a \\ a \\ \vdots \\ \vdots \\ a \\ a-1 \end{pmatrix}.$$

This system was studied by Brown [18] and, when $a = 0$, a breakdown occurs for every odd step in Lanczos' method. For small values of a, the convergence has a sawtooth behavior. For $n = 300, a = 10^{-10}, x_0 = 0, y = r_0$, and $\varepsilon_1 = 10^{-200}$, there is no jump, the norm of the residual oscillates (between 10^5 and 10^{13} for the CGS, and between 10^0 and 10^7 for the Lanczos/Orthosym), and an overflow occurs at $k = 19$.

When $\varepsilon_1 = 10^{-7}$, we have jumps of length 2 from the beginning, and we obtain $n_{150} = 300$, $\|r_{\text{CGS}}\| = 0.16 \cdot 10^{-8}$, and $\|r_{\text{O/Sym}}\| = 0.26 \cdot 10^{-10}$, which both coincide with the actual residuals.

References

[1] E.H. Ayachour, *Avoiding look-ahead in Lanczos method and Padé approximation*, Appl. Math. (Warsaw), to appear.

[2] R.E. Bank and T.F. Chan, *An analysis of the composite step biconjugate gradient method*, Numer. Math., **66** (1993), 295–319.

[3] R.E. Bank and T.F. Chan, *A composite step bi-conjugate gradient algorithm for nonsymmetric linear systems*, Numer. Algorithms, **7** (1994), 1–16.

[4] C. Brezinski, *Padé-type approximation and general orthogonal polynomials*, Internat. Ser. Numer. Math. **50**, Birkhäuser, Basel, 1980.

[5] C. Brezinski, *CGM: a whole class of Lanczos-type solvers for linear systems*, Note ANO, Laboratoire d'Analyse Numérique et d'Optimisation, Université des Sciences et Technologies de Lille, **253** (1991).

[6] C. Brezinski, *Krylov subspace methods, biorthogonal polynomials and Padé-type approximants*, Numer. Algorithms, to appear.

[7] C. Brezinski and M. Redivo Zaglia, *Hybrid procedures for solving linear systems*, Numer. Math., **67** (1994), 1–19.

[8] C. Brezinski and M. Redivo Zaglia, *Breakdowns in the computation of orthogonal polynomials*, in: A. Cuyt, Ed., *Nonlinear numerical methods and rational approximation. II*, Kluwer, Dordrecht, 1994, 49–59.

[9] C. Brezinski and M. Redivo Zaglia, *Treatment of near-breakdown in the CGS algorithm*, Numer. Algorithms, **7** (1994), 33–73.

[10] C. Brezinski and M. Redivo-Zaglia, *Look-ahead in Bi-CGSTAB and other product methods for linear systems*, BIT, **35** (1995), 169–201.

[11] C. Brezinski and M. Redivo-Zaglia, *Transpose-free Lanczos-type algorithms for nonsymmetric linear systems*, Numer. Algorithms, **17** (1998), 67–103.

[12] C. Brezinski, M. Redivo Zaglia, and H. Sadok, *Avoiding breakdown and near-breakdown in Lanczos type algorithms*, Numer. Algorithms, **1** (1991), 261–284.

[13] C. Brezinski, M. Redivo Zaglia, and H. Sadok, *A breakdown-free Lanczos type algorithm for solving linear systems*, Numer. Math., **63** (1992), 29–38.

[14] C. Brezinski, M. Redivo-Zaglia, and H. Sadok, *New look-ahead Lanczos-type algorithms for linear systems*, Numer. Math., to appear.

[15] C. Brezinski and H. Sadok, *Avoiding breakdown in the CGS algorithms*, Numer. Algorithms, **1** (1991), 199–206.

[16] C. Brezinski and H. Sadok, *Some vector sequence transformations with applications to systems of equations*, Numer. Algorithms, **3** (1992), 75–80.

[17] C. Brezinski and H. Sadok, *Lanczos-type algorithms for solving systems of linear equations*, Appl. Numer. Math., **11** (1993), 443–473.

[18] P.N. Brown, *A theoretical comparison of the Arnoldi and GMRES algorithms*, SIAM J. Sci. Statist. Comput., **12** (1991), 58–78.

[19] T.F. Chan and T. Szeto, *A composite step conjugate gradients squared algorithm for solving nonsymmetric linear systems*, Numer. Algorithms, **7** (1994), 17–32.

[20] T.F. Chan, L. de Pillis, and H. van der Vorst, *Transpose-free formulations of Lanczos-type methods for nonsymmetric linear systems*, Numer. Algorithms, **17** (1998), 51–66.

[21] A. Draux, *Polynômes orthogonaux formels. Applications*, Lecture Notes in Mathematics **974**, Springer, Berlin, 1983.

[22] R. Fletcher, *Conjugate gradient methods for indefinite systems*, in: G.A. Watson, Ed., *Numerical analysis, Dundee 1975*, Lecture Notes in Mathematics **506**, Springer, Berlin, 1976, 73–89.

[23] R.W. Freund, M.H. Gutknecht, and N.M. Nachtigal, *An implementation of the look-ahead Lanczos algorithm for non-Hermitian matrices*, SIAM J. Sci. Comput., **14** (1993), 137–158.

[24] R.W. Freund and N.M. Nachtigal, *Implementation details of the coupled QMR algorithm*, in: L. Reichel, A. Ruttan, and R.S. Varga, Eds., *Numerical linear algebra*, de Gruyter, Berlin, 1993, 123–140.

[25] R.W. Freund and N.M. Nachtigal, *An implementation of the QMR method based on coupled two-term recurrences*, SIAM J. Sci. Comput., **15** (1994), 313–337.

[26] P.R. Graves-Morris, *A "look-around Lanczos" algorithm for solving a system of linear equations*, Numer. Algorithms, **15** (1997), 247–274.

[27] P.R. Graves-Morris and A. Salam, *Avoiding breakdown in van der Vorst's method*, Numer. Algorithms, to appear.

[28] M.H. Gutknecht, *The unsymmetric Lanczos algorithms and their relations to Padé approximation, continued fractions, and the qd algorithm*, in: Preliminary Proceedings of the Copper Mountain Conference on Iterative Methods, April 1–5, 1990.

[29] M.H. Gutknecht, *A completed theory of the unsymmetric Lanczos process and related algorithms. I*, SIAM J. Matrix Anal. Appl., **13** (1992), 594–639.

[30] M.H. Gutknecht, *Variants of BiCGSTAB for matrices with complex spectrum*, SIAM J. Sci. Comput., **14** (1993), 1020–1033.

[31] M.H. Gutknecht, *A completed theory of the unsymmetric Lanczos process and related algorithms. II*, SIAM J. Matrix Anal. Appl., **15** (1994), 15–58.

[32] M.H. Gutknecht, *Lanczos-type solvers for nonsymmetric linear systems of equations*, in: A. Iserles, Ed., *Acta numerica 1997*, Acta Numer. **6**, Cambridge Univ. Press, Cambridge, 1997, 271–397.

[33] M.H. Gutknecht and K.J. Ressel, *Look-ahead procedures for Lanczos-type product methods based on three-term recurrences*, Tech. Report TR-96-19, Swiss Center for Scientific Computing, Zürich, 1996.

[34] M.H. Gutknecht and Z. Strakoš, *Accuracy of three-term and two-term recurrences for Krylov space solvers*, Tech. Report TR-97-21, Swiss Center for Scientific Computing, Zürich, 1997.

[35] N.J. Higham, *The test matrix toolbox for MATLAB (Version 3.0)*, Numerical Analysis Report, Department of Mathematics, The University of Manchester, **276**, 1995.

[36] M. Hochbruck, *The Padé table and its relation to certain numerical algorithms*, Habilitationsschrift, Universität Tübingen, 1996.

[37] C. Lanczos, *An iteration method for the solution of the eigenvalue problem of linear differential and integral operators*, J. Research Nat. Bur. Standards, **45** (1950), 255–282.

[38] C. Lanczos, *Solution of systems of linear equations by minimized-iterations*, J. Research Nat. Bur. Standards, **49** (1952), 33–53.

[39] B.N. Parlett, D.R. Taylor, and Z.A. Liu, *A look-ahead Lánczos algorithm for unsymmetric matrices*, Math. Comp., **44** (1985), 105–124.

[40] P. Sonneveld, *CGS, a fast Lanczos-type solver for nonsymmetric linear systems*, SIAM J. Sci. Statist. Comput., **10** (1989), 36–52.

[41] H.A. van der Vorst, *Bi-CGSTAB: a fast and smoothly converging variant of Bi-CG for the solution of nonsymmetric linear systems*, SIAM J. Sci. Statist. Comput., **13** (1992), 631–644.

[42] P.K.W. Vinsome, *Orthomin, an iterative method for solving sparse sets of simultaneous linear equations* in: Proceedings 4th Symposium on Reservoir Simulation, Society of Petroleum Engineers of AIME, 1976, 149–159.

[43] D.M. Young and K.C. Jea, *Generalized conjugate-gradient acceleration of nonsymmetrizable iterative methods*, Linear Algebra Appl., **34** (1980), 159–194.

Claude Brezinski
Laboratoire d'Analyse Numérique et d'Optimisation
Université des Sciences et Technologies de Lille
F-59655 Villeneuve d'Ascq Cedex, France
E-mail address: `Claude.Brezinski@univ-lille1.fr`

Michela Redivo-Zaglia
Dipartimento di Elettronica e Informatica
Università degli Studi di Padova
Via Gradenigo 6/a
I-35131 Padova, Italy
E-mail address: `michela@dei.unipd.it`

International Series of Numerical Mathematics
Vol. 131, © 1999 Birkhäuser Verlag Basel/Switzerland

Applications of Anti-Gauss Quadrature Rules in Linear Algebra

Daniela Calvetti, Lothar Reichel, and Fiorella Sgallari

Abstract. The need to inexpensively determine upper and lower bounds for matrix functionals of the form $F(A) := u^T f(A)u$ arises in a large number of applications. Here A denotes a large symmetric matrix and u is a vector. Golub and collaborators have described how such bounds can be computed by using Gauss and Gauss-Radau quadrature rules when the derivatives $\frac{d^j}{dt^j} f(t)$, $j = 1, 2, \ldots$, are of constant sign in an interval that contains the spectrum of A. However, many matrix functionals of interest in applications are defined by functions f whose derivatives do not have constant sign on the spectrum of A. We describe a new method for inexpensively computing candidates for upper and lower bounds of $F(A)$ based on the application of pairs of Gauss and anti-Gauss quadrature rules. This method does not require the sign of the derivates of f to be constant on an interval that contains the spectrum of A. Anti-Gauss rules are modifications of Gauss rules recently introduced by Laurie. We also discuss applications to matrix functionals with nonsymmetric matrices.

1. Introduction

In the last few years there has been considerable interest in the inexpensive computation of upper and lower bounds for matrix functionals of the form

$$F(A) := u^T f(A)u, \qquad A \in \mathbb{R}^{n\times n}, \qquad u \in \mathbb{R}^n, \tag{1}$$

where A is a large, possibly sparse, symmetric matrix, and u is a given vector. For notational simplicity, we will assume that $\|u\| = 1$. Throughout this paper, $\|\cdot\|$ denotes the Euclidean vector norm or the associated induced matrix norm. Functions f discussed in the literature include $f(t) := 1/t$, $f(t) := 1/(\nu + t^2)$ and $f(t) := \exp(\alpha t)$, where ν and α are given constants.

In a sequence of papers Golub and collaborators, see, e.g., [2, 1, 3, 4, 5, 7], have exploited the connection between matrix functionals of the form (1), Stieltjes integrals, Gauss-type quadrature rules, and the Lanczos process to derive powerful inexpensive algorithms for the computation of upper and lower bounds for $F(A)$. For definiteness, introduce the spectral factorization

$$A = S\Lambda S^{-1}, \qquad \Lambda = \mathrm{diag}[\lambda_1, \lambda_2, \ldots , \lambda_n], \tag{2}$$

with $\lambda_1 \leq \lambda_2 \leq \cdots \leq \lambda_n$ and $\|S\| = 1$. We let $\lambda(A)$ denote the spectrum of A. Define

$$f(A) := Sf(\Lambda)S^{-1} \tag{3}$$

and introduce the vector $[\mu_1, \mu_2, \ldots, \mu_n] := u^T S$. It follows from the symmetry of A that $S^{-1} = S^T$, and therefore

$$F(A) = u^T Sf(\Lambda)S^T u = \sum_{j=1}^{n} f(\lambda_j)\mu_j^2. \tag{4}$$

We write the right-hand side of (4) as a Stieltjes integral

$$\mathcal{I}f := \int_{-\infty}^{\infty} f(t)d\mu(t). \tag{5}$$

The nonnegative measure $d\mu(t)$ is such that $\mu(t)$ is a nondecreasing step function defined on $\mathbb{R}$ with jumps of height μ_j^2 at $t = \lambda_j$. It follows from $\|u\| = 1$ that the measure $d\mu(t)$ is of total mass one. The m-point Gauss quadrature rule associated with the measure $d\mu(t)$ is of the form

$$\mathcal{G}_m f := \sum_{j=1}^{m} f(\theta_j)\gamma_j^2 \tag{6}$$

and is characterized by the fact that

$$\mathcal{I}f = \mathcal{G}_m f, \qquad \forall f \in \mathbb{P}_{2m-1}, \tag{7}$$

where $\mathbb{P}_{2m-1}$ denotes the set of polynomials of degree at most $2m-1$. It is assumed that the matrix A has at least m distinct eigenvalues. The nodes θ_j of the quadrature rule are known to be the zeros of an mth-degree orthogonal polynomial with respect to the inner product

$$(f, g) := \mathcal{I}(fg). \tag{8}$$

Assume that the nodes are ordered so that

$$\theta_1 < \theta_2 < \cdots < \theta_m.$$

If f is a $2m$ times continuously differentiable function, then the error in the quadrature rule can be expressed as

$$\mathcal{E}_m f := (\mathcal{I} - \mathcal{G}_m)f = \frac{f^{(2m)}(\lambda_{\mathcal{G}})}{(2m)!} \cdot \int_{-\infty}^{\infty} \prod_{j=1}^{m} (t - \theta_j)^2 d\mu(t) \tag{9}$$

for some $\lambda_{\mathcal{G}} \in [\lambda_1, \lambda_n]$. Note that if the derivative $f^{(2m)}(t)$ is of known constant sign in the interval $[\lambda_1, \lambda_n]$, then the sign of the error in $\mathcal{I}f$ is known. For instance, if $f^{(2m)}(t) > 0$ for $\lambda_1 \leq t \leq \lambda_n$, then $\mathcal{E}_m f > 0$ and therefore $\mathcal{G}_m f < \mathcal{I}f$. If in addition $f^{(2m+1)}(t) < 0$ for $\lambda_1 \leq t \leq \lambda_n$, then an upper bound for $\mathcal{I}f$ can be determined analogously by use of an $(m+1)$-point Gauss-Radau quadrature rule with a fixed node at $t = \lambda_1$; see [3]. However, if $f^{(2m)}(t)$ or $f^{(2m+1)}(t)$ change sign in $[\lambda_1, \lambda_m]$, then the evaluation of pairs of Gauss and Gauss-Radau rules might not yield lower and upper bounds for $\mathcal{I}f$.

Recently, Laurie [8] introduced quadrature rules referred to as anti-Gauss rules. Let

$$\tilde{\mathcal{G}}_{m+1}f = \sum_{j=1}^{m+1} f(\tilde{\theta}_j)\tilde{\gamma}_j^2, \tag{10}$$

be the $(m+1)$-point anti-Gauss rule associated with the measure $d\mu(t)$. This rule is determined by the requirement that

$$(\mathcal{I} - \tilde{\mathcal{G}}_{m+1})f = -\mathcal{E}_m f, \qquad \forall f \in \mathbb{P}_{2m+1}, \tag{11}$$

where $\mathcal{E}_m$ is the quadrature error (9) obtained with the Gauss rule $\mathcal{G}_m$.

We propose to evaluate pairs of Gauss and anti-Gauss rules to determine candidates for upper and lower bounds of $F(A)$. These do not require that the derivatives $f^{(2m)}(t)$ and $f^{(2m+1)}(t)$ be of constant sign on the interval $[\lambda_1, \lambda_n]$. On the other hand, we cannot guarantee that the candidates are in fact upper and lower bounds. Details are provided in §2. We conclude this section with a discussion on how to evaluate the Gauss rule (6) and introduce notation that will be used later in the paper.

Gauss quadrature rules with respect to the measure $d\mu(t)$ can conveniently be determined by the Lanczos process applied to the matrix A with initial vector $v_1 := u$, where u is of unit norm; see, e.g., [6, Chapter 9] or [9, Chapter 6] for a detailed discussion of the Lanczos process for symmetric matrices. Application of m steps of the Lanczos process yields the decomposition

$$AV_m = V_m T_m + \beta_m v_{m+1} e_m^T, \tag{12}$$

where $V_m = [v_1, v_2, \ldots, v_m] \in \mathbb{R}^{n\times m}$ and $v_{m+1} \in \mathbb{R}^n$ satisfy $V_m^T V_m = I_m$, $v_{m+1}^T v_{m+1} = 1$, $V_m^T v_{m+1} = 0$ and $\beta_m \in \mathbb{R}$. Moreover,

$$T_m := \begin{bmatrix} \alpha_1 & \beta_1 & & & & 0 \\ \beta_1 & \alpha_2 & \beta_2 & & & \\ & \beta_2 & \alpha_3 & & & \\ & & & \ddots & & \\ & & \ddots & \ddots & \beta_{m-2} & \\ & & & \beta_{m-2} & \alpha_{m-1} & \beta_{m-1} \\ 0 & & & & \beta_{m-1} & \alpha_m \end{bmatrix} \in \mathbb{R}^{m\times m}$$

is a symmetric tridiagonal matrix. Throughout this paper, I_m denotes the $m \times m$ identity matrix, and e_j the jth axis vector. The relation (12) between the columns v_j of V_m shows that

$$v_j = p_{j-1}(A)v_1, \qquad 1 \le j \le m+1, \tag{13}$$

for certain polynomials p_{j-1} of degree $j-1$. It follows from the orthonormality of the vectors v_j that

$$\begin{aligned}(p_{j-1},p_{k-1}) &= \int_{-\infty}^{\infty} p_{j-1}(t)p_{k-1}(t)d\mu(t) = u^T S p_{j-1}(\Lambda)p_{k-1}(\Lambda)S^T u \\ &= u^T p_{j-1}(A)p_{k-1}(A)u = v_1^T p_{j-1}(A)p_{k-1}(A)v_1 \\ &= v_j^T v_k = \begin{cases} 0, & j \neq k, \\ 1, & j = k. \end{cases}\end{aligned} \tag{14}$$

Thus, the polynomials p_j are orthogonal with respect to the inner product (8). The relation (12) for the columns v_j of V_m yields, in view of (13), the recurrence relation

$$\begin{aligned}&\beta_1 p_1(t) = (t-\alpha_1)p_0(t), \qquad p_0(t) = 1, \\ &\beta_j p_j(t) = (t-\alpha_j)p_{j-1}(t) - \beta_{j-1}p_{j-2}(t), \qquad 2 \leq j \leq m.\end{aligned} \tag{15}$$

It follows from (15) and (14) that the recurrence coefficients can be expressed as

$$\begin{aligned}\alpha_j &= (p_{j-1}, tp_{j-1}), \\ \beta_j^2 &= ((t-\alpha_j)p_{j-1} - \beta_{j-1}p_{j-2}, (t-\alpha_j)p_{j-1} - \beta_{j-1}p_{j-2}).\end{aligned} \tag{16}$$

The recurrence relation (15) can be written in the form

$$\begin{aligned}[p_0(t), p_1(t), \ldots, p_{m-1}(t)]T_m = t[&p_0(t), p_1(t), \ldots, p_{m-1}(t)] \\ &-\beta_m[0, \ldots, 0, p_m(t)],\end{aligned}$$

which shows that the zeros of p_m are the eigenvalues of T_m. Introduce the spectral decomposition

$$T_m = Q_m D_m Q_m^T,$$

where $D_m = \mathrm{diag}[\theta_1, \theta_2, \ldots, \theta_m]$ and $Q_m^T Q_m = I_m$. It is well known that the weights of the Gauss rule (6) are given by $\gamma_j^2 = (e_1^T Q_m e_j)^2$, $1 \leq j \leq m$. It follows that the Gauss rule (6) can be written in the form

$$\mathcal{G}_m f = e_1^T Q_m f(D_m) Q_m^T e_1 = e_1^T f(T_m) e_1. \tag{17}$$

Thus, the value $\mathcal{G}_m f$ can be determined by first computing the Lanczos decomposition (12) and then evaluating one of the expressions in (17).

This paper is organized as follows. Section 2 reviews properties of anti-Gauss quadrature rules and discusses their application to the determination of candidates for bounds of the functional $F(A)$ given by (1) when A is a symmetric matrix. A generalization that allows A to be nonsymmetric is described in §3, and a few numerical examples are presented in §4.

2. Anti-Gauss Rules and Symmetric Matrix Functionals

This section discusses the application of anti-Gauss rules to the computation of candidates for upper and lower bounds for symmetric matrix functionals, i.e., functionals of the form (1) with a symmetric matrix A. Anti-Gauss rules were introduced by Laurie in [8] as a tool to estimate the error in Gauss quadrature formulas. Compared with Gauss-Kronrod quadrature rules, which are used for the same purpose, anti-Gauss rules have the advantage that they always exist for nonnegative measures.

This section reviews some of the properties of anti-Gauss rules for nonnegative measures with support on the real axis shown by Laurie [8]. In the next section, we consider more general measures. It follows from (9) and (11) that

$$\tilde{\mathcal{G}}_{m+1}f = (2\mathcal{I} - \mathcal{G}_m)f, \qquad \forall f \in \mathbb{P}_{2m+1}, \tag{18}$$

i.e., $\tilde{\mathcal{G}}_{m+1}$ is the $(m+1)$-point Gauss rule for the functional

$$\mathcal{J}f := (2\mathcal{I} - \mathcal{G}_m)f.$$

Introduce the inner product

$$< f, g > := \mathcal{J}(fg),$$

and let $\{\tilde{p}_j\}_{j=0}^{m+1}$ be the first $m+2$ orthogonal polynomials with respect to $< \cdot, \cdot >$ normalized so that $< \tilde{p}_j, \tilde{p}_j > = 1$ for all j. These polynomials satisfy the recurrence relation

$$\begin{aligned} &\tilde{\beta}_1\tilde{p}_1(t) = (t - \tilde{\alpha}_1)\tilde{p}_0(t), \qquad \tilde{p}_0(t) = 1, \\ &\tilde{\beta}_j\tilde{p}_j(t) = (t - \tilde{\alpha}_j)\tilde{p}_{j-1}(t) - \tilde{\beta}_{j-1}\tilde{p}_{j-2}(t), \qquad 2 \le j \le m+1, \end{aligned} \tag{19}$$

where

$$\begin{aligned} \tilde{\alpha}_j &= < \tilde{p}_{j-1}, t\tilde{p}_{j-1} >, \\ \tilde{\beta}_j^2 &= < (t - \tilde{\alpha}_j)\tilde{p}_{j-1} - \tilde{\beta}_{j-1}\tilde{p}_{j-2}, (t - \tilde{\alpha}_j)\tilde{p}_{j-1} - \tilde{\beta}_{j-1}\tilde{p}_{j-2} > . \end{aligned} \tag{20}$$

It follows from (18) and (7) that

$$< f, g > = (f, g) = \mathcal{I}(fg), \qquad f \in \mathbb{P}_{m-1}, \quad g \in \mathbb{P}_m,$$

and this equality applied to (16) and (20) yields

$$\begin{aligned} \tilde{\alpha}_j &= \alpha_j, \qquad && 1 \le j \le m, \\ \tilde{\beta}_j^2 &= \beta_j^2, \qquad && 1 \le j < m. \end{aligned} \tag{21}$$

We may choose $\tilde{\beta}_j = \beta_j$ for $1 \le j < m$, and then $\tilde{p}_j = p_j$, $0 \le j < m$. Consider the polynomial

$$\hat{p}_m(t) := (t - \tilde{\alpha}_m)\tilde{p}_{m-1}(t) - \tilde{\beta}_{m-1}\tilde{p}_{m-2}(t), \tag{22}$$

which also can be written as

$$\hat{p}_m(t) = (t - \alpha_m)p_{m-1}(t) - \beta_{m-1}p_{m-2}(t). \tag{23}$$

Since this polynomial is a multiple of p_m, we have $\mathcal{G}_m \hat{p}_m^2 = 0$. It follows from (16), (20), (22) and (23) that

$$\tilde{\beta}_m^2 = < \hat{p}_m, \hat{p}_m > = 2\mathcal{I}(\hat{p}_m^2) - \mathcal{G}_m(\hat{p}_m^2) = 2\mathcal{I}(\hat{p}_m^2) = 2(\hat{p}_m, \hat{p}_m) = 2\beta_m^2.$$

We may choose $\tilde{\beta}_m = \sqrt{2}\beta_m$. Since, furthermore, $\tilde{p}_m(t) = p_m(t)/\sqrt{2}$, we have $\tilde{\alpha}_{m+1} = < \tilde{p}_m, t\tilde{p}_m > = \frac{1}{2} < p_m, tp_m > = \frac{1}{2} \cdot 2\mathcal{I}(tp_m^2) = \alpha_{m+1}$. It follows that the symmetric tridiagonal matrix associated with the quadrature rule $\tilde{\mathcal{G}}_{m+1}$ is given by

$$\tilde{T}_{m+1} = \begin{bmatrix} \alpha_1 & \beta_1 & & & & & 0 \\ \beta_1 & \alpha_2 & \beta_2 & & & & \\ & \beta_2 & \alpha_3 & & & & \\ & & & \ddots & & & \\ & & \ddots & \ddots & \beta_{m-1} & & \\ & & & \beta_{m-1} & \alpha_m & \sqrt{2}\beta_m & \\ 0 & & & & \sqrt{2}\beta_m & \alpha_{m+1} \end{bmatrix} \in \mathbb{R}^{(m+1)\times(m+1)}.$$

The entries of $\tilde{T}_{m+1}$ can be computed by $m+1$ steps of the Lanczos process, cf. (12). Similarly to the formula (17), the anti-Gauss quadrature rule (10) can be evaluated according to

$$\tilde{\mathcal{G}}_{m+1} f = e_1^T f(\tilde{T}_{m+1}) e_1. \tag{24}$$

We are now in a position to discuss how to determine an approximation of the symmetric matrix functional $F(A)$ as well as candidates for upper and lower bounds. Assume for the moment that we can carry out n steps of the Lanczos process without breakdown. This yields an orthonormal basis $\{v_j\}_{j=1}^n$ of $\mathbb{R}^n$ and an associated sequence of polynomials $\{p_j\}_{j=0}^{n-1}$ defined by (13) that satisfy (14). Since the vectors $[p_0(t), p_1(t), \ldots, p_{n-1}(t)]$ for $t \in \lambda(A)$ form an (orthogonal) basis in $\mathbb{R}^n$, we have

$$f(t) = \sum_{j=0}^{n-1} \eta_j p_j(t), \qquad t \in \lambda(A). \tag{25}$$

In view of $\mathcal{I}p_0 = 1$ and $\mathcal{I}p_j = 0$ for $j > 0$, application of the integral operator $\mathcal{I}$ to (25) yields $\mathcal{I}f = \eta_0$. Therefore, applying the Gauss rule $\mathcal{G}_m$ and anti-Gauss rule

$\tilde{\mathcal{G}}_{m+1}$ to (25), using property (7), yields for $n \geq 2m+2$ that

$$\mathcal{G}_m f = \mathcal{I}f + \sum_{j=2m}^{n-1} \eta_j \mathcal{G}_m p_j, \tag{26}$$

$$\begin{aligned} \tilde{\mathcal{G}}_{m+1} f &= \sum_{j=0}^{n-1} \eta_j \tilde{\mathcal{G}}_{m+1} p_j = \sum_{j=0}^{2m+1} \eta_j (2\mathcal{I} - \mathcal{G}_m) p_j + \sum_{j=2m+2}^{n-1} \eta_j \tilde{\mathcal{G}}_{m+1} p_j \\ &= \sum_{j=0}^{2m+1} \eta_j 2\mathcal{I} p_j - \sum_{j=0}^{2m+1} \eta_j \mathcal{G}_m p_j + \sum_{j=2m+2}^{n-1} \eta_j \tilde{\mathcal{G}}_{m+1} p_j \\ &= \mathcal{I}f - \eta_{2m} \mathcal{G}_m p_{2m} - \eta_{2m+1} \mathcal{G}_m p_{2m+1} + \sum_{j=2m+2}^{n-1} \eta_j \tilde{\mathcal{G}}_{m+1} p_j, \end{aligned} \tag{27}$$

where in the last step we used $\mathcal{I}p_0 = \mathcal{G}_m p_0 = 1$, $\mathcal{I}p_j = 0$, $j > 0$, and $\mathcal{G}_m p_j = 0$, $0 < j < 2m$.

Assume that the coefficients η_j converge rapidly to zero with increasing index. Then the leading terms in the expansions (26) and (27) dominate the error, i.e.,

$$\begin{aligned} \mathcal{E}_m f &:= (\mathcal{I} - \mathcal{G}_m) f \approx -\eta_{2m} \mathcal{G}_m p_{2m} - \eta_{2m+1} \mathcal{G}_m p_{2m+1}, \\ \tilde{\mathcal{E}}_{m+1} f &:= (\mathcal{I} - \tilde{\mathcal{G}}_{m+1}) f \approx \eta_{2m} \mathcal{G}_m p_{2m} + \eta_{2m+1} \mathcal{G}_m p_{2m+1}, \end{aligned} \tag{28}$$

where $\approx$ stands for "approximately equal to". Thus, the errors $\mathcal{E}_m f$ and $\tilde{\mathcal{E}}_{m+1} f$ are of opposite sign and of roughly the same magnitude. Numerous numerical experiments indicate that for many integrands f, symmetric matrices A and vectors u, the quadrature rules $\mathcal{G}_m f$ and $\tilde{\mathcal{G}}_{m+1} f$ provide accurate candidates for upper and lower bounds of $\mathcal{I}f$, and therefore of $F(A)$. We illustrate their accuracy in computed examples in §4. Note that an estimate of the error in $\mathcal{G}_m f$ can be computed as

$$\frac{1}{2}(\mathcal{E}_m f - \tilde{\mathcal{E}}_{m+1} f) \approx -\eta_{2m} \mathcal{G}_m p_{2m} - \eta_{2m+1} \mathcal{G}_m p_{2m+1}.$$

The formulas (28) suggest that the quadrature rule

$$\mathcal{L}_{2m+1} f := \frac{1}{2}(\mathcal{G}_m + \tilde{\mathcal{G}}_{m+1}) f \tag{29}$$

yields a good approximation of $\mathcal{I}f$. Indeed, it follows from (26) and (27) that

$$\mathcal{L}_{2m+1} f = \mathcal{I}f + \sum_{j=2m+2}^{n-1} \eta_j \mathcal{L}_{2m+1} p_j, \tag{30}$$

which shows that the quadrature rule (29) is exact for polynomials of degree $2m+1$. Laurie [8] refers to $\mathcal{L}_{2m+1}$ as the average integration rule.

3. Anti-Gauss Rules and Nonsymmetric Matrix Functionals

We extend the discussion of the previous sections to nonsymmetric matrix functionals of the form

$$F(A) := u^T f(A)u', \qquad A \in \mathbb{R}^{n\times n}, \qquad u, u' \in \mathbb{R}^n, \tag{31}$$

where A is a large, possibly nonsymmetric matrix, and u and u' are given vectors scaled so that $u^T u' = 1$. Assume that A is diagonalizable, i.e., A can be factored according to (2), and let $f(A)$ be defined according to (3). In the present section, generally, $S^{-1} \neq S^T$. Introduce the vectors $[\mu_1, \mu_2, \dots, \mu_n] := u^T S$ and $[\mu'_1, \mu'_2, \dots, \mu'_n]^T := S^{-1}u'$. Then

$$F(A) = u^T S f(\Lambda) S^{-1} u' = \sum_{j=1}^{n} f(\lambda_j)\mu_j \mu'_j. \tag{32}$$

We can represent the right-hand side of (32) by a Stieltjes integral in terms of a measure $d\omega(z)$ with support in the complex plane

$$\mathcal{I}f := \int f(z)d\omega(z). \tag{33}$$

Note that the measure $d\omega(z)$ may be indefinite. It follows from $u^T u' = 1$ that $\int d\omega(z) = 1$.

The m-point Gauss quadrature rule associated with the measure $d\omega(z)$, if it exists, is of the form

$$\mathcal{G}_m f := \sum_{j=1}^{m} f(\zeta_j)\gamma_j \gamma'_j \tag{34}$$

and is characterized by

$$\mathcal{I}f = \mathcal{G}_m f, \qquad \forall f \in \mathbb{P}_{2m-1}. \tag{35}$$

The nodes ζ_j of the quadrature rule (34) are known to be the zeros of an mth-degree orthogonal polynomial with respect to the bilinear form

$$[f, g] := \mathcal{I}(fg). \tag{36}$$

We also define the $(m+1)$-point anti-Gauss rule

$$\tilde{\mathcal{G}}_{m+1} f := \sum_{j=1}^{m+1} f(\tilde{\zeta}_j)\tilde{\gamma}_j \tilde{\gamma}'_j \tag{37}$$

associated with $d\omega(z)$. It is determined by the requirement that

$$(\mathcal{I} - \tilde{\mathcal{G}}_{m+1})f = -(\mathcal{I} - \mathcal{G}_m)f, \qquad \forall f \in \mathbb{P}_{2m+1}. \tag{38}$$

Gauss and anti-Gauss quadrature rules with respect to the measure $d\omega(z)$ can be computed conveniently by the nonsymmetric Lanczos process applied to the matrix A with initial vectors $v_1 := u'$ and $w_1 := u$; see, e.g., [6, Chapter 9] or [9, Chapter 7] for details on the nonsymmetric Lanczos process. It is well known that the nonsymmetric Lanczos process can break down, however; in the present section we assume that m is small enough for this not to take place within $m+1$ steps of the Lanczos process.

Application of m steps of the nonsymmetric Lanczos process yields the decompositions

$$\begin{aligned} AV_m &= V_m T_m + \beta_m v_{m+1} e_m^T, \\ A^T W_m &= W_m T_m^T + \beta'_m w_{m+1} e_m^T, \end{aligned} \tag{39}$$

where $V_m = [v_1, v_2, \dots, v_m] \in \mathbb{R}^{n\times m}$, $W_m = [w_1, w_2, \dots, w_m] \in \mathbb{R}^{n\times m}$, and $v_{m+1}, w_{n+1} \in \mathbb{R}^n$ satisfy $V_m^T W_m = I_m$, $v_{m+1}^T w_{m+1} = 1$, $V_m^T w_{m+1} = 0$, $W_m^T v_{m+1} = 0$, and $\beta_m, \beta'_m \in \mathbb{R}$. Moreover,

$$T_m := \begin{bmatrix} \alpha_1 & \beta'_1 & & & & & 0 \\ \beta_1 & \alpha_2 & \beta'_2 & & & & \\ & \beta_2 & \alpha_3 & & & & \\ & & & \ddots & & & \\ & & \ddots & \ddots & \beta'_{m-2} & & \\ & & & \beta_{m-2} & \alpha_{m-1} & \beta'_{m-1} \\ 0 & & & & \beta_{m-1} & \alpha_m \end{bmatrix} \in \mathbb{R}^{m\times m} \tag{40}$$

is a nonsymmetric tridiagonal matrix. The relations (39) between the columns v_j of V_m and w_j of W_m show that

$$\begin{aligned} v_j &= p_{j-1}(A)v_1, \\ w_j &= q_{j-1}(A^T)w_1, \end{aligned} \qquad 1 \le j \le m+1, \tag{41}$$

for certain polynomials p_{j-1} and q_{j-1} of degree $j-1$. It follows from the biorthogonality of the vectors v_j and w_k that

$$\begin{aligned} [q_{k-1}, p_{j-1}] &= \int q_{k-1}(z)p_{j-1}(z)d\omega(z) = u^T S q_{k-1}(\Lambda)p_{j-1}(\Lambda)S^{-1}u' \\ &= u^T q_{k-1}(A)p_{j-1}(A)u' = w_1^T q_{k-1}(A)p_{j-1}(A)v_1 \\ &= w_k^T v_j = \begin{cases} 0, & j \ne k, \\ 1, & j = k. \end{cases} \end{aligned} \tag{42}$$

Thus, the polynomials q_k and p_j are biorthogonal with respect to the bilinear form (36). The relations between the columns v_j of V_m and w_k of W_m expressed by (39) yield, in view of (41), the recurrence relations

$$\begin{aligned} \beta_1 p_1(z) &= (z-\alpha_1)p_0(z), \qquad p_0(z) = 1, \\ \beta_j p_j(z) &= (z-\alpha_j)p_{j-1}(z) - \beta'_{j-1}p_{j-2}(z), \qquad 2 \le j \le m, \end{aligned} \tag{43}$$

and

$$\begin{aligned}\beta_1' q_1(z) &= (z-\alpha_1)q_0(z), \qquad q_0(z)=1,\\ \beta_j' q_j(z) &= (z-\alpha_j)q_{j-1}(z) - \beta_{j-1}q_{j-2}(z), \qquad 2\le j\le m.\end{aligned} \tag{44}$$

The recurrence formulas (43) and (44), and the biorthogonality (42), allow us to express the recurrence coefficients in the following manner:

$$\begin{aligned}\alpha_j &= [q_{j-1}, zp_{j-1}],\\ \beta_j\beta_j' &= [(z-\alpha_j)p_{j-1} - \beta_{j-1}'p_{j-2}, (z-\alpha_j)q_{j-1} - \beta_{j-1}q_{j-2}].\end{aligned} \tag{45}$$

The recurrence relations (43) and (44) can be written as

$$\begin{aligned}[p_0(z), p_1(z), \ldots, p_{m-1}(z)]\, T_m &= z\,[p_0(z), p_1(z), \ldots, p_{m-1}(z)]\\ &\quad -\beta_m\,[0,\ldots,0,p_m(z)],\\ [q_0(z), q_1(z), \ldots, q_{m-1}(z)]\, T_m^T &= z\,[q_0(z), q_1(z), \ldots, q_{m-1}(z)]\\ &\quad -\beta_m'\,[0,\ldots,0,q_m(z)],\end{aligned} \tag{46}$$

which shows that the eigenvalues of T_m (and of T_m^T) are zeros of p_m (and of q_m). We will assume that T_m has m distinct eigenvalues $\zeta_1, \zeta_2, \ldots, \zeta_m$. Define the matrices

$$P_m = \begin{bmatrix} p_0(\zeta_1) & p_1(\zeta_1) & \cdots & p_{m-1}(\zeta_1)\\ p_0(\zeta_2) & p_1(\zeta_2) & \cdots & p_{m-1}(\zeta_2)\\ \vdots & \vdots & & \vdots\\ p_0(\zeta_m) & p_1(\zeta_m) & \cdots & p_{m-1}(\zeta_m)\end{bmatrix},$$

$$Q_m = \begin{bmatrix} q_0(\zeta_1) & q_1(\zeta_1) & \cdots & q_{m-1}(\zeta_1)\\ q_0(\zeta_2) & q_1(\zeta_2) & \cdots & q_{m-1}(\zeta_2)\\ \vdots & \vdots & & \vdots\\ q_0(\zeta_m) & q_1(\zeta_m) & \cdots & q_{m-1}(\zeta_m)\end{bmatrix},$$

and $D_m = \mathrm{diag}[\zeta_1, \zeta_2, \ldots, \zeta_m]$. It follows from (46) that

$$T_m P_m^{-1} = P_m^{-1} D_m, \qquad T_m Q_m^T = Q_m^T D_m.$$

Thus, both P_m^{-1} and Q_m^T are eigenvector matrices of T_m. Therefore, there is a diagonal matrix $\Gamma_m = \mathrm{diag}[\hat\gamma_1, \hat\gamma_2, \ldots, \hat\gamma_m]$ such that

$$P_m^{-1} = Q_m^T \Gamma_m. \tag{47}$$

Since all entries of the first column of Q_m are unity, the entries $\hat\gamma_j$ of Γ_m are the entries of the first row of P_m^{-1}. These entries can be determined by writing (47) in the form $\Gamma_m^{-1} = P_m Q_m^T$, i.e.,

$$\hat\gamma_k = \left(\sum_{j=0}^{m-1} p_j(\zeta_k) q_j(\zeta_k)\right)^{-1}, \qquad 1\le k\le m. \tag{48}$$

Moreover, writing (47) as $Q_m^T \Gamma_m P_m = I_m$ shows that, for $0 \le j, k < m$,

$$\sum_{\ell=1}^{m} q_j(\zeta_\ell) p_k(\zeta_\ell) \hat{\gamma}_\ell = \begin{cases} 1, & j = k, \\ 0, & j \neq k. \end{cases} \tag{49}$$

Proposition 3.1. *The zeros of p_m are the nodes, and the $\hat{\gamma}_k$ defined by (48) are the weights, of the Gauss quadrature rule (34). Let*

$$T_m = Z_m D_m Z_m^{-1} \tag{50}$$

be a spectral factorization of the matrix (40) and define the vectors $[\gamma_1, \gamma_2, \dots, \gamma_m] := e_1^T Z_m$ and $[\gamma_1', \gamma_2', \dots, \gamma_m']^T := Z_m^{-1} e_1$. Then the Gauss weights are given by $\hat{\gamma}_k = \gamma_k \gamma_k'$ and the Gauss rule (34) can be expressed as

$$\mathcal{G}_m f = e_1^T Z_m f(D_m) Z_m^{-1} e_1 = e_1^T f(T_m) e_1. \tag{51}$$

Proof. We show that the quadrature rule (34) satisfies (35) and therefore is a Gauss rule. Let $f \in \mathbb{P}_{2m-1}$. Then for certain polynomials

$$\begin{aligned} r(z) &:= \textstyle\sum_{j=0}^{m-1} \delta_j q_j(z), \\ s(z) &:= \textstyle\sum_{j=0}^{m-1} \delta_j' q_j(z), \end{aligned} \tag{52}$$

we can represent f as

$$f(z) = p_m(z) s(z) + r(z). \tag{53}$$

Integration of (53), using (36), (42), (52) and the fact that $q_0(z) = 1$, yields

$$\mathcal{I} f = [p_m, s] + [r, 1] = [r, 1] = \delta_0.$$

Moreover, since the nodes ζ_j of $\mathcal{G}_m$ are the zeros of p_m, we have

$$\mathcal{G}_m f = \mathcal{G}_m(p_m s) + \mathcal{G}_m r = \mathcal{G}_m r,$$

and by (49) and the fact that $p_0(z) = 1$, we obtain

$$\mathcal{G}_m r = \sum_{j=0}^{m-1} \delta_j \left(\sum_{k=1}^{m} q_j(\zeta_k) p_0(\zeta_k) \hat{\gamma}_k \right) = \delta_0.$$

This shows (35).

We now establish the formulas (51). Spectral factorizations (50) may differ in the scaling of the columns of the eigenvector matrix Z_m. However, the products $\gamma_j \gamma_j'$, $1 \le j \le m$, are invariant under this scaling. This shows (51). □

We turn to anti-Gauss rules (37) associated with the measure $d\omega(z)$. It follows from (38) that

$$\tilde{\mathcal{G}}_{m+1} f = (2\mathcal{I} - \mathcal{G}_m) f, \qquad \forall f \in \mathbb{P}_{2m+1}, \tag{54}$$

i.e., $\tilde{\mathcal{G}}_{m+1}$ is the $(m+1)$-point Gauss rule for the functional

$$\mathcal{J} f := (2\mathcal{I} - \mathcal{G}_m) f.$$

Introduce the bilinear form

$$\{f,g\} := \mathcal{J}(fg),$$

and let $\{\tilde{p}_j\}_{j=0}^{m+1}$ and $\{\tilde{q}_j\}_{j=0}^{m+1}$ be families of biorthogonal polynomials with respect to $\{\cdot,\cdot\}$ normalized so that $\{\tilde{q}_j,\tilde{p}_j\} = 1$ for all j. These polynomials satisfy the recurrence relations

$$\begin{aligned} \tilde{\beta}_1\tilde{p}_1(z) &= (z-\tilde{\alpha}_1)\tilde{p}_0(z), \qquad \tilde{p}_0(z) = 1, \\ \tilde{\beta}_j\tilde{p}_j(z) &= (z-\tilde{\alpha}_j)\tilde{p}_{j-1}(z) - \tilde{\beta}'_{j-1}\tilde{p}_{j-2}(z), \qquad 2 \le j \le m+1, \end{aligned} \tag{55}$$

and

$$\begin{aligned} \tilde{\beta}'_1\tilde{q}_1(z) &= (z-\tilde{\alpha}_1)\tilde{q}_0(z), \qquad \tilde{q}_0(z) = 1, \\ \tilde{\beta}'_j\tilde{q}_j(z) &= (z-\tilde{\alpha}_j)\tilde{q}_{j-1}(z) - \tilde{\beta}_{j-1}\tilde{q}_{j-2}(z), \qquad 2 \le j \le m+1, \end{aligned} \tag{56}$$

where

$$\begin{aligned} \tilde{\alpha}_j &= \{\tilde{q}_{j-1}, z\tilde{p}_{j-1}\}, \\ \tilde{\beta}_j\tilde{\beta}'_j &= \{(z-\tilde{\alpha}_j)\tilde{p}_{j-1} - \tilde{\beta}'_{j-1}\tilde{p}_{j-2}, (z-\tilde{\alpha}_j)\tilde{q}_{j-1} - \tilde{\beta}_{j-1}\tilde{q}_{j-2}\}. \end{aligned} \tag{57}$$

It follows from (54) and (35) that

$$\{f,g\} = [f,g] = \mathcal{I}(fg), \qquad f \in \mathbb{P}_{m-1}, \quad g \in \mathbb{P}_m,$$

and application of these equalities to (45) and (57) yields

$$\begin{aligned} \tilde{\alpha}_j &= \alpha_j, & 1 \le j \le m, \\ \tilde{\beta}_j\tilde{\beta}'_j &= \beta_j\beta'_j, & 1 \le j < m. \end{aligned} \tag{58}$$

We may choose $\tilde{\beta}_j = \beta_j$ and $\tilde{\beta}'_j = \beta'_j$ for $1 \le j < m$, and then $\tilde{p}_j = p_j$ and $\tilde{q}_j = q_j$ for $0 \le j < m$. Consider the polynomials

$$\begin{aligned} \hat{p}_m(z) &:= (z-\tilde{\alpha}_m)\tilde{p}_{m-1}(z) - \tilde{\beta}'_{m-1}\tilde{p}_{m-2}(z), \\ \hat{q}_m(z) &:= (z-\tilde{\alpha}_m)\tilde{q}_{m-1}(z) - \tilde{\beta}_{m-1}\tilde{q}_{m-2}(z), \end{aligned}$$

which also can be written as

$$\begin{aligned} \hat{p}_m(z) &= (z-\alpha_m)p_{m-1}(z) - \beta'_{m-1}p_{m-2}(z), \\ \hat{q}_m(z) &= (z-\alpha_m)q_{m-1}(z) - \beta_{m-1}q_{m-2}(z). \end{aligned}$$

These polynomials are multiples of p_m and q_m, and therefore, the zeros of p_m being the Gauss nodes, $\mathcal{G}_m(\hat{p}_m\hat{q}_m) = 0$. Thus, by (45) and (57),

$$\begin{aligned} \tilde{\beta}_m\tilde{\beta}'_m &= \{\hat{p}_m,\hat{q}_m\} = 2\mathcal{I}(\hat{p}_m\hat{q}_m) - \mathcal{G}_m(\hat{p}_m\hat{q}_m) \\ &= 2\mathcal{I}(\hat{p}_m\hat{q}_m) = 2[\hat{p}_m,\hat{q}_m] = 2\beta_m\beta'_m. \end{aligned}$$

We may choose $\tilde{\beta}_m = \sqrt{2}\beta_m$ and $\tilde{\beta}'_m = \sqrt{2}\beta'_m$. It follows as in §2 that the nonsymmetric tridiagonal matrix associated with the quadrature rule $\tilde{\mathcal{G}}_{m-1}$ is given by

$$\tilde{T}_{m+1} = \begin{bmatrix} \alpha_1 & \beta'_1 & & & & 0 \\ \beta_1 & \alpha_2 & \beta'_2 & & & \\ & \beta_2 & \alpha_3 & & & \\ & & & \ddots & & \\ & & \ddots & \ddots & \beta'_{m-1} & \\ & & & \beta_{m-1} & \alpha_m & \sqrt{2}\beta'_m \\ 0 & & & & \sqrt{2}\beta_m & \alpha_{m+1} \end{bmatrix} \in \mathbb{R}^{(m+1)\times(m+1)}.$$

The entries of $\tilde{T}_{m+1}$ can be computed by $m+1$ steps of the Lanczos process, cf. (39). Analogously to formula (51), the anti-Gauss quadrature rule (37) can be evaluated according to

$$\tilde{\mathcal{G}}_{m+1}f = e_1^T f(\tilde{T}_{m+1})e_1. \tag{59}$$

Similarly as in §2, we define the average integration rule $\mathcal{L}_{2m+1}$ as the average of $\mathcal{G}_m$ and $\tilde{\mathcal{G}}_{m+1}$.

4. Computed Examples

In this section we describe three computed examples which illustrate the performance of the method proposed in the paper. The examples were generated by using MATLAB 5.01 on a MICRON personal computer, i.e., with approximately 15 significant digits. We remark that since the matrices in the following examples are either symmetric indefinite or nonsymmetric, the bounds based on Gauss and Gauss-Radau quadrature rules do not apply.

Example 4.1. *We would like to compute an approximation of the functional*

$$F(A) := u^T \sin(A)u,$$

where A is a 100×100 real symmetric matrix with randomly generated uniformly distributed eigenvalues in the interval $[-\frac{1}{2}, \frac{1}{2}]$ and u is a random vector of unit length. Table 1 shows the exact value of $F(A)$ and approximations obtained by the Gauss, anti-Gauss, and average quadrature rules with integrand $f(t) := \sin(t)$. As we would expect, the exact value of $F(A)$ lies inside the intervals determined by the estimates obtained by the Gauss and the anti-Gauss rules. We remark that the approximations of $F(A)$ obtained by the average rule agree with the exact values in the first 10 and 13 significant digits for $m = 3, 4$, respectively. The agreeing digits in the approximations determined by the average rule are underlined.

TABLE 1. $F(A) = u^T \sin(A)u$, A symmetric indefinite

$n = 100$	$m = 3$	$m = 4$
$F(A)$	$7.388323113654 \cdot 10^{-2}$	$7.388323113654 \cdot 10^{-2}$
$\mathcal{G}_m f$	$7.388322938159 \cdot 10^{-2}$	$7.388323113673 \cdot 10^{-2}$
$\tilde{\mathcal{G}}_{m+1} f$	$7.388323289305 \cdot 10^{-2}$	$7.388323113635 \cdot 10^{-2}$
$\mathcal{L}_{2m+1} f$	$\underline{7.388323113}732 \cdot 10^{-2}$	$\underline{7.388323113654} \cdot 10^{-2}$

TABLE 2. $F(A) = u^T(I + A^2)^{-1}u$, A symmetric indefinite

$n = 200$	$m = 6$	$m = 11$	$m = 12$	$m = 14$
$F(A)$	0.2816	0.2816	0.2816	0.2816
$\mathcal{G}_m f$	0.2424	0.2883	0.2779	0.2798
$\tilde{\mathcal{G}}_{m+1} f$	0.3350	0.2762	0.2858	0.2833
$\mathcal{L}_{2m+1} f$	$\underline{0.28}87$	$\underline{0.282}2$	$\underline{0.281}9$	$\underline{0.2816}$

Example 4.2. *We would like to determine an approximation of the functional*

$$F(A) := u^T(I + A^2)^{-1}u,$$

where A is a 200×200 real symmetric matrix with randomly generated uniformly distributed eigenvalues in the interval $[-5, 5]$ and u is a random vector of unit length. Table 2 displays the value of $F(A)$ and approximations obtained by the Gauss, anti-Gauss, and average quadrature rules with integrand $f(t) := 1/(1 + t^2)$. For each value of m, the value of $F(A)$ is in an interval with endpoints $\mathcal{G}_m f$ and $\tilde{\mathcal{G}}_{m+1} f$; however, for some values of m, $\mathcal{G}_m f > F(A) > \tilde{\mathcal{G}}_{m+1} f$ and for other values $\mathcal{G}_m f < F(A) < \tilde{\mathcal{G}}_{m+1} f$; see, e.g., $m = 11, 12$. The average rule yields approximations of $F(A)$ with agreement in the first two to four significant digits. The agreeing digits are underlined in the table. The fact that the average rule in this example yields less accuracy than in Example 4.1 stems from the fact that the integrand in the present example has poles in the complex plane at $\pm i$, where $i = \sqrt{-1}$, while the integrand in Example 4.1 is an entire function.

The following two examples are concerned with functionals of real nonsymmetric matrices.

Example 4.3. *We would like to compute an approximation of*

$$F(A) := u^T \exp(A)u',$$

where A is a 200×200 real nonsymmetric matrix with random entries generated with the MATLAB command `A=rand(200)/100`*, and u, u' are random vectors scaled so that $u^T u' = 1$. Table 3 displays the exact value of $F(A)$ and three approximations determined by the Gauss, anti-Gauss, and average quadrature rules with integrand $f(t) = \exp(t)$ and $m = 3, 4$. The value of $F(A)$ lies between the*

TABLE 3. $F(A) = u^T \exp(A)u'$, A nonsymmetric

$n = 200$	$m = 3$	$m = 4$
$F(A)$	2.68128047787425	2.68128047787425
$\mathcal{G}_m f$	2.68128047607348	2.68128047787426
$\tilde{\mathcal{G}}_{m+1} f$	2.68128047966772	2.68128047787424
$\mathcal{L}_{2m+1} f$	2.68128047787060	2.68128047787425

TABLE 4. $F(A) = u^T A^{-1} u'$, A nonsymmetric

$n = 300$	$m = 2$	$m = 3$	$m = 4$
$F(A)$	4.517591178156344	4.517591178156344	4.517591178156344
$\mathcal{G}_m f$	4.517578214459956	4.517591178488559	4.517591178156347
$\tilde{\mathcal{G}}_{m+1} f$	4.517605348649931	4.517591177824231	4.517591178156343
$\mathcal{L}_{2m+1} f$	4.517591178155443	4.517591178156395	4.517591178156345

values $\mathcal{G}_m f$ and $\tilde{\mathcal{G}}_{m+1} f$. The average rule yields a remarkably accurate approximation of $F(A)$ with agreement in the first 12 and 15 significant digits for $m = 3, 4$, respectively.

Example 4.4. *Let $f(t) := 1/t$. We would like to determine an approximation of $F(A) := u^T f(A)u'$, where A is a 300×300 real, nonsymmetric random matrix generated similarly as in Example 4.3, and u, u' are random vectors scaled so that $u^T u' = 1$. The value of $F(A)$ and approximations thereof determined by Gauss, anti-Gauss, and average quadrature rules are reported in Table 4. The estimates provided by the average rule and the exact value of $F(A)$ agree in the first 11, 13, and 14 significant digits for $m = 2, 3, 4$, respectively.*

The above examples illustrate that the evaluation of pairs of Gauss and anti-Gauss quadrature rules can be applied to inexpensively compute candidates for upper and lower bounds of matrix functionals.

Acknowledgements

We would like to thank Walter Gautschi and Gene Golub for comments. Work by D.C. was supported in part by NSF grant DMS-9896073, work by L.R. was supported in part by NSF grant DMS-9404706, and work by F.S. was supported in part by University of Bologna funds for selected research topics.

References

[1] Z. Bai, M. Fahey, and G. Golub, *Some large-scale matrix computation problems*, J. Comput. Appl. Math., **74** (1996), 71–89.

[2] Z. Bai, M. Fahey, G. Golub, M. Menon, and E. Richter, *Computing partial eigenvalue sum in electronic structure calculations*, Report SCCM-98-03, Computer Science Department, Stanford University, Stanford, 1998.

[3] G.H. Golub and G. Meurant, *Matrices, moments and quadrature*, in: D.F. Griffiths and G.A. Watson, Eds., *Numerical analysis 1993*, Pitman Res. Notes Math. Ser. **303**, Longman, Harlow, 1994, 105–156.

[4] G.H. Golub and G. Meurant, *Matrices, moments and quadrature. II. How to compute the norm of the error in iterative methods*, BIT, **37** (1997), 687–705.

[5] G.H. Golub and Z. Strakoš, *Estimates in quadratic formulas*, Numer. Algorithms, **8** (1994), 241–268.

[6] G.H. Golub and C.F. Van Loan, *Matrix computations*, 3d ed., Johns Hopkins University Press, Baltimore, 1996.

[7] G.H. Golub and U. von Matt, *Generalized cross-validation for large-scale problems*, J. Comput. Graph. Statist., **6** (1997), 1–34.

[8] D.P. Laurie, *Anti-Gaussian quadrature formulas*, Math. Comp., **65** (1996), 739–747.

[9] Y. Saad, *Iterative methods for sparse linear systems*, PWS Publ. Co., Boston, 1996.

Daniela Calvetti
Department of Mathematics
Case Western Reserve University
Cleveland, OH 44106-7058, USA
E-mail address: `dxc57@po.cwru.edu`

Lothar Reichel
Department of Mathematics and Computer Science
Kent State University
Kent, OH 44242-0001, USA
E-mail address: `reichel@mcs.kent.edu`

Fiorella Sgallari
Dipartimento di Matematica
Università di Bologna
Piazza P.ta S. Donato 5
I-40127 Bologna, Italy
E-mail address: `sgallari@dm.unibo.it`

International Series of Numerical Mathematics
Vol. 131, © 1999 Birkhäuser Verlag Basel/Switzerland

Stieltjes Polynomials and the Error of Gauss-Kronrod Quadrature Formulas

Sven Ehrich

Abstract. The Gauss-Kronrod quadrature scheme, which is based on the zeros of Legendre polynomials and Stieltjes polynomials, is the standard method for automatic numerical integration in mathematical software libraries. For a long time, very little was known about the error of the Gauss-Kronrod scheme. Essential progress was made only recently, based on new bounds and asymptotic properties for the Stieltjes polynomials. The purpose of this paper is to give a survey on these results. In particular, the quality of the Gauss-Kronrod formula for smooth and for nonsmooth functions is investigated and compared with other quadrature formulas.

1. Introduction

The numerical evaluation of definite integrals is an important step in many applications of mathematics. From a practical point of view, great interest lies in automatic routines, whose only input are the integrand function, the domain of integration, and a tolerance for the error. In applications, such routines are expected to give quick and reliable approximations. Among the most often used routines for this purpose are D01AJF from the NAG library [38], QDAG from the IMSL library [30], and NIntegrate from the Mathematica software package [54]. All these routines use adaptive subdivisions of the integration domain and local error estimators which are based on the Gauss-Kronrod quadrature scheme. This is one of the most important quadrature methods for practical applications. It seems surprising that until recently only very little was known about the error of Gauss-Kronrod quadrature methods. In contrast, very detailed investigations exist for all other quadrature formulas of practical interest, in particular for the rectangular, midpoint, trapezoidal and Simpson formulas, and those of Gauss, Clenshaw-Curtis, and Romberg (see [4, 9, 20] and the literature cited therein). The purpose of this paper is to give a survey on recent results concerning the error of the Gauss-Kronrod quadrature formula.

The efficiency of the Gauss-Kronrod quadrature method is due to a nested sequence of nodes and the use of existing information. The basic concept, which consists in "refining" the information, is characteristic for the general class of multiscale methods, which presently attracts much interest in numerical analysis.

Gauss-Kronrod formulas were introduced in 1964 by A.S. Kronrod [31, 32] in order to estimate the error of Gaussian quadrature formulas. Based on the n nodes $x_{1,n}, \ldots, x_{n,n}$ of the Gaussian formula Q_n^G, the Gauss-Kronrod formula

$$Q_{2n+1}^{GK}[f] = \sum_{\nu=1}^{n} A_{\nu,n} f(x_{\nu,n}) + \sum_{\mu=1}^{n+1} B_{\mu,n+1} f(\xi_{\mu,n+1})$$

is constructed by adding $n+1$ additional nodes $\xi_{1,n+1}, \ldots, \xi_{n+1,n+1}$ which are chosen in such a way that polynomials of degree as high as possible are integrated exactly. The weights $A_{\nu,n}$ and $B_{\mu,n+1}$ are also uniquely defined by this requirement. Contrary to Kronrods original idea, in the abovementioned automatic quadrature routines, Q_{2n+1}^{GK} most often is used to compute the quadrature approximation, and the error estimate uses the difference $Q_{2n+1}^{GK} - Q_n^G$. Therefore, it is important to study the error of Q_{2n+1}^{GK}.

The additional nodes $\xi_{1,n+1}, \ldots, \xi_{n+1,n+1}$ are precisely the zeros of the Stieltjes polynomials E_{n+1}, defined up to a multiplicative constant by the orthogonality relations (see §2)

$$\int_{-1}^{1} P_n(x) E_{n+1}(x)\, x^k \, dx = 0, \qquad k = 0, 1, \ldots, n.$$

Here, P_n is the nth Legendre polynomial whose zeros are the nodes of Q_n^G. Since P_n changes sign n times in the interval $[-1,1]$, the associated inner product is not definite. Hence, one of the basic assumptions in the theory of orthogonal polynomials is violated. The polynomials E_{n+1} were first studied by T.J. Stieltjes in 1894 [2, Lettre 429], who conjectured that its zeros, for all $n \in \mathbb{N}$, are real, distinct, and inside $(-1,1)$, and that they interlace with the zeros of P_n. G. Szegö proved these properties in 1935 [52] and extended them to generalised Stieltjes polynomials E_{n+1}^{λ} and ultraspherical polynomials $P_n^{(\lambda)}$, with the parameter λ in the ultraspherical weight function $w_\lambda(x) = (1-x^2)^{\lambda-\frac{1}{2}}$ restricted to $\lambda \in (0,2]$ (for $\lambda = 0$, two zeros are in the endpoints ± 1, but all other properties are still valid). The generalised Stieltjes polynomials E_{n+1}^{λ} are defined by

$$\int_{-1}^{1} P_n^{(\lambda)}(x) E_{n+1}^{\lambda}(x)\, x^k \, dx = 0, \qquad k = 0, 1, \ldots, n.$$

After Szegö's paper, for a long time no new results on the Stieltjes polynomials appeared in the literature. Also Kronrod's work contains no references to Stieltjes' and Szegö's work. The connection was independently pointed out in the Eastern literature by I.P. Mysovskih in 1964 [37] and in the Western literature by P. Barrucand in 1970 [3]. G. Monegato proved in 1976 [33] that the positivity of the quadrature weights $B_{\mu,n+1}$ associated with the additional nodes $\xi_{\mu,n+1}$ is equivalent to the interlacing property of the nodes. The positivity of all quadrature weights was proved by Monegato in 1978 [34]. Many authors have considered the location of the zeros and the positivity of the quadrature weights for more general weight functions. In particular, cf. Gautschi and Notaris [23], Gautschi and Rivlin

[24], Monegato [35], Peherstorfer [45], as well as the survey papers of Monegato [36], Gautschi [21], and Notaris [42]. Until recently, only few error bounds existed in the literature, and nothing was known about their sharpness. Of special interest is the comparison with the Gaussian quadrature formula Q^G_{2n+1}, since the new approach of Peano stopping functionals in [19] allows one to estimate the quadrature error, unlike in Kronrod's approach, essentially without adding further nodes.

A major drawback in the investigation of Gauss-Kronrod formulas was a lack of precise knowledge on the Stieltjes polynomials E_{n+1} and its zeros. While Szegö proved the interlacing property of the nodes in [52], for a long time no sharper results on the asymptotic behaviour of the zeros, or lower bounds for the differences, were known. Peherstorfer in [45] considered Stieltjes polynomials for weight functions $w(x) = \frac{W(x)}{\sqrt{1-x^2}}$, where $W \in C^2[-1,1]$ and $W \geq C > 0$ for some constant C, and proved results on the asymptotic behaviour of the associated Stieltjes polynomials. The paper [45] generalised earlier works on Bernstein-Szegö weight functions in [40, 44]. However, these problems remained open for the most important case of the Legendre weight function (see also [45, p. 186]).

Essential progress was made in the paper [14], which contains results on the asymptotic behaviour of Stieltjes polynomials for ultraspherical weight functions w_λ, $\lambda \in [0,1]$. This includes the Legendre case for $\lambda = \frac{1}{2}$. The asymptotic formulas have been used to obtain error estimates for Gauss-Kronrod quadrature formulas in many important function spaces and comparisons with other quadrature formulas. Later in [18], also pointwise bounds for the Stieltjes polynomials, which are precise in the whole interval $[-1,1]$, and lower bounds for the distances of the zeros have been proved. For completeness, we also include these results in this paper.

These new results provide a detailed description of the error of the Gauss-Kronrod quadrature formula. They motivate a survey in which both the fundamental facts about Gauss-Kronrod formulas and pointers to the original literature, as well as possible consequences for practical calculations are outlined. In §2, we state the fundamental properties of Stieltjes polynomials as well as asymptotic relations and inequalities. Section 3 contains an overview of results on the error of the Gauss-Kronrod quadrature formula and comparisons with other quadrature formulas. Some numerical examples are given in §4.

2. Stieltjes Polynomials

Let P_n be the Legendre polynomial defined by

$$\int_{-1}^{1} P_n(x)\, x^k\, dx = 0, \qquad k = 0, 1, \dots, n-1, \tag{1}$$

and $P_n(1) = 1$. For $n \geq 0$, the Stieltjes polynomial E_{n+1} is defined by

$$\int_{-1}^{1} P_n(x) E_{n+1}(x) x^k\, dx = 0, \qquad k = 0, 1, \dots, n, \tag{2}$$

and the normalisation

$$E_{n+1}(x) = \frac{2^n}{\gamma_n} x^{n+1} + p(x), \qquad \gamma_n = \frac{2^{2n}\, n!^2}{(2n+1)!}, \quad p \in \mathbb{P}_n. \tag{3}$$

Here and in the following, $\mathbb{P}_n$ is the space of all algebraic polynomials of degree $\leq n$. Up to a multiplicative constant, the polynomial E_{n+1} is defined uniquely by (2). The conditions (2) are nonstandard orthogonality relations that differ from (1) in two respects: the weight function P_n changes sign in $(-1,1)$, and this weight depends on n. Hence, the underlying inner product is not definite. T.J. Stieltjes posed the following two problems in 1894 [2, vol. 2, p. 440]:

(I) Are all zeros of E_{n+1} real, distinct, and in $(-1,1)$ for all $n \in \mathbb{N}$?
(II) If (I) is true, do the zeros of E_{n+1} interlace for all $n \in \mathbb{N}$ with the zeros of P_n?

These conjectures were proved in 1935 by G. Szegö [52]. However, no sharper results on Stieltjes polynomials and its zeros have been known for a long time.

A classical approximation to the Legendre polynomials is given by Laplace's formula,

$$P_n(\cos\theta) = \sqrt{\frac{2}{\pi\, n\, \sin\theta}}\, \cos\left\{\left(n+\frac{1}{2}\right)\theta - \frac{\pi}{4}\right\} + \mathcal{O}(n^{-3/2}), \tag{4}$$

which holds uniformly for $\epsilon \leq \theta \leq \pi - \epsilon$, $\epsilon \in (0, \frac{\pi}{2})$ arbitrary but fixed. The asymptotic behaviour of the Stieltjes polynomials was studied numerically in [35, p. 235]. This reference contains the observation that $P_n E_{n+1}$ numerically behaves like the Chebyshev polynomial of the first kind T_{2n+1}. In [45], Peherstorfer proved asymptotic formulas for the case of weight functions of the kind $w(x) = \frac{W(x)}{\sqrt{1-x^2}}$, $W \in C^2[-1,1]$, and $W \geq C > 0$ (cf. [45, §4]), but the question remained open for the Legendre weight [45, p. 186]. This problem was solved in [14].

Theorem 2.1. *We have*

$$E_{n+1}(\cos\theta) = 2\sqrt{\frac{2n\sin\theta}{\pi}}\, \cos\left\{\left(n+\frac{1}{2}\right)\theta + \frac{\pi}{4}\right\} + \mathcal{O}(1), \tag{5}$$

uniformly for $\epsilon < \theta < \pi - \epsilon$, $\epsilon \in (0, \frac{\pi}{2})$ *fixed.*

While Theorem 2.1 gives the precise behaviour inside of the interval $(-1,1)$, one cannot deduce the boundary behaviour near the endpoints ± 1. For Legendre polynomials, a well-known and more precise result is (cf., [53, Theorem 8.21.13])

$$P_n(\cos\theta) = \sqrt{\frac{2}{\pi\, n\, \sin\theta}} \left\{\cos\left[\left(n+\frac{1}{2}\right)\theta - \frac{\pi}{4}\right] + (n\sin\theta)^{-1}\mathcal{O}(1)\right\}, \tag{6}$$

uniformly for $cn^{-1} \leq \theta \leq \pi - cn^{-1}$, $c > 0$ arbitrary but fixed. The following theorem contains the analogous result for Stieltjes polynomials (cf. [17, Lemma 1]).

Theorem 2.2. *We have*

$$E_{n+1}(\cos\theta) = C_n(\theta)\sqrt{n\sin\theta}\left\{\cos\left[\left(n+\frac{1}{2}\right)\theta+\frac{\pi}{4}\right]+(n\sin\theta)^{-1}A_n(\theta)\right\}+B_n(\theta),$$

where for every c which is independent of n, there exists $\mathcal{C}=\mathcal{C}(c)$ such that $\max(|A_n(\theta)|,|B_n(\theta)|,|C_n(\theta)|)<\mathcal{C}$ *for* $\theta\in[cn^{-1},\pi-cn^{-1}]$. *For $c\geq\pi$, we have* $1\leq\frac{1}{2}\sqrt{\frac{\pi}{2}}C_n(\theta)\leq 1.0180\ldots$.

In [14, Theorem (ii)], an asymptotic approximation was also given for the derivative of Stieltjes polynomials.

Theorem 2.3. *We have*

$$E'_{n+1}(\cos\theta)=2n\sqrt{\frac{2n}{\pi\sin\theta}}\,\sin\left\{\left(n+\frac{1}{2}\right)\theta+\frac{\pi}{4}\right\}+\mathcal{O}(\sqrt{n}),$$

uniformly for $\varepsilon<\theta<\pi-\varepsilon$, $\varepsilon\in(0,\frac{\pi}{2})$.

Theorem 2.1 and (4) show that the product P_nE_{n+1} behaves asymptotically like a constant multiple of the Chebyshev polynomial T_{2n+1},

$$P_n(x)E_{n+1}(x)=\frac{2}{\pi}\,T_{2n+1}(x)+o(1),\qquad x\in[-1+\delta,1-\delta],\tag{7}$$

$\delta>0$ independent of n.

In the following, for the zeros $\xi_{1,n+1},\ldots,\xi_{n+1,n+1}$ of E_{n+1}, ordered increasingly, the cos-arguments will be denoted by $\theta_{\mu,n+1}$,

$$\xi_{\mu,n+1}=\cos\theta_{\mu,n+1},\qquad \mu=1,2,\ldots,n+1.$$

The zeros of P_nE_{n+1}, ordered increasingly, will be denoted by

$$y_{\nu,2n+1}=\cos\psi_{\nu,2n+1},\qquad \nu=1,2,\ldots,2n+1.$$

As an application of the formula (5), the paper [14] also contains results on the asymptotic distribution of the zeros of E_{n+1}.

Corollary 2.4. *Let $\epsilon>0$ be fixed. For $n\to\infty$, we have uniformly for all* $\theta_{n+2-\mu,n+1}\in[\epsilon,\pi-\epsilon]$

$$\theta_{n+2-\mu,n+1}=\frac{\mu-\frac{3}{4}+o(1)}{n+\frac{1}{2}}\,\pi.$$

An analogous result for the zeros of Legendre polynomials is well known (cf. [53, Theorem 8.9.1]):

$$\phi_{n+1-\nu,n}=\frac{\nu-\frac{1}{4}+o(1)}{n+\frac{1}{2}}\,\pi\tag{8}$$

uniformly for $x_{\nu,n}\in[-1+\epsilon,1-\epsilon]$, $\epsilon>0$ fixed.

There are further open problems on Stieltjes polynomials to which Theorem 2.1 can be applied. For example, it is conjectured in [36] that the zeros of E_{n+1} interlace with the zeros of E_n.

Theorem 2.5. *For every $\varepsilon \in (0,1)$, there is a $N \in \mathbb{N}$, such that for every $n \geq N$ those zeros of E_n and E_{n+1} which are inside $[-1+\varepsilon, 1-\varepsilon]$ interlace.*

The following uniform upper bound on Stieltjes polynomials is given in [36],

$$|E_{n+1}(x)| \leq \frac{4}{\gamma_n}, \qquad x \in [-1,1],$$

with γ_n as in (3). The asymptotic formula (5) indicates that this bound is not sharp near the endpoints ± 1. A bound which gives the precise order was proved in [18].

Theorem 2.6. *For $n \geq 1$ we have*

$$|E_{n+1}(x)| \leq 2\,C^* \sqrt{\frac{2n+1}{\pi}}\ \sqrt[4]{1-x^2} + 55, \qquad x \in [-1,1], \tag{9}$$

where $C^* = 1.0180\ldots$, and

$$E_{n+1}(1) \geq \frac{2}{3\sqrt{\pi}}, \qquad n \geq 1.$$

The second assertion in Theorem 2.6 shows that the first assertion is of precise order also in the endpoints ± 1. Similarly as in (7), we obtain from Theorem 2.6 an upper bound for the product $P_n E_{n+1}$ which has the precise order in the whole interval $[-1,1]$.

Corollary 2.7. *For $n \geq 0$ we have*

$$|P_n(x) E_{n+1}(x)| \leq \mathcal{C}, \qquad -1 \leq x \leq 1,$$

where $\mathcal{C}$ is a positive constant which is independent of n.

Corollary 2.4 implies that the zeros of Stieltjes polynomials have a very regular distribution which is typical for orthogonal polynomials. Furthermore, Corollary 2.4 states that the zeros of E_{n+1} are also distributed very regularly with respect to the zeros of P_n. In the asymptotic sense of Theorem 2.2, they lie midway between two successive zeros of P_n. However, these statements only follow for the zeros which are inside closed subintervals of $(-1,1)$. The following theorem from [18, Theorem 2.4] improves the interlacing result of Szegö by stating lower bounds for the distances of all zeros.

Theorem 2.8. *We have*

$$\liminf_{n\to\infty} \inf_{0\leq\nu\leq 2n+1} (2n+1)\,(\psi_{\nu,2n+1} - \psi_{\nu+1,2n+1}) > \mathcal{C} > 0,$$

and

$$\liminf_{n\to\infty} \inf_{0\leq\mu\leq n+1} (n+1)\,(\theta_{\mu,n+1} - \theta_{\mu+1,n+1}) > \mathcal{C} > 0,$$

where $\psi_{0,2n+1} = \theta_{0,n+1} = \pi$, $\psi_{2n+2,2n+1} = \theta_{n+2,n+1} = 0$, and $\mathcal{C}$ is a positive constant which is independent of n, ν and μ.

The following important result on the derivatives of Stieltjes polynomials was proved in [18].

Theorem 2.9. *There exists a positive constant C such that for all $n \in \mathbb{N}$*

$$\frac{C^{-1}}{n\sqrt{n}}\,(1-(\xi_{\mu,n+1})^2)^{\frac{1}{4}} \le \frac{1}{|E'_{n+1}(\xi_{\mu,n+1})|} \le \frac{C}{n\sqrt{n}}\,(1-(\xi_{\mu,n+1})^2)^{\frac{1}{4}}. \tag{10}$$

3. Error of the Gauss-Kronrod Quadrature Formula

In automatic quadrature routines, the user needs to supply the integrand function, but no further information about it. In applications, both smooth functions and those with discontinuities, oscillations, or other types of singularities may occur. In general, the user expects that the method converges very rapidly for smooth functions, while for nonsmooth functions the method should still give a reliable quadrature approximation. For the routines mentioned in the introduction, the accuracy of the quadrature approximation depends on the accuracy of the Gauss-Kronrod formula. This motivates investigating the error of the Gauss-Kronrod formula for many classes of functions with different smoothness properties. Results in that direction were not available in the literature until recently.

3.1. Basic facts

For the Gauss-Kronrod formula

$$Q_{2n+1}^{GK}[f] = \sum_{\nu=1}^{n} A_{\nu,n}\, f(x_{\nu,n}) + \sum_{\mu=1}^{n+1} B_{\mu,n+1} f(\xi_{\mu,n+1}),$$

the new nodes $\xi_{1,n+1}, \dots, \xi_{n+1,n+1}$ and the new weights $A_{1,n}, \dots, A_{n,n}$ and $B_{1,n+1}$, $\dots, B_{n+1,n+1}$ are chosen such that the degree of polynomial exactness is as high as possible. A necessary and sufficient condition is that the nodes $\xi_{\mu,n+1}$ are the zeros of the Stieltjes polynomial E_{n+1}. The precise degree of exactness is $3n+1$ if n is even, and $3n+2$ if n is odd [48]. By adding further nodes, one can construct the so-called Patterson extensions, cf. [43, 26].

One of the most important properties of Q_{2n+1}^{GK} is the positivity of the weights $A_{\nu,n}$ and $B_{\mu,n+1}$, which was proved by Monegato [33, 34]. More precise information on the size of the weights was stated in [14, Corollary 3].

Theorem 3.1. *We have*

$$A_{\nu,n} = \frac{\pi}{2n+\frac{3}{2}}\, \sin\phi_{\nu,n}\,(1+o(1)),$$

$$B_{\mu,n+1} = \frac{\pi}{2n+\frac{3}{2}}\, \sin\theta_{\nu,n+1}\,(1+o(1)),$$

uniformly for those weights $A_{\nu,n}$, $B_{\mu,n+1}$ for which the associated nodes satisfy $\varepsilon < \phi_{\nu,n} < \pi-\varepsilon$ and $\varepsilon < \theta_{\mu,n+1} < \pi-\varepsilon$.

It is interesting to compare these results with well-known results for the weights of the Gaussian quadrature formulas. In the asymptotic sense, the Gauss-Kronrod weights associated with the Gaussian nodes are asymptotically half as large as the respective weights of the Gaussian formula.

A very important tool for obtaining error bounds for quadrature formulas is Peano's Theorem, which gives a representation of linear functionals (cf., e.g., [4, 7, 28]).

Theorem 3.2 (Peano, cf. [7]). *Let R be a linear functional on $C[a,b]$ with $R[\mathbb{P}_{s-1}] = 0$. If $f^{(s-1)}$ is absolutely continuous in $[a,b]$, then we have*

$$R[f] = \int_a^b f^{(\nu)}(x)\, K_\nu(x)\, dx, \qquad \nu = 1, 2, \ldots, s, \tag{11}$$

where

$$K_1(x) := \begin{cases} 0, & x = a, \\ \lim_{\nu\to\infty} R[s_{x,\nu}], & a < x \leq b, \end{cases}$$

$$s_{x,\nu}(t) := \begin{cases} 0, & t < x, \\ \nu(t-x), & t \in [x, x+\nu^{-1}], \\ 1, & t > x + \nu^{-1}, \end{cases}$$

and

$$K_\nu(x) := \int_x^b K_{\nu-1}(u)\, du, \qquad \nu = 2, 3, \ldots, s.$$

The functions $K_1, \ldots, K_s$ are called Peano kernels. A simple representation is (cf. [7])

$$K_\nu(x) = R\left[\frac{(\cdot - x)_+^{\nu-1}}{(\nu-1)!}\right], \qquad \nu = 2, 3, \ldots, s.$$

If $f^{(\nu)} \in L^p[a,b]$, $1 \leq p \leq \infty$, then we obtain from (11) with Hölder's inequality

$$|\, R[f]\, | \leq \|\, K_\nu \,\|_{p'} \; \|\, f^{(\nu)} \,\|_p, \qquad \frac{1}{p} + \frac{1}{p'} = 1. \tag{12}$$

In practical applications, bounds for norms of derivatives are rarely available. But the inequality (12) is very useful, since it is nonimprovable (cf. [7]),

$$\|\, K_s \,\|_{p'} = \sup_{\|\, f^{(s)} \,\|_p \leq 1} |\, R[f]\, |, \tag{13}$$

and the constant $\|\, K_s \,\|_{p'}$ is thus the largest value which $|R[f]|$ can have for f from the function class

$$A_{s,p}(I) = \{f \mid \|f^{(s)}\|_{L^p(I)} \leq 1\}, \qquad I = [a,b].$$

This is a classical measure of quality (cf., e.g., Gautschi [20, p. 115]). The most often used constants are

$$c_s(R) := c_{s,1}(R) := \|\, K_s \,\|_1$$

for $A_{s,\infty}$, which shall be investigated here for the Gauss-Kronrod formula.

3.2. Error for smooth functions

By definition, the Gauss-Kronrod formula has a high degree of exactness which increases with the number of nodes. In the class of arbitrarily often differentiable functions, we can consider error constants with index $3n+2+\kappa$ for Q_{2n+1}^{GK}, where $\kappa = 0$ if n is even, and $\kappa = 1$ if n is odd. Monegato proved in [34]

$$c_{3n+2+\kappa}(R_{2n+1}^{GK}) \leq \mathcal{C}\, \frac{2^{-3n}\sqrt{n}}{(3n+2+\kappa)!},$$

where $\mathcal{C}$ is a positive constant which is independent of n. More general results of Brass and Förster [6] give the improvement

$$c_{3n+2+\kappa}(R_{2n+1}^{GK}) \leq \mathcal{C}\, \frac{2^{-3n}}{(3n+2+\kappa)!}.$$

In [11, Theorem], error bounds of the order

$$\mathcal{C}_1\, \frac{2^{-3n}\; n^{-11/2}}{(3n+2+\kappa)!} \leq c_{3n+2+\kappa}(R_{2n+1}^{GK}) \leq \mathcal{C}_2\, \frac{2^{-3n}\; n^{-1/2}}{(3n+2+\kappa)!} \tag{14}$$

are proved with explicit constants $\mathcal{C}_1, \mathcal{C}_2$ which do not depend on n. The following result from [16, Theorem 1] states the precise order of $c_{3n+2+\kappa}(R_{2n+1}^{GK})$.

Theorem 3.3. *There holds*

$$c_{3n+2+\kappa}(R_{2n+1}^{GK}) \sim \frac{2^{-3n}\; n^{-5/2}}{(3n+2+\kappa)!}. \tag{15}$$

The notation "$\sim$" here and in the following will be used if there exists a positive constant $\mathcal{C}$, which is independent of the parameters in the respective context (here n), such that

$$A \sim B \iff \max\{|A/B|, |B/A|\} \leq \mathcal{C}.$$

The proof of Theorem 3.3 uses results on quadrature errors for Chebyshev polynomials of the first kind. These results are of interest in their own right and will be discussed later in this section. For the error constants $c_{3n+2+\kappa-s}(R_{2n+1}^{GK})$, where $s = s(n)$ is chosen such that $\lim_{n\to\infty} \frac{3n+2+\kappa-s}{n} = A$ holds with $A > 0$, upper and lower bounds are proved in [10, Satz 1]. For constant s, the upper and lower bounds differ by a factor of polynomial order in n, see [10, Korollar].

For functions which are analytic in a certain region containing the domain of integration, error bounds can be given which do not use derivatives. Many results are known, in particular for the Gaussian formula (cf. Hämmerlin [27], Gautschi and Varga [25]). In view of the high degree of exactness, such error bounds are also interesting for Gauss-Kronrod formulas. For functions f which are analytic in the open circle $K_R = \{z \in \mathbb{C} : |z| < R\}$, $R > 1$, Notaris [41, (2.18)] proved the error bound

$$|R_{2n+1}^{GK}[f]| \le \frac{(n!)^2(d_n - \iota_n + 1)!}{2^{n-2}(2n)!(d_n + 1)!} \cdot \inf_{1<r<R} \left\{ \left[\frac{r}{(r-1)^{d_n-\iota_n+2}} + (-1)^{\iota_n} \frac{r}{(r+1)^{d_n-\iota_n+2}} \right] \max_{|z|=r} |f(z)| \right\}. \tag{16}$$

Here $d_n = 3n+1+\kappa$, and ι_n are certain integers which depend on n. They have to be computed numerically, and tables have been given in [41] for $n = 2, 3, \ldots, 30$. In general, only $\iota_n \ge 0$ is known. A simpler bound was proved in [12, Theorem 1].

Theorem 3.4. *Let C be a closed path in the complex plane with length $L(C)$ which contains $[-1,1]$ in its interior. Let f be analytic in a simply connected domain which contains C in its interior, and let $\delta > 0$ be a lower bound for the distance of the points of C to $[-1,1]$. Then*

$$|R_{2n+1}^{GK}[f]| \le \left\{ \frac{17}{10\sqrt{3n-3(2+\kappa)}} + \frac{4\sqrt{2}}{\sqrt{(6n+3+\kappa)(6n+5+2\kappa)}} \right\} \frac{L(C)\max_{z\in C}|f(z)|}{\pi\, 2^{3n+2+\kappa}\, \delta^{3n+3+\kappa}}.$$

In the special case $C = K_r$ we have

$$|R_{2n+1}^{GK}[f]| \le \left\{ \frac{17}{10\sqrt{3n-3(2+\kappa)}} + \frac{4\sqrt{2}}{\sqrt{(6n+3+\kappa)(6n+5+2\kappa)}} \right\} \frac{r\, \max_{z\in K_r}|f(z)|}{2^{3n+1}(r-1)^{3n+3}}.$$

The comparably simple bound for $\iota_n = 0$ in (16) is worse by a factor of order $\mathcal{O}(n)$. In particular, if f is analytic in the whole complex plane, then for every $\varepsilon > 0$ there is an $r > 1$ such that

$$|R_{2n+1}^{GK}[f]| \le \frac{34}{5\sqrt{3n-3(2+\kappa)}} \frac{2(r^2+1)}{r^{3n+4+\kappa}} \max_{z\in\mathcal{C}_r} |f(z)|\ (1+\varepsilon),$$

where $\mathcal{C}_r$ is the ellipse centered at the origin with semiaxes of length $r + r^{-1}$ and $r - r^{-1}$ (cf. [12, Remark]).

The Chebyshev polynomials T_n have many interesting properties for numerical methods (cf., e.g., [50]). In particular, the investigation of quadrature errors for Chebyshev polynomials is fundamental for error bounds and has attracted special interest in the literature (cf., e.g., [39]). In particular, Peano kernels K_m of a continuous linear functional R on $C[-1,1]$ can often be well approximated for $m > 1$ by

$$K_{m,s}(x) = \frac{2}{\pi} \frac{m!\,2^m}{(2m)!} \sum_{\mu=0}^{s-1} (1-x^2)^{m-1/2}\, R[T_{m+\mu}]\, \frac{P_\mu^{(m)}(x)}{P_\mu^{(m)}(1)},$$

where $P_\mu^{(m)}$ is the μth ultraspherical polynomial with respect to the ultraspherical weight function w_m. Indeed, if

$$\int_{-1}^{1} |K_m(x)|\, dx = \int_{-1}^{1} |K_{m,s}(x)|\, dx \; + \; \rho_s,$$

then

$$|\rho_s| \le \frac{1}{m! 2^{m-1}} \left(\frac{(2m)!\; s!}{(2m+s)!} \, \frac{2m+s}{2m-1} \right)^{\frac{1}{2}} \sup_{\mu \ge s} |\, R[T_{\mu+m}]\,|.$$

This is an important result of Brass and Förster [6]. Since $R_{2n+1}^{GK}[T_\nu] = 0$ for $\nu = 0, 1, \ldots, 3n+1+\kappa$, and

$$\sup_{\mu \ge s} |\, R_{2n+1}^{GK}[T_{\mu+m}]\,| \le \|\, R_{2n+1}^{GK} \,\| \le 4,$$

the errors of the first Chebyshev polynomials which are not integrated exactly are particularly important. The following result is from [16, Theorem 3].

Theorem 3.5. *Let $k \in \{0,1,2,3\}$. Then there exist positive constants C_1, C_2, which are independent of n, such that*

$$\frac{C_1}{n^2\sqrt{n}} \le |R_{2n+1}^{GK}[T_{3n+2+\kappa+2k}]| \le \frac{C_2}{n^2\sqrt{n}} \,.$$

In view of symmetry, we have $R_{2n+1}^{GK}[T_{3n+1+\kappa+2k}] = 0$.

3.3. Error for nonsmooth functions

It is interesting to study how the error of the Gauss-Kronrod formula depends on the smoothness of the function f. From more general results for positive formulas we obtain for fixed $s \in \mathbb{N}$

$$|R_{2n+1}^{GK}[f]| \le C\, n^{-s},$$

if $f \in C^s[-1,1]$. However, this statement cannot be used for comparisons with other quadrature formulas, for which at least the asymptotic behaviour of the best constant is known. For the Gauss-Kronrod formula, the precise asymptotic limit is given by the following theorem from [15, Theorem].

Theorem 3.6. *Let $s \in \mathbb{N}$ be fixed. We have*

$$\lim_{n\to\infty} \; (2n+1)^s \; c_s(R_{2n+1}^{GK}) = \pi^s \int_0^1 |B_s(x)|\, dx \; B\left(\frac{1}{2}\, s+1, \frac{1}{2}\right),$$

where B_s is the sth Bernoulli polynomial and

$$B(x,y) = \int_0^1 t^{x-1}(1-t)^{y-1} dt = \frac{\Gamma(x)\,\Gamma(y)}{\Gamma(x+y)}$$

is the Beta function.

For many applications, it is not realistic to assume that the function is differentiable. It is known that Peano's Theorem can also be used for the class of functions of bounded variation (cf. [5]). Hence it is interesting to obtain results for

$$\rho_V(R_{2n+1}^{GK}) := \sup\{|R_{2n+1}^{GK}[f]| \mid \operatorname{Var} f \le 1\},$$

where Var$f$ denotes the total variation of f on $[-1,1]$. In view of

$$|R_{2n+1}^{GK}[f]| \le \rho_V(R_{2n+1}^{GK}) \operatorname{Var} f,$$

bounds for $\rho_V(R_{2n+1}^{GK})$ lead to error bounds for Q_{2n+1}^{GK}. The following result on the constants $\rho_V(R_{2n+1}^{GK})$ is from [13, Theorem 1.6].

Theorem 3.7. *There holds*

$$\lim_{n\to\infty} (2n+1)\,\rho_V(R_{2n+1}^{GK}) = \frac{\pi}{2}\,.$$

As Gautschi points out in [20, p. 116], "It appears therefore, contrary to widespread belief, that Gauss-Christoffel formulae are not only effective for highly regular functions, but also handle functions of low-order continuity at least as well as other common quadrature rules". A natural question is if this is also true for the Gauss-Kronrod formula. In particular, an interesting question is if for functions of bounded variation, other choices of additional nodes lead to better quadrature formulas. The following answer from [13, Corollary 1.7] is essentially negative.

Theorem 3.8. *Among all quadrature formulas which use the Gaussian nodes $x_{1,n}, \ldots, x_{n,n}$ and $n+1$ additional nodes which interlace with the $x_{\nu,n}$, the Gauss-Kronrod formula is asymptotically optimal with respect to ρ_V.*

The statement of Theorem 3.8 is interesting, since Gauss-Kronrod formulas are constructed with a high degree of exactness and hence for smooth functions. However, Theorem 3.8 states that also for nonsmooth functions we cannot construct an essentially better extension of the Gaussian quadrature formula.

3.4. Comparison with other quadrature formulas

One of the basic assumptions for the use of Gauss-Kronrod formulas in automatic quadrature routines is that Q_{2n+1}^{GK} gives a better approximation than Q_n^G. Brass and Förster have shown in [6] that this assumption is justified for arbitrarily often differentiable functions, in the sense that $c_{2n}(R_{2n+1}^{GK})$ is much smaller than $c_{2n}(R_n^G)$,

$$\frac{c_{2n}(R_{2n+1}^{GK})}{c_{2n}(R_n^G)} < C\,\sqrt[4]{n}\left(\frac{16}{25\sqrt{5}}\right)^n.$$

For the construction of efficient algorithms, it is important to know how much better Q_{2n+1}^{GK} is than Q_n^G, depending on the smoothness of f. The following corollary follows from Theorem 3.6 and known properties of the Gaussian formula [46].

Corollary 3.9. *For every $s \in \mathbb{N}$ we have*

$$\lim_{n\to\infty} \frac{c_s(R_{2n+1}^{GK})}{c_s(R_n^G)} = 2^{-s}.$$

For less smooth functions f one thus can only expect a slight superiority of Q_{2n+1}^{GK} over Q_n^G. But for the class of functions of bounded variation, Q_{2n+1}^{GK} is still twice as good as Q_n^G, with respect to the constant ρ_V.

Corollary 3.10. *We have*

$$\lim_{n\to\infty} \frac{\rho_V(R_{2n+1}^{GK})}{\rho_V(R_n^G)} = \frac{1}{2}.$$

However, the most important question is whether the Gauss-Kronrod scheme is a good method for automatic quadrature. Since Q_{2n+1}^{GK} is used for approximating the integral and the difference $Q_{2n+1}^{GK} - Q_n^G$ is used for approximating the error, a natural question is whether the $2n+1$ nodes could have been chosen better from the beginning. Using the concept of Peano stopping functionals from [19], one can construct stopping functionals for any quadrature formula essentially without adding extra nodes. Thus the Gauss-Kronrod formula Q_{2n+1}^{GK} has to compete in particular with the Gaussian formula Q_{2n+1}^G. The following result from [11, Corollary] shows that for arbitrarily often differentiable functions the Gaussian formula Q_{2n+1}^G is significantly better than Q_{2n+1}^{GK}.

Theorem 3.11. *For all $n \geq 1$ we have*

$$c_{3n+2+\kappa}(R_{2n+1}^G) < c_{3n+2+\kappa}(R_{2n+1}^{GK}),$$

and for $n \geq 15$ we have

$$\frac{c_{3n+2+\kappa}(R_{2n+1}^G)}{c_{3n+2+\kappa}(R_{2n+1}^{GK})} < 3^{-n+1}.$$

Asymptotically we have

$$\lim_{n\to\infty} \left(\frac{c_{3n+2+\kappa}(R_{2n+1}^G)}{c_{3n+2+\kappa}(R_{2n+1}^{GK})}\right)^{\frac{1}{n}} = \sqrt{\frac{6^6}{7^7}} = \frac{1}{4.2013\ldots}.$$

However, this disadvantage could be disregarded if for most applications one would have nonsmooth integrands, and if the Gauss-Kronrod formula would do much better for these. Hence it is important to compare the two formulas also for nonsmooth functions. From Theorem 3.6 and known facts about the Gaussian quadrature formula [46] we obtain the following corollary.

Corollary 3.12. *Let $s \in \mathbb{N}$ be fixed. Then we have*

$$\lim_{n\to\infty} \frac{c_s(R_{2n+1}^G)}{c_s(R_{2n+1}^{GK})} = 1, \qquad \lim_{n\to\infty} \frac{\rho_V(R_{2n+1}^G)}{\rho_V(R_{2n+1}^{GK})} = 1.$$

In combination with suitable stopping rules, Gaussian quadrature formulas are thus a very interesting alternative to Gauss-Kronrod formulas.

Nevertheless, the Gauss-Kronrod formula is attractive if the nesting of nodes is of importance, as in the global quadrature routines in [38, 47]. The only quadrature method of practical importance which has the same property is the Clenshaw-Curtis method. While nodes and weights of the Clenshaw-Curtis method are much easier to compute, Clenshaw-Curtis formulas are significantly worse for arbitrarily often differentiable functions,

$$\lim_{n\to\infty}\left(\frac{c_{2n+1}(R^{GK}_{2n+1})}{c_{2n+1}(R^{CC}_{2n+1})}\right)^{\frac{1}{n}} \le \frac{16}{25\sqrt{5}}.$$

In this context, it would be interesting to obtain further results for the Patterson extensions [43, 26].

3.5. Gauss-Kronrod product integration

In the final part of this section, we study a further application of quadrature formulas that involve the zeros of Stieltjes polynomials, namely Gauss-Kronrod methods for product integration. If the integrand contains singularities or is oscillating, then often product integration methods of the type

$$\int_{-1}^{1} k(x)f(x)\,dx$$
$$= \sum_{\nu=1}^{n} a_{\nu,n}[k]\, f(x_{\nu,n}) + R_n[k,f] =: Q_n[k,f] + R_n[k,f], \qquad k \in L^1,$$

are used, where f is assumed to be smooth. By this approach, a high order of convergence can be obtained under weak assumptions on the function k [51]. A standard method for product integration is based on Clenshaw-Curtis formulas. The main reason is the important result of Sloan and Smith [51] that product integration methods for the Clenshaw-Curtis nodes $x^{CC}_{1,n}, \ldots, x^{CC}_{n,n}$ converge if $k \in L^p$ for some $p > 1$ and f is Riemann integrable, and that under the same assumptions the quadrature is stable,

$$\lim_{n\to\infty} \sum_{\nu=1}^{n} |a^{CC}_{\nu,n}[k]| = \int_{-1}^{1} |k(x)|\,dx < \infty, \tag{17}$$

or, more generally,

$$\lim_{n\to\infty} \sum_{\nu=1}^{n} |a^{CC}_{\nu,n}[k]||f(x^{CC}_{\nu,n}) = \int_{-1}^{1} |k(x)|f(x)\,dx < \infty \tag{18}$$

for every Riemann integrable function f. As pointed out in [51], the limits in (17) and in (18) cannot be replaced by smaller constants for any quadrature formula.

Many quadrature formulas, in particular those with Gauss nodes, do not have these properties. For the product integration method $Q^{GK}_{2n+1}[k,f]$ based on the zeros of $P_n E_{n+1}$, for a long time convergence and stability was known only under much more restrictive assumptions [8]. But the above assertions hold true also for the Gauss-Kronrod method [17, Theorem 1].

Theorem 3.13. *Let $k \in L^p$ for some $p > 1$. Then the product integration method $Q^{GK}_{2n+1}[k,f]$ converges for every Riemann integrable function f. Furthermore, for* *every Riemann integrable* f,

$$\lim_{n\to\infty} \sum_{\nu=1}^{2n+1} |a_{\nu,2n+1}^{GK}[k]|\, f(y_{\nu,2n+1}) = \int_{-1}^{1} |k(x)| f(x)\, dx,$$

where $y_{\nu,2n+1}$ *are the nodes of the Gauss-Kronrod formula* Q_{2n+1}^{GK}.

The product integration formulas $Q_{n+1}^{K}[k,f]$ based on the zeros of the Stieltjes polynomials have similarly good properties (cf. [17, Theorem 2]).

Theorem 3.14. *Let* $\varphi(x) = \sqrt{1-x^2}$. *If* $\sqrt{\varphi}k \in L^p$ *for some* $p > 1$, *then* $Q_{n+1}^{K}[k,f]$ *converges for every Riemann integrable function* f. *If additionally* $\varphi k \in L^{2p}$, *then for every Riemann integrable* f,

$$\lim_{n\to\infty} \sum_{\nu=1}^{n+1} |a_{\nu,n+1}^{K}[k]|\, f(\xi_{\nu,n+1}) = \int_{-1}^{1} |k(x)| f(x)\, dx,$$

where $\xi_{\nu,n+1}$ *are the zeros of the Stieltjes polynomial* E_{n+1}.

These results can be used to obtain uniform convergence of the product integration method for weakly singular integral equations of the type

$$Y(t) = f(t) + \int_a^b k(t,s)\, Y(s)\, ds, \tag{19}$$

where a and b are finite, f is continuous, and k is given by

$$k(t,s) = h(t,s)\, r(t,s),$$

with continuous r and possibly weakly singular h [17]. Replacing the integration in (19) by a product quadrature method, we obtain an approximate solution Y_m which satisfies

$$Y_m(t) = f(t) + \sum_{i=1}^{m} a_{i,m}[h(t,\,\cdot\,)]\, r(t, s_{i,m})\, Y_m(s_{i,m})$$

($m = 2n+1$ for Q_{2n+1}^{GK} and $m = n+1$ for Q_{n+1}^{K}). The following theorems from [17, Corollary 1, Corollary 2] guarantee the uniform convergence of the approximate solutions under suitable assumptions on h.

Theorem 3.15. *If the function* h *satisfies*

$$\int_a^b |h(t,s)|^p\, ds < \infty \qquad \textit{for all} \quad t \in [a,b],$$

and

$$\lim_{t'\to t} \int_a^b |h(t',s) - h(t,s)|^p\, ds = 0$$

for some $p > 1$, *then the solution of the product integration method* Q_{2n+1}^{GK} *based on the Gauss-Kronrod nodes converges uniformly in* $[a,b]$ *to the solution of* (19).

Theorem 3.16. *If the function h satisfies*

$$\int_a^b \sqrt{\varphi(s)}\, |h(t,s)|^p \, ds < \infty \qquad \text{for all} \quad t \in [a,b],$$

with φ as in Theorem 3.14, *and*

$$\lim_{t' \to t} \int_a^b \sqrt{\varphi(s)}\, |h(t',s) - h(t,s)|^p \, ds = 0$$

for some $p > 1$, then the solution of the product integration method Q_{n+1}^K based on the zeros of the Stieltjes polynomials converges uniformly in $[a,b]$ to the solution of (19).

4. Numerical Examples

In this section, we give three numerical examples which illustrate the theoretical results of §3. The functions f are

$$f_1(x) = e^{-\left(\frac{x}{c}\right)^2}, \qquad f_2(x) = \frac{x+1}{c+(x+1)^2}, \qquad f_3(x) = \begin{cases} 0, & x < c, \\ 1, & x \geq c. \end{cases}$$

We compare the Gauss formula Q_n^G, the Gauss-Kronrod formula Q_{2n+1}^{GK} and the Gauss formula Q_{2n+1}^G with respect to their errors in integrating these functions over $[-1,1]$ for specific parameters c.

The function f_1 is "smooth", i.e., arbitrarily often differentiable in $[-1,1]$. The results in Table 1 for $c = \frac{1}{3}$ show that Q_{2n+1}^{GK} is more accurate than Q_n^G by several orders of magnitude. Hence, the standard assumption for the Gauss-Kronrod scheme is fulfilled in this case. The numerical results also show that the Gauss formula Q_{2n+1}^G is better than Q_{2n+1}^{GK} for this function, and its superiority grows exponentially for increasing n as predicted by Theorem 3.11.

The function f_2 is also arbitrarily often differentiable in $[-1,1]$ for $c > 0$, but has poles in the complex plane close to the interval of integration, which have a negative influence on the error. Still, in Table 2, Q_{2n+1}^{GK} is much better than Q_n^G, but here Q_{2n+1}^{GK} and Q_{2n+1}^G could be considered equally good.

The function f_3 is a simple prototype of a function of bounded variation. In Table 3, the ratio of the errors of Q_{2n+1}^{GK} and Q_n^G for odd n approximates $\frac{1}{2}$, as predicted by Corollary 3.10. The formulas Q_{2n+1}^{GK} and Q_{2n+1}^G are equally good, as predicted by Corollary 3.12. Note that for $c = 0$, as in Table 3, the error of a symmetric quadrature formula is equal to half of the weight which is associated with the node 0 (resp. equal to 0 if 0 is not a node).

TABLE 1. Errors for integrating f_1 over $[-1, 1]$, $c = \frac{1}{3}$

n	Abs. Error Q_n^G	Abs. Error Q_{2n+1}^{GK}	Abs. Error Q_{2n+1}^G
1	1.41	3.03 (−1)	3.03 (−1)
2	4.91 (−1)	8.05 (−2)	4.88 (−2)
3	3.03 (−1)	9.67 (−3)	4.62 (−3)
4	1.29 (−1)	4.78 (−4)	2.79 (−4)
5	4.88 (−2)	1.48 (−5)	1.15 (−5)
6	1.60 (−2)	1.37 (−6)	3.46 (−7)
7	4.62 (−3)	6.01 (−8)	7.87 (−9)
8	1.19 (−3)	1.61 (−9)	1.40 (−10)
9	2.79 (−4)	8.44 (−11)	1.99 (−12)
10	5.92 (−5)	4.15 (−12)	2.33 (−14)

TABLE 2. Errors for integrating f_2 over $[-1, 1]$, $c = 0.02$

n	Abs. Error Q_n^G	Abs. Error Q_{2n+1}^{GK}	Abs. Error Q_{2n+1}^G
1	6.91 (−1)	2.99 (−1)	2.99 (−1)
2	1.05 (−1)	1.02 (−2)	5.80 (−2)
3	2.99 (−1)	1.47 (−2)	1.46 (−2)
4	1.90 (−1)	2.29 (−3)	3.23 (−3)
5	5.80 (−2)	9.68 (−4)	5.96 (−4)
6	2.98 (−3)	1.33 (−4)	1.77 (−4)
7	1.46 (−2)	7.00 (−5)	2.23 (−5)
8	9.37 (−3)	3.17 (−6)	9.30 (−6)
9	3.23 (−3)	4.46 (−6)	7.34 (−7)
10	1.68 (−4)	2.32 (−7)	4.72 (−7)
11	5.96 (−4)	2.19 (−7)	1.79 (−8)
12	4.47 (−4)	3.43 (−8)	2.32 (−8)
13	1.77 (−4)	6.19 (−9)	4.69 (−12)
14	2.39 (−5)	2.13 (−9)	1.11 (−9)
15	2.23 (−5)	1.24 (−10)	4.06 (−11)
16	2.06 (−5)	4.04 (−11)	5.16 (−11)
17	9.30 (−6)	2.28 (−11)	3.84 (−12)
18	1.89 (−6)	5.97 (−12)	2.32 (−12)
19	7.34 (−7)	3.25 (−13)	2.69 (−13)
20	9.19 (−7)	6.44 (−13)	1.01 (−13)

TABLE 3. Errors for integrating f_3 over $[-1,1]$, $c=0$

n	Abs. Error Q_n^G	Abs. Error Q_{2n+1}^{GK}	Abs. Error Q_{2n+1}^G
1	1.00	4.44 (−1)	4.44 (−1)
2	0.00	3.11 (−1)	2.84 (−1)
3	4.44 (−1)	2.25 (−1)	2.09 (−1)
4	0.00	1.73 (−1)	1.65 (−1)
5	2.84 (−1)	1.41 (−1)	1.36 (−1)
6	0.00	1.21 (−1)	1.16 (−1)
7	2.09 (−1)	1.05 (−1)	1.01 (−1)
8	0.00	9.22 (−2)	8.97 (−2)
9	1.65 (−1)	8.24 (−2)	8.05 (−2)
10	0.00	7.47 (−2)	7.30 (−2)
11	1.36 (−1)	6.83 (−2)	6.68 (−2)
12	0.00	6.28 (−2)	6.16 (−2)
13	1.16 (−1)	5.81 (−2)	5.71 (−2)
14	0.00	5.41 (−2)	5.32 (−2)
15	1.01 (−1)	5.07 (−2)	4.97 (−2)
16	0.00	4.76 (−2)	4.69 (−2)
17	8.97 (−2)	4.48 (−2)	4.42 (−2)
18	0.00	4.24 (−2)	4.19 (−2)
19	8.05 (−2)	4.03 (−2)	3.98 (−2)
20	0.00	3.83 (−2)	3.78 (−2)

References

[1] M. Abramowitz and I.A. Stegun, Eds., *Handbook of mathematical functions*, Dover Publications, New York, 1965.

[2] B. Baillaud and H. Bourget, *Correspondance d'Hermite et de Stieltjes I,II*, Gauthier-Villars, Paris, 1905.

[3] P. Barrucand, *Intégration numérique, abscisse de Kronrod-Patterson et polynômes de Szegö*, C.R. Acad. Sci. Paris Sér. A-B, **270** (1970), A336–A338.

[4] H. Brass, *Quadraturverfahren*, Studia Mathematica, Skript 3, Vandenhoeck & Ruprecht, Göttingen, 1977.

[5] H. Brass, *Bounds for Peano kernels*, in: H. Brass and G. Hämmerlin, Eds., *Numerical integration IV*, Internat. Ser. Numer. Math. **112**, Birkhäuser, Basel, 1993.

[6] H. Brass and K.-J. Förster, *On the estimation of linear functionals*, Analysis, **7** (1987), 237–258.

[7] H. Brass and K.-J. Förster, *On the application of the Peano representation of linear functionals in numerical analysis*, in: G.V. Milovanović, Ed., *Recent progress in inequalities*, Kluwer, Dordrecht, 1998, 175–202.

[8] C. Dagnino, *Extended product integration rules*, BIT, **23** (1983), 488–499.

[9] P.J. Davis and P. Rabinowitz, *Methods of numerical integration*, 2d ed., Computer Science and Applied Mathematics, Academic Press, Orlando, FL, 1984.

[10] S. Ehrich, *Einige neue Ergebnisse zu den Fehlerkonstanten der Gauß-Kronrod-Quadraturformel*, Z. Angew. Math. Mech., **73** (1993), T882–T886.

[11] S. Ehrich, *Error bounds for Gauss-Kronrod quadrature formulae*, Math. Comp., **62** (1994), 295–304.

[12] S. Ehrich, *Gauss-Kronrod quadrature error estimates for analytic functions*, Z. Angew. Math. Mech., **74** (1994), T691–T693.

[13] S. Ehrich, *On the error of extended Gaussian quadrature formulae for functions of bounded variation*, in: R.V.M. Zahar, Ed., *Approximation and Computation, a festschrift in honor of Walter Gautschi*, Internat. Ser. Numer. Math. **119**, Birkhäuser, Basel, 1994, 169–178.

[14] S. Ehrich, *Asymptotic properties of Stieltjes polynomials and Gauss-Kronrod quadrature formulae*, J. Approx. Theory, **82** (1995), 287–303.

[15] S. Ehrich, *A note on Peano constants of Gauss-Kronrod quadrature schemes*, J. Comput. Appl. Math., **66** (1996), 177–183.

[16] S. Ehrich, *High order error constants of Gauss-Kronrod quadrature formulas*, Analysis, **16** (1996), 335–345.

[17] S. Ehrich, *On product integration with Gauss-Kronrod nodes*, SIAM J. Numer. Anal., **35** (1998), 78–92.

[18] S. Ehrich and G. Mastroianni, *Stieltjes polynomials and Lagrange interpolation*, Math. Comp., **66** (1997), 311–331.

[19] K.-J. Förster, *A survey of stopping rules in quadrature based on Peano kernel methods*, Rend. Circ. Mat. Palermo (2) Suppl. No. **33** (1993), 311–330.

[20] W. Gautschi, *A survey of Gauss-Christoffel quadrature formulae*, in: P.L. Butzer and F. Fehér, Eds., *E.B. Christoffel*, Birkhäuser, Basel, 1981, 72–147.

[21] W. Gautschi, *Gauss-Kronrod quadrature — a survey*, in: G.V. Milovanović, Ed., *Numerical methods and approximation theory III*, Univ. Niš, Niš, 1988, 39–66.

[22] W. Gautschi, *On mean convergence of extended Lagrange interpolation*, J. Comput. Appl. Math., **43** (1992), 19–35.

[23] W. Gautschi and S.E. Notaris, *An algebraic study of Gauss-Kronrod quadrature formulae for Jacobi weight functions*, Math. Comp., **51** (1988), 231–248.

[24] W. Gautschi and T.J. Rivlin, *A family of Gauss-Kronrod quadrature formulae*, Math. Comp., **51** (1988), 749–754.

[25] W. Gautschi and R.S. Varga, *Error bounds for Gaussian quadrature of analytic functions*, SIAM J. Numer. Anal., **20** (1983), 1170–1186.

[26] Th. Gerstner and M. Griebel, *Numerical integration using sparse grids*, manuscript (1998).

[27] G. Hämmerlin, *Fehlerabschätzung bei numerischer Integration nach Gauss*, in: B. Brosowski and E. Martensen, Eds., *Methoden und Verfahren der mathematischen Physik*, Band 6, B.I.-Hochschultaschenbücher, No. **725**, Bibliographisches Inst., Mannheim, 1972, 153–163.

[28] G. Hämmerlin and K.-H. Hoffmann, *Numerische Mathematik*, 4th ed., Springer-Lehrbuch, Grundwissen Mathematik **7**, Springer, Berlin, 1994.

[29] E. Heine, *Handbuch der Kugelfunktionen. Theorie und Anwendungen I,II*, 2d ed., Reimer, Berlin, 1878, 1881.

[30] IMSL Fortran Numerical Libraries, Version 3.0, Visual Numerics, 1998.

[31] A.S. Kronrod, *Nodes and weights of quadrature formulas. Sixteen-place tables*, Nauka, Moscow, 1964; translation by Consultants Bureau, New York, 1965.

[32] A.S. Kronrod, *Integration with control of accuracy*, Soviet Physics Dokl., **9** (1964), 17–19.

[33] G. Monegato, *A note on extended Gaussian quadrature rules*, Math. Comp., **30** (1976), 812–817.

[34] G. Monegato, *Positivity of the weights of extended Gauss-Legendre quadrature rules*, Math. Comp., **32** (1978), 243–245.

[35] G. Monegato, *An overview of results and questions related to Kronrod schemes*, in: G. Hämmerlin, Ed., *Numerische Integration*, Internat. Ser. Numer. Math. **45**, Birkhäuser, Basel, 1979, 231–240.

[36] G. Monegato, *Stieltjes polynomials and related quadrature rules*, SIAM Rev., **24** (1982), 137–158.

[37] I.P. Mysovskih, *A special case of quadrature formulae containing preassigned nodes* (Russian), Vesci Akad. Navuk BSSR Ser. Fiz.-Tehn. Navuk, No. **4** (1964), 125–127.

[38] NAG Fortran Library, Mark 18, 1998.

[39] D. Nicholson, P. Rabinowitz, N. Richter-Dyn, and D. Zeilberger, *On the error in the numerical integration of Chebyshev polynomials*, Math. Comp., **25** (1971), 79–86.

[40] S.E. Notaris, *Gauss-Kronrod quadrature formulae for weight functions of Bernstein-Szegö type II*, J. Comput. Appl. Math., **29** (1990), 161–169.

[41] S.E. Notaris, *Error bounds for Gauss-Kronrod quadrature formulae of analytic functions*, Numer. Math., **64** (1993), 371–380.

[42] S. E. Notaris, *An overview of results on the existence or nonexistence and the error term of Gauss-Kronrod quadrature formulae*, in: R.V.M. Zahar, Ed., *Approximation and computation, a festschrift in honor of Walter Gautschi*, Internat. Ser. Numer. Math. **119**, Birkhäuser, Boston, MA, 1994, 485–496.

[43] T.N.L. Patterson, *The optimum addition of points to quadrature formulae*, Math. Comp., **22** (1968), 847–856; loose microfiche suppl. C1–C11; errata, ibid., **23** (1969), 892.

[44] F. Peherstorfer, *Weight functions admitting repeated positive Kronrod quadrature*, BIT, **30** (1990), 145–151.

[45] F. Peherstorfer, *On the asymptotic behaviour of functions of the second kind and Stieltjes polynomials and on the Gauss-Kronrod quadrature formulas*, J. Approx. Theory, **70** (1992), 156–190.

[46] K. Petras, *Asymptotic behaviour of Peanokernels of fixed order*, in: H. Brass and G. Hämmerlin, Eds., *Numerical integration III*, Internat. Ser. Numer. Math. **85**, Birkhäuser, Basel, 1988, 186–198.

[47] R. Piessens, E. de Doncker-Kapenga, C.W. Überhuber, and D.K. Kahaner, *QUADPACK – a subroutine package for automatic integration*, Springer Ser. Comput. Math. **1**, Springer, Berlin, 1983.

[48] P. Rabinowitz, *The exact degree of precision of generalized Gauss-Kronrod integration rules*, Math. Comp., **35** (1980), 1275–1283. [Corrigendum in [49]]

[49] P. Rabinowitz, *On the definiteness of Gauss-Kronrod integration rules*, Math. Comp., **46** (1986), 225–227.

[50] T.J. Rivlin, *The Chebyshev polynomials. From approximation theory to algebra and number theory*, 2d ed., Pure and Applied Mathematics, Wiley, New York, 1990.

[51] I.H. Sloan and W.E. Smith, *Product-integration with the Clenshaw-Curtis points: implementation and error estimates*, Numer. Math., **34** (1980), 387–401.

[52] G. Szegö, *Über gewisse orthogonale Polynome, die zu einer oszillierenden Belegungsfunktion gehören*, Math. Ann., **110** (1935), 501–513 [cf. also R. Askey, Ed., *Collected Works*, vol. 2, 545–558].

[53] G. Szegö, *Orthogonal polynomials*, Amer. Math. Soc. Colloq. Publ. **23**, Amer. Math. Soc., Providence, RI, 1939.

[54] S. Wolfram, *The mathematica book*, Wolfram Media and Cambridge University Press, Cambridge, 1996.

Sven Ehrich
GSF-National Research Center for Environment and Health
Institute of Biomathematics and Biometrics
Ingolstädter Landstr. 1
D-85764 Neuherberg, Germany
E-mail address: `ehrich@gsf.de`

International Series of Numerical Mathematics
Vol. 131,

Fast Solution of Confluent Vandermonde-Like Linear Systems Using Polynomial Arithmetic

Hans-Jürgen Fischer

Abstract. In the present paper we propose a method for the fast solution of some confluent Vandermonde-like systems by polynomial arithmetic. The results (though less general than known results of Hao Lu) can be formulated explicitly in terms of numerator polynomials. In order to avoid numerical difficulties with the canonical (monomial) basis, algorithms for multiplication and division in arbitrary orthogonal bases are developed. We discuss applications to the numerical computation of Gauss-rules from given modified moments and to the solution of Gram systems.

1. Introduction

Let σ be a given positive measure with infinite support $S(\sigma)$ and finite moments of all orders. Then there exists a unique family of monic polynomials π_j with

$$\int \pi_l(x)\pi_j(x)\,d\sigma(x) = 0 \quad \text{for } l < j \text{ and } \quad \int \pi_j^2(x)\,d\sigma(x) = \gamma_j^{-2} > 0 \qquad (1)$$

for $j = 0, 1, \ldots$ (γ_j is the leading coefficient of the *orthonormal* polynomial of degree j). They satisfy a three-term recurrence relation

$$\pi_{j+1}(x) = (x - \alpha_j)\pi_j(x) - \beta_j\pi_{j-1}(x) \quad \text{for} \quad j = 0, 1, \ldots, \qquad (2)$$

if we set $\pi_{-1}(x) \equiv 0$.

From the standard theory of orthogonal polynomials we know that there is an n-point quadrature formula with nodes $\{\tau_i\}_{i=1,\ldots,n}$ and weights $\{\sigma_i\}_{i=1,\ldots,n}$ integrating exactly all polynomials of degree $\leq 2n-1$:

$\{\tau_i\}_{i=1,\ldots,n}$ are the zeros of π_n and the weights σ_i are defined by

$$\sigma_i^{-1} = \gamma_{n-1}^2\pi_{n-1}(\tau_i)\pi_n'(\tau_i) \quad \text{for} \quad i = 1, \ldots, n$$

(for the last equation, compare Exercise 6.2 and Equation (4.10) in Chapter I of Chihara's book [1]).

If we are given modified moments $m_j = \int p_j(x)\,d\sigma(x)$ for some appropriate set of polynomials $\{p_j\}$, then our orthogonal polynomials and thus the recurrence coefficients α_j and β_j are well defined in terms of them. Usually, the polynomials p_j are classical orthogonal polynomials with known recurrence coefficients. Then, the recurrence coefficients α_j and β_j can be computed with the well-known method of modified moments. In order to study the numerical condition of this algorithm, Gautschi considered the map from modified moments m_j to $\{\sigma_i\}, \{\tau_i\}$ instead (see [5], [6], and [7]). In this analysis, he used known results on inverses of confluent Vandermonde-like matrices. The (well-conditioned) map from $\{\alpha_j\}, \{\beta_j\}$ to $\{\sigma_i\}, \{\tau_i\}$ is an eigenvalue problem for the tridiagonal Jacobi matrix. These problems can be solved very efficiently, see [8], [10], and [9].

However, we can try to calculate directly the nodes $(\tau_i)_{i=1,\dots,n}$ and the weights $(\sigma_i)_{i=1,\dots,n}$ of a Gaussian quadrature formula for the measure σ from our modified moments m_j. Since the quadrature formula is exact for polynomials of degree $2n-1$, we obtain the nonlinear system of equations

$$\sum_{i=1}^{n} \sigma_i\, p_j(\tau_i) = m_j, \qquad j = 0, \dots, 2n-1.$$

This can be written as

$$F(\sigma, \tau) = m$$

with $\sigma = [\sigma_1, \dots, \sigma_n]^T$, $\tau = [\tau_1, \dots, \tau_n]^T$, $m = [m_0, \dots, m_{2n-1}]^T$, and the vector-valued function

$$F(\sigma, \tau) = \sum_{i=1}^{n} \sigma_i\, p(\tau_i),$$

where $p = [p_0, \dots, p_{2n-1}]^T$. The method of Newton leads to an iteration

$$\mathbf{P}_c\mathbf{D} \begin{bmatrix} \sigma^{(k+1)} - \sigma^{(k)} \\ \tau^{(k+1)} - \tau^{(k)} \end{bmatrix} = m - F(\sigma^{(k)}, \tau^{(k)}),$$

where $\mathbf{P}_c = \mathbf{P}_c(\tau^{(k)})$ will be defined in the next section and

$$\mathbf{D} = \mathbf{D}(\sigma^{(k)}) = \operatorname{diag}(1, \dots, 1, \sigma_1^{(k)}, \dots, \sigma_n^{(k)}).$$

The calculation of the residual vector can be avoided. We have

$$F(\sigma, \tau) = \mathbf{P}_c(\tau) \begin{bmatrix} \sigma \\ \mathbf{0} \end{bmatrix} = \mathbf{P}_c(\tau)\mathbf{D}(\sigma) \begin{bmatrix} \sigma \\ \mathbf{0} \end{bmatrix},$$

and this implies that we have to solve the system

$$\mathbf{P}_c\mathbf{D} \begin{bmatrix} \sigma^{(k+1)} \\ \tau^{(k+1)} - \tau^{(k)} \end{bmatrix} = m$$

in each iteration.

The solution of linear systems with a general $n \times n$ matrix, of course, requires $O(n^3)$ arithmetic operations. Fortunately, there are known fast algorithms for this special structure of matrices with only $O(n^2)$ operations if the polynomials p_j satisfy a three-term recurrence relation; see the bibliography in [17]. In this 1994 paper of Hao Lu an even faster algorithm is proposed for confluent Vandermonde matrices, which is based on fast polynomial arithmetic and requires only $O(n \log^2 n)$ arithmetic operations. Our results, Theorem 1 and Theorem 2, were inspired by this article. In the meantime, Hao Lu generalized his results considerably in the article [18] (of which the author became aware only in February 1998, after finishing the investigations reported here). In fact, Hao Lu is able to handle the case of polynomials satisfying a k-term recurrence relation. Thus, our result is much weaker — it is just the identification of what Hao Lu calls "J-match" in terms of numerator polynomials in the special case $k = 3$. Some drawbacks and advantages of our formulation will be discussed below. Moreover, we handle here only the special case of confluent Vandermonde-like matrices (n double nodes) necessary for our iterative algorithm above. The general case could be formulated in the same way as in [17] and [18].

2. Vandermonde-Like Matrices and Their Inversion

Many problems in interpolation or construction of quadrature formulas require the solution of linear systems with (possibly confluent) Vandermonde-like matrices or their transposes. If we are given a set of polynomials $\{p_j\}_{j=0,\dots,n-1}$, where p_j has degree j, and a set of distinct nodes $\{\tau_i\}_{i=1,\dots,n}$, then the $n \times n$ Vandermonde-like matrix is defined by

$$\mathbf{P} = \begin{bmatrix} p_0(\tau_1) & p_0(\tau_2) & \cdots & p_0(\tau_n) \\ p_1(\tau_1) & p_1(\tau_2) & \cdots & p_1(\tau_n) \\ \vdots & \vdots & & \vdots \\ p_{n-1}(\tau_1) & p_{n-1}(\tau_2) & \cdots & p_{n-1}(\tau_n) \end{bmatrix}. \tag{3}$$

We consider here only the simplest case of a $2n \times 2n$ confluent Vandermonde-like matrix related to the set of polynomials $\{p_j\}_{j=0,\dots,2n-1}$ and the set of nodes $\{\tau_i\}_{i=1,\dots,n}$ defined by

$$\mathbf{P}_c = \begin{bmatrix} p_0(\tau_1) & \cdots & p_0(\tau_n) & p_0'(\tau_1) & \cdots & p_0'(\tau_n) \\ p_1(\tau_1) & \cdots & p_1(\tau_n) & p_1'(\tau_1) & \cdots & p_1'(\tau_n) \\ \vdots & & \vdots & \vdots & & \vdots \\ p_{2n-1}(\tau_1) & \cdots & p_{2n-1}(\tau_n) & p_{2n-1}'(\tau_1) & \cdots & p_{2n-1}'(\tau_n) \end{bmatrix}. \tag{4}$$

Results on the structure of the inverses of (confluent) Vandermonde-like matrices have been known for some time. Usually, they are formulated in terms of interpolation polynomials at the nodes τ_i. If we write interpolation polynomials in Newton's form, we obtain algorithms due to Higham; see his papers [14] and [15]. Kailath and Olshevsky use another approach in their paper [16], where many bibliographical references are given.

3. The Method of Hao Lu

In this section, we assume that the polynomials p_j are orthogonal with respect to some measure s. There are several normalizations due to historical reasons, so we do not require them to be monic or orthonormal. Instead, we assume the leading coefficient of p_j to be K_j. The polynomials then satisfy a three-term recurrence

$$xp_j(x) = \frac{K_j}{K_{j+1}} p_{j+1}(x) + a_j p_i(x) + \frac{K_j}{K_{j-1}} b_j p_{j-1}(x) \quad \text{for} \quad j = 0, 1, \dots , \tag{5}$$

where $p_{-1} \equiv 0$ and $p_0 \equiv K_0$. Their orthogonality relation is

$$\int p_j(x)\, p_k(x)\, ds(x) = \frac{K_j^2}{c_j^2}\, \delta_{kj} = H_j\, \delta_{kj}, \tag{6}$$

where the leading coefficients c_j of *orthonormal* polynomials can be found from

$$c_j^{-2} = c_0^{-2} \prod_{k=1}^{j} b_k$$

($c_0^{-2} = \int 1\, ds(x)$ is the full weight of the measure). It is well known that the polynomials

$$q_j(x) = \int \frac{p_j(x) - p_j(t)}{x - t}\, ds(t) \tag{7}$$

for $j > 0$ satisfy the same recurrence (5) with the initial values $q_0 \equiv 0$ and $q_1 \equiv c_0^{-2} K_1$ (and this means that the degree of q_j is $j - 1$; the polynomials are called *associated polynomials of the second kind* or *numerator polynomials*). We can use them to formulate the following identity:

Lemma 1. *Let p_j be the orthogonal polynomials with respect to some measure s satisfying Equation* (5) *and the orthogonality conditions* (6), *and let the numerator polynomials be defined by* (7). *Then we have*

$$\frac{p_n(x) - p_n(\tau)}{x - \tau} = \sum_{j=0}^{n-1} \frac{1}{H_j} \left[p_j(x)\, q_n(x) - p_n(x)\, q_j(x) \right] p_j(\tau). \tag{8}$$

Proof. We follow closely the proof of [19, Chapter 2, Lemma 7.4]. The expression

$$\frac{p_n(x) - p_n(\tau)}{x - \tau}$$

is a polynomial of degree $n - 1$ in τ and thus can be written as

$$\frac{p_n(x) - p_n(\tau)}{x - \tau} = \sum_{j=0}^{n-1} d_j\, p_j(\tau).$$

Multiplying both sides by $p_j(\tau)$ and integrating, we obtain from (6)

$$\int \frac{p_n(x) - p_n(\tau)}{x - \tau}\, p_j(\tau)\, ds(\tau) = H_j\, d_j.$$

The proposition now follows from the identity

$$\frac{p_n(x)-p_n(\tau)}{x-\tau}\,p_j(\tau) = \frac{p_n(x)-p_n(\tau)}{x-\tau}\,p_j(x) - \frac{p_j(x)-p_j(\tau)}{x-\tau}\,p_n(x) + \frac{p_j(x)-p_j(\tau)}{x-\tau}\,p_n(\tau),$$

since

$$\int \frac{p_n(x)-p_n(\tau)}{x-\tau}\,p_j(x)\,ds(\tau) = q_n(x)\,p_j(x),$$

$$\int \frac{p_j(x)-p_j(\tau)}{x-\tau}\,p_n(x)\,ds(\tau) = q_j(x)\,p_n(x),$$

and

$$\int \frac{p_j(x)-p_j(\tau)}{x-\tau}\,p_n(\tau)\,ds(\tau) = 0,$$

owing to orthogonality (observe that $\frac{p_j(x)-p_j(\tau)}{x-\tau}$ is a polynomial of degree $j-1 < n$ in τ). □

This auxiliary result enables us to formulate the method of Hao Lu for arbitrary orthogonal polynomials. First, we introduce some notations.

Definition 1. *Let q and r be polynomials. There are unique polynomials s and t with $\deg(t) < \deg(r)$ and $q(x) = r(x)\,s(x) + t(x)$, and we will write $s = q \div r$ and $t = q \bmod r$.*

Now we are in a position to formulate a variant of Theorem 1.1 of [17].

Theorem 1. *The solution $\mathbf{y} = [y_1, \dots, y_n]^T$ of the Vandermonde-like system*

$$\mathbf{Py} = \mathbf{r}$$

with $\mathbf{r} = [r_0, \dots, r_{n-1}]^T$ is given by

$$y_i = \frac{g(\tau_i)}{\omega_n'(\tau_i)},$$

where

$$\omega_n(x) = (x-\tau_1)\cdots(x-\tau_n), \quad g(x) = \omega_n(x)r(x) \div p_n(x)$$

and

$$r(x) = q_n(x)\sum_{j=0}^{n-1}\frac{r_j}{H_j}\,p_j(x) \bmod p_n(x).$$

Proof. We multiply the equation

$$\sum_{i=1}^{n} y_i p_j(\tau_i) = r_j$$

by $\dfrac{1}{H_j}\,[p_j(x)\,q_n(x) - p_n(x)\,q_j(x)]$ and sum over j. This (in view of the identity (8)) gives the equation

$$\sum_{i=1}^{n} y_i \frac{p_n(x) - p_n(\tau_i)}{x - \tau_i} = \sum_{j=0}^{n-1} \frac{r_j}{H_j}\,[p_j(x)\,q_n(x) - p_n(x)\,q_j(x)]$$

$$= q_n(x) \sum_{j=0}^{n-1} \frac{r_j}{H_j}\,p_j(x) - p_n(x) \sum_{j=0}^{n-1} \frac{r_j}{H_j}\,q_j(x).$$

Since the left-hand side (and thus the right-hand side) is a polynomial of degree $\leq n-1$, it coincides with $r(x)$. If we multiply this equation by $\omega_n(x)$, we have

$$p_n(x) \sum_{i=1}^{n} y_i \frac{\omega_n(x)}{x - \tau_i} - \sum_{i=1}^{n} y_i\, p_n(\tau_i) \frac{\omega_n(x)}{x - \tau_i} = \omega_n(x) r(x).$$

Now the polynomial $\displaystyle\sum_{i=1}^{n} y_i\, p_n(\tau_i) \frac{\omega_n(x)}{x - \tau_i}$ has degree $\leq n-1$, consequently

$$\sum_{i=1}^{n} y_i \frac{\omega_n(x)}{x - \tau_i} = \omega_n(x) r(x) \div p_n(x) = g(x).$$

Substituting $x = \tau_i$, we obtain $y_i\, \omega_n'(\tau_i) = g(\tau_i)$, and this proves our theorem. □

We now formulate the result for the confluent case.

Theorem 2. *The solution*

$$\begin{bmatrix} \mathbf{y} \\ \mathbf{z} \end{bmatrix} = [y_1, \dots, y_n, z_1, \dots, z_n]^T$$

of the confluent Vandermonde-like system

$$\mathbf{P}_c \begin{bmatrix} \mathbf{y} \\ \mathbf{z} \end{bmatrix} = \mathbf{r}$$

with $\mathbf{r} = [r_0, \dots, r_{2n-1}]^T$ *is given by*

$$y_i = \frac{g'(\tau_i)}{\omega_n'^2(\tau_i)} - 2\frac{\omega_n''(\tau_i)\, g(\tau_i)}{\omega_n'^3(\tau_i)} \quad \textit{and} \quad z_i = \frac{g(\tau_i)}{\omega_n'^2(\tau_i)},$$

where

$$\omega_n(x) = (x - \tau_1) \cdots (x - \tau_n), \quad g(x) = \omega_n^2(x) r(x) \div p_{2n}(x)$$

and

$$r(x) = q_{2n}(x) \sum_{j=0}^{2n-1} \frac{r_j}{H_j}\, p_j(x) \bmod p_{2n}(x).$$

Proof. From the identity (8) with $2n$ instead of n we obtain by differentiating with respect to τ

$$\frac{p_{2n}(x)-p_{2n}(\tau)}{(x-\tau)^2}-\frac{p_{2n}'(\tau)}{x-\tau}=\sum_{j=0}^{2n-1}\frac{1}{H_j}\,[p_j(x)\,q_{2n}(x)-p_{2n}(x)\,q_j(x)]\,p_j'(\tau).$$

Together with the original identity we obtain

$$\sum_{i=1}^{n}\left[y_i\,\frac{p_{2n}(x)-p_{2n}(\tau_i)}{x-\tau_i}+z_i\left\{\frac{p_{2n}(x)-p_{2n}(\tau_i)}{(x-\tau_i)^2}-\frac{p_{2n}'(\tau_i)}{x-\tau_i}\right\}\right]=r(x)$$

as above. Multiplying by $\omega_n^2(x)$, we have

$$p_{2n}(x)\sum_{i=1}^{n}\left[y_i\,\frac{\omega_n^2(x)}{x-\tau_i}+z_i\,\frac{\omega_n^2(x)}{(x-\tau_i)^2}\right]+R_{2n-1}(x)=r(x)\,\omega_n^2(x),$$

where

$$R_{2n-1}(x)=-\sum_{i=1}^{n}\left[(y_i\,p_{2n}(\tau_i)+z_i\,p_{2n}'(\tau_i))\,\frac{\omega_n^2(x)}{x-\tau_i}+z_i\,p_{2n}(\tau_i)\,\frac{\omega_n^2(x)}{(x-\tau_i)^2}\right]$$

is a polynomial of degree $\leq 2n-1$. This gives the equation

$$\sum_{i=1}^{n}\left[y_i\,\frac{\omega_n^2(x)}{x-\tau_i}+z_i\,\frac{\omega_n^2(x)}{(x-\tau_i)^2}\right]=r(x)\,\omega_n^2(x)\div p_{2n}(x)=g(x).$$

Now substituting $x=\tau_i$, we obtain

$$z_i\,\omega_n'^2(\tau_i)=g(\tau_i),$$

and differentiating with respect to x and substituting $x=\tau_i$, we have

$$y_i\,\omega_n'^2(\tau_i)+2z_i\,\omega_n''(\tau_i)\,\omega_n'(\tau_i)=g'(\tau_i).$$

Our proposition follows immediately from these two equations. □

We return now to the general results in [18]. As we mentioned above, they are valid for polynomials p_j satisfying some k-term recurrence relation. The polynomial r of our theorems is obtained there in terms of a J-match without a polynomial multiplication and division (see Theorems 3.2 and 3.3 in [18]). For numerical computations, however, it has to be converted into the polynomial basis used for the polynomial arithmetic. This requires some effort, too, since the J-match is defined by another recurrence (see Theorem 2.3 in [18]). Hao Lu proposed the use of the canonical basis $\{x^j\}$ for computations, allowing fast multiplications/divisions by FFT. We suspect that this could lead to serious numerical difficulties in some cases. If the p_j are Chebyshev polynomials of the first kind, the J-match consists of Chebyshev polynomials of the second kind. The coefficients of these polynomials in the canonical basis are prohibitively large. The sum of their absolute values is of order $(1+\sqrt{2})^j$ for U_j (see Lemma 2(ii) in our paper [4]). Already for $n=1000$ this number is too large to be representable as a floating-point number on many computers. Moreover, the error of the final result will be influenced by the *absolute* error of these coefficients, owing to cancellation of leading digits. We only mention here that this exponential growth of the relevant condition number was found already in Gautschi's paper [5] (compare his Equation (3.10) for $\omega = 1$).

For these reasons we feel that the arithmetic operations with polynomials should be performed in the more natural basis of the p_j themselves. Algorithms for the arithmetic operations are developed in §5 below. As we will see, the algorithms use only the recurrence coefficients of the basis. Then, the identification of the J-match in terms of the numerator polynomials is an advantage, because they satisfy essentially the same recurrence relation.

4. Some Numerical Experiments

In this section, we report the results of some numerical experiments with the iterative method mentioned in the Introduction. More precisely, we consider a variant, namely the system

$$\mathbf{P}_c\mathbf{D}(\sigma^{(k+1)}) \begin{bmatrix} \sigma^{(k+1)} \\ \tau^{(k+1)} - \tau^{(k)} \end{bmatrix} = m.$$

We assume that we are given Chebyshev moments, i.e., $p_k = T_k$. This makes our algorithm very efficient: Up to a numerical factor, the U_n are the second-kind polynomials for T_n. Division by T_n is trivial in view of the identities

$$2T_n(x)T_m(x) = T_{n+m}(x) + T_{|n-m|}(x)$$

and

$$2U_n(x)T_m(x) = \begin{cases} U_{n+m}(x) + U_{n-m}(x) & , n \geq m, \\ U_{2m-1} & , n = m-1, \\ U_{n+m}(x) + U_{m-n-2}(x), & n \leq m-2, \end{cases}$$

and the product of two polynomials of degree n can be calculated by the fast DCT (discrete cosine transform) in $O(n \log n)$ arithmetic operations. This is interpolation at the zeros of Chebyshev polynomials, and it was also used for the computation of the polynomial ω_n in Theorem 2.

As starting values $\tau^{(0)}$ we used the zeros of T_n; the vector $\sigma^{(0)}$ is not needed. We did not attempt to find conditions for convergence. The measure σ in our experiments was the measure of orthogonality of Jacobi polynomials in the following three cases:

1. $\alpha = \beta = 0$ — Legendre polynomials,
2. $\alpha = \beta = \frac{1}{2}$ — Chebyshev polynomials of the second kind,
3. $\alpha = -\beta = -\frac{1}{2}$ — Chebyshev polynomials of the third kind.

As a defect criterion, we used the largest (in absolute value) component of the computed correction $\tau^{(k+1)} - \tau^{(k)}$. The results are displayed in the Tables 1–3.

TABLE 1. Defect after k iterations, Case 1, $n = 500$

k	defect	k	defect
1	4.22E–04	5	5.94E–08
2	1.44E–04	6	4.95E–12
3	2.87E–05	7	1.65E–16
4	4.67E–06		

TABLE 2. Defect after k iterations, Case 2, $n = 51$

k	defect	k	defect
1	8.65E–03	6	1.29E–04
2	6.03E–03	7	2.32E–06
3	5.06E–03	8	5.62E–10
4	2.73E–03	9	1.76E–15
5	9.90E–04		

TABLE 3. Defect after k iterations, Case 3, $n = 25$

k	defect	k	defect
1	3.33E–02	6	9.11E–05
2	2.29E–02	7	2.40E–07
3	1.78E–02	8	1.14E–12
4	6.55E–03	9	1.17E–15
5	1.61E–03		

Surprisingly, in the first case our iteration converged for all values of n we tried, up to $n = 500$. However, to obtain a defect of the order of the machine precision, we needed 7 iterations. The CPU-time required was 7.09 seconds, whereas the traditional method (solving the eigenvalue problem for the Jacobi matrix) required only 0.88 seconds. The zeros of Chebyshev polynomials of the second and third kinds, of course, are not so near to the starting values. The values of n displayed in these cases were the largest for which the iteration converged at all.

We see that our direct method is far inferior to traditional methods, even for very large n. However, the explicit formulas for the solution of (confluent) Vandermonde-like systems are interesting in their own right. Moreover, the resulting iterative procedure is interesting from a theoretical point of view, since it shows that the time complexity can be reduced to $O(n \log^2 n)$.

The implementation of the algorithm in the case of Chebyshev moments uses fast polynomial arithmetic (the algorithms for fast discrete cosine transforms are taken from the paper of Steidl and Tasche [20]). This makes our algorithm faster than that of Gohberg and Olshevsky in [12], even though their method is based essentially on the same identity (8).

The idea of using nested polynomial division (and FFT) to speed up the evaluation of a polynomial in n given points (this gives the operation count of $O(n \log^2 n)$ in [17]) was used already in the paper of Finck, Heinig and Rost [3] in an algorithm for Cauchy-Vandermonde systems (with the same operation count). We add the remark that the evaluation of a polynomial in the Chebyshev basis at n given points can be speeded up (asymptotically) by another method based on interpolation and FFT; see the article of Dutt and Rokhlin [2].

5. The Arithmetic of Orthogonal Polynomials

To implement our algorithms in the general case, we must have algorithms for arithmetic operations with polynomials. Of course, we assume that any polynomial is given as a linear combination of our orthogonal polynomials p_j. Thus, the polynomials are encoded as vectors, and it is clear how to add or subtract or multiply by a scalar factor. Multiplication by a polynomial is far more complicated. In the general case, the products $p_n p_m$ cannot be written easily as linear combinations of p_j. Even if the respective coefficients were known, this approach would require $O(n^3)$ operations for the computation of the product of two polynomials of degree n.

Before we explain how to multiply, or divide by, polynomials, we look at the problem of evaluation in some point t. The starting point is the recurrence relation (5). In this section we assume that our polynomials are *monic* (i.e., $K_j \equiv 1$) — this assumption saves many multiplications. Now for the evaluation of the polynomial

$$q(x) = \sum_{j=0}^{n} q_j p_j(x)$$

at $x = t$ we have Clenshaw's algorithm (see Algorithm 1).

$u_n \leftarrow q_n, \quad u_{n+1} \leftarrow 0,$
for $j = n-1$ downto 0 **do**
  $u_j \leftarrow (t - a_j) \times u_{j+1} - b_{j+1} \times u_{j+2} + q_j$
end for
$q(t) \leftarrow u_0$

Algorithm 1: Calculate $q(t)$

Since we assumed $K_j \equiv 1$, the recurrence (5) can be written in the form

$$xp_j(x) = p_{j+1}(x) + a_j p_j(x) + b_j p_{j-1}(x)$$

for $j = 0, 1, \ldots$. As we can easily see, this equation allows us to multiply a polynomial by the independent variable x. Obviously, the corresponding transformation of the coefficient vector is just a multiplication by a tridiagonal matrix. We will denote this transformation by Ψ (for a polynomial q of degree n, $\Psi(q)$ can be computed in $O(n)$ operations).

Thus, we can use Clenshaw's algorithm not only with some numerical value t, but with the independent variable x, assuming the quantities u_j to be polynomials. This seems to be a rather strange idea, since we just reconstruct the polynomial $q(x)$. Consequently, this is an implementation of the identity map in $O(n^2)$ arithmetical operations. However, this can be very practical!

First, we can perform the operations of the algorithm in *another* orthogonal basis satisfying another recurrence of type (5). This gives a conversion algorithm from one orthogonal basis to another in $O(n^2)$ arithmetic operations.

Second, we can assume that instead of the numerical values q_j we have the polynomials $q_j r(x)$, where $r(x)$ is some given polynomial. Then the algorithm obviously calculates

$$\sum_{j=0}^{n} q_j r(x) p_j(x) = q(x) r(x).$$

Again, we formulate this as Algorithm 2 (where we simply write $c \times u$ for the multiplication of a vector u by a number c). If the degree of $r(x)$ is m, then the

$u_n \leftarrow q_n \times r, \quad u_{n+1} \leftarrow 0,$ $\{u_j$ and r are polynomials$\}$
for $j = n-1$ downto 0 **do**
  $u_j \leftarrow \Psi(u_{j+1}) - a_j \times u_{j+1} - b_{j+1} \times u_{j+2} + q_j \times r$ $\{\Psi$ multiplies by $x\}$
end for
$s \leftarrow u_0$ {the result is a polynomial}

Algorithm 2: Calculate $s(x) = q(x)r(x)$

operation count of our algorithm obviously is $O((m + n/2)n)$. Here the order of factors is crucial — we should guarantee $n \le m$ for efficiency (if at all possible, since of course $r(x)$ and thus the result could be represented in another orthogonal basis).

Interestingly, the computation of $q \div r$ and $q \bmod r$ can be performed along the same lines. In fact, any polynomial u_j in the polynomial version of Clenshaw's Algorithm 1 can be witten uniquely as

$$u_j = v_j r + w_j \quad \text{with} \quad \deg(w_j) < \deg(r).$$

For $\deg(u_j) < \deg(r)$ we have $v_j \equiv 0$ and $w_j = u_j$; this is true for the initial values u_{n+1} and u_n. Obviously, addition, subtraction, and multiplication by a constant of the representations (v_j, w_j) of u_j can be performed componentwise. The multiplication by the variable x is not difficult if we first multiply the "lower" component w_j by x and then reduce the product mod r to obtain the "carry" for the higher part v_j. Since $\deg(\Psi(w_j)) = \deg(w_j) + 1 \le \deg(r)$, this carry is always a number d_j. Now, a careful analysis of the operations we perform with the polynomials v_j shows that this is exactly the symbolic version of Clenshaw's algorithm. Thus, it reconstructs the polynomial from its coefficients, and this means that the numbers d_j are the coefficients of $q \div r$. These observations give an elegant algorithm for polynomial division in the basis $\{p_j\}$:

$w_n \leftarrow q_n, \quad w_{n+1} \leftarrow 0$, $\{w_j$ are polynomials$\}$
for $j = n-1$ downto 0 **do**
  $w_j \leftarrow \Psi(w_{j+1}) - a_j \times w_{j+1} - b_{j+1} \times w_{j+2} + q_j$ $\{\Psi$ multiplies by $x\}$
  $d_j \leftarrow p^*_m(w_j)/p^*_m(r)$ {quotient of highest coefficients, $m = \deg(r)$}
  $w_j \leftarrow w_j - d_j \times r$
end for
$s \leftarrow [d_0, \dots, d_{n-m}]$
$t \leftarrow w_0$ {the result is a polynomial}

Algorithm 3: Calculate $s = q \div r$ and $t = q \bmod r$

6. Another Application of the Method of Hao Lu

Some problems in weighted L^2-approximation require the solution of linear systems with Gram matrices. The $n \times n$ Gram matrix is defined by

$$\mathbf{G} = [h_{ij}]_{i,j=1}^n \text{ with } h_{ij} = \int p_{i-1}(x)p_{j-1}(x)\, d\sigma(x),$$

where σ is some measure. Of course, all integrals have to be finite. Furthermore, we assume that the support $S(\sigma)$ of our measure contains at least n points. The monic orthogonal polynomials π_j with respect to σ satisfy the three-term recurrence relation (2).

The Gaussian quadrature formula for the measure σ gives a factorization of $\mathbf{G}$ (we omit the elementary proof):

Lemma 2. *With the notations defined before and* $\mathbf{D} = \operatorname{diag}(\sigma_1, \dots, \sigma_n)$ *we have the representation*

$$\mathbf{G} = \mathbf{P}\mathbf{D}\mathbf{P}^T.$$

Now we can formulate our result for Gram systems.

Theorem 3. *Let* $\mathbf{y} = [y_0, \dots, y_{n-1}]^T$ *be the solution of the linear system*

$$\mathbf{G}\mathbf{y} = \mathbf{r},$$

where $\mathbf{G}$ *is defined as above and* $\mathbf{r} = [r_0, \dots, r_{n-1}]^T$. *Then the polynomial*

$$y(x) = \sum_{j=0}^{n-1} y_j p_j(x)$$

can be computed from

$$y(x) = \gamma_{n-1}^2 g(x)\, \pi_{n-1}(x) \bmod \pi_n(x),$$

with

$$g(x) = r(x)\,\pi_n(x) \div p_n(x)$$

and

$$r(x) = q_n(x) \sum_{j=0}^{n-1} \frac{r_j}{H_j}\, p_j(x) \bmod p_n(x).$$

Proof. The solution $\mathbf{g} = [g_1, \dots, g_n]^T$ of $\mathbf{Pg} = \mathbf{r}$ is given by

$$g_i = \frac{g(\tau_i)}{\pi_n'(\tau_i)}$$

(since we have $\omega_n(x) = \pi_n(x)$), and multiplying by $\mathbf{D}^{-1}$, using the formula for σ_i^{-1} in the Introduction, we obtain

$$\mathbf{P}^T\mathbf{y} = [\gamma_{n-1}^2 \pi_{n-1}(\tau_1) g(\tau_1), \dots, \gamma_{n-1}^2 \pi_{n-1}(\tau_n) g(\tau_n)]^T.$$

But we have

$$\mathbf{P}^T\mathbf{y} = [y(\tau_1), \dots, y(\tau_n)]^T$$

by the definition of the polynomial y, and if the polynomial y of degree $\leq n-1$ coincides with $\gamma_{n-1}^2 \pi_{n-1}(x) g(x)$ in all zeros of π_n, then this implies

$$y(x) = \gamma_{n-1}^2 \pi_{n-1}(x) g(x) \bmod \pi_n(x). \qquad \square$$

This result was inspired by an article of Gemignani [11], where he re-invents a known method (see [13]).

References

[1] T.S. Chihara, *An introduction to orthogonal polynomials*, Gordon and Breach, New York, 1978.

[2] A. Dutt and V. Rokhlin, *Fast Fourier transforms for nonequispaced data*, SIAM J. Sci. Comput., **14** (1993), 1368–1393.

[3] T. Finck, G. Heinig, and K. Rost, *An inversion formula and fast algorithms for Cauchy-Vandermonde matrices*, Linear Algebra Appl., **183** (1993), 179–191.

[4] H.-J. Fischer, *On the condition of orthogonal polynomials via modified moments*, Z. Anal. Anwendungen, **15** (1996), 223–244.

[5] W. Gautschi, *Questions of numerical condition related to polynomials*, in: C. de Boor and G.H. Golub, Eds., *Recent advances in numerical analysis*, Academic Press, New York, 1978, 45–72.

[6] W. Gautschi, *On generating orthogonal polynomials*, SIAM J. Sci. Statist. Comput., **3** (1982), 289–317.

[7] W. Gautschi, *On the sensitivity of orthogonal polynomials to perturbations in the moments*, Numer. Math., **48** (1986), 369–382.

[8] W. Gautschi. *Computational aspects of orthogonal polynomials*, in: P. Nevai, Ed., *Orthogonal polynomials, theory and practice*, NATO Adv. Sci. Inst. Ser. C Math. Phys. Sci. **294**, Kluwer, Dordrecht, 1990, 181–216.

[9] W. Gautschi, *Algorithm 726: ORTHPOL — a package of routines for generating orthogonal polynomials and Gauss-type quadrature rules*, ACM Trans. Math. Software, **20** (1994), 21–62.

[10] W. Gautschi, *Orthogonal polynomials: applications and computation*, in: A. Iserles, Ed., *Acta numerica 1996*, Acta Numer. **5**, Cambridge Univ. Press, Cambridge, 1996, 45–119.

[11] L. Gemignani. *A fast algorithm for generalized Hankel matrices arising in finite-moment problems*, Linear Algebra Appl., **267** (1997), 41–52.

[12] I. Gohberg and V. Olshevsky, *Fast inversion of Chebyshev-Vandermonde matrices*, Numer. Math., **67** (1994), 71–92.

[13] G. Heinig, W. Hoppe, and K. Rost, *Structured matrices in interpolation and approximation problems*, Wiss. Z. Tech. Univ. Karl-Marx-Stadt, **31** (1989), 196–202.

[14] N.J. Higham, *Fast solution of Vandermonde-like systems involving orthogonal polynomials*, IMA J. Numer. Anal., **8** (1988), 473–486.

[15] N.J. Higham, *Stability analysis of algorithms for solving confluent Vandermonde-like systems*, SIAM J. Matrix Anal. Appl., **11** (1990), 23–41.

[16] T. Kailath and V. Olshevsky, *Displacement-structure approach to polynomial Vandermonde and related matrices*, Linear Algebra Appl., **261** (1997), 49–90.

[17] H. Lu, *Fast solution of confluent Vandermonde linear systems*, SIAM J. Matrix Anal. Appl., **15** (1994), 1277–1289.

[18] H. Lu, *Solution of Vandermonde-like systems and confluent Vandermonde-like systems*, SIAM J. Matrix Anal. Appl., **17** (1996), 127–138.

[19] E.M. Nikishin and V.N. Sorokin, *Rational approximations and orthogonality*, Translations of Mathematical Monographs **92**, Amer. Math. Soc., Providence, RI, 1991.

[20] G. Steidl and M. Tasche, *A polynomial approach to fast algorithms for discrete Fourier-cosine and Fourier-sine transforms*, Math. Comp., **56** (1991), 281–296.

Hans-Jürgen Fischer
Fakultät für Mathematik
Technische Universität Chemnitz
D-09107 Chemnitz, Germany
E-mail address: hfischer@mathematik.tu-chemnitz.de

International Series of Numerical Mathematics
Vol. 131, © 1999 Birkhäuser Verlag Basel/Switzerland

On Discrete Polynomial Least-Squares Approximation in Moving Time Windows

Erich Fuchs

Abstract. The paper presents a modification algorithm for least-squares approximations in moving time windows for equidistant discrete data. Properties of discrete shift and difference operators are used to derive fast square-root- and division-free algorithms, especially suitable for implementation on DSPs, for some particular discrete weights. It can be shown that a modification of an equally weighted least-squares fit needs not more than $5n + 2$ multiplications and $5n - 2$ additions for polynomial degrees $n \leq 5$.

1. Introduction

Time series methods for signal trend analysis are widely used to provide the possibility of a trend correction as a signal preprocessing step, or to extrapolate the received signal a small time period into the future. With high sampling rates, such methods often fail under real-time conditions, owing to the computational costs of generating a single solution, even if the underlying model is kept simple. However, many real-time applications within monitoring systems controlling industrial production processes require that a signal trend based on the last, say m, measured (discrete) values be available after each receipt of a new measurement.

Therefore, the practical needs in the area of least-squares problems triggered the development of techniques modifying an existing solution by adding (or deleting) a data value with less effort than a computation from scratch (so-called modified least-squares problems). Expressed in mathematical terms, the factorization of an $m \times n$ matrix $A = QR$ with $m > n$ in an orthogonal factor Q and an upper triangular factor R can be updated by adding a new row in $O(n^2)$, or deleting an old one in $O(mn)$, operations while the original decomposition requires $O(mn^2)$ operations (see [1] and [7]).

If polynomials are used for approximating (not necessarily equidistant) discrete data, it is useful to represent the least-squares solution in polynomials orthogonal on that data. Updating (adding a data point) and downdating (deleting a data point) techniques for such polynomial least-squares fits were analyzed by Elhay, Golub, and Kautsky in [2] making use of special properties of orthogonal polynomials; several modification algorithms for adding or deleting a value y at abscissa x with a certain weight w were derived and compared. The most economical method of this type needs $14n + O(1)$ floating-point operations and $2n$ square roots to modify a polynomial fit of degree n by computing a new set of orthogonal polynomials and the respective least-squares fit coefficients. These very general and efficient algorithms can be used for creating "moving" least-squares fits by updating (include a new measurement) and downdating (discard an old value) an existing solution. Visually, this can be imagined as a window (with variable size depending on the abscissas) moving over the received signal data.

Most of the sensors used, e.g. in process monitoring systems, however, generate equidistant measured values. This fact allows us to change the point of view in the following way: instead of using a window moving over the data (and changing the orthogonal polynomial basis with every step) one can consider a fixed window corresponding to a fixed coordinate system and a signal passing by that window. Translating this line of thinking into mathematics leads to the following advantage: one can keep the same orthogonal polynomial basis for every least-squares fit, and therefore it is only required to modify the orthogonal expansion coefficients representing the least-squares solution in that basis.

It is the purpose of this paper to derive and analyze iterative algorithms modifying polynomial least-squares fits by taking advantage of equidistant measuring data. On the one hand, the methods developed in this paper are not applicable in the general sense considered in [2], but on the other hand, the specialization to equidistant data reveals new possibilities to modify an approximating polynomial by exploiting the simple "shifted" structure of the new data to be processed. By restriction to equidistant data it will be shown that for some suitable discrete weights square-root-free algorithms can be derived which are just right for the implementation on digital signal processors (DSPs) often used for signal processing under real-time conditions. For a polynomial degree $n \leq 5$ a modification of an equally weighted least-squares fit requires only $\leq 5n + 2$ multiplications and $\leq 5n - 2$ additions.

The main results presented here are based on properties of discrete shift and difference operators, which will be introduced in §2. After deriving some basic formulas for orthogonal polynomials shifted along the real axis in §2, modification algorithms for least-squares approximations with equal weights in moving time windows are presented in §3 together with some numerical results. Section 4 analyzes the question as to which other discrete weights are suitable for these fast iterative approximation updates. Some application examples in §5 conclude the paper.

2. Notation and Basic Relations

Since we want to shift a signal with every step it passes by a certain window, we need suitable mathematical definitions covering these aspects; the linear operators $\mathrm{T}, \nabla : \mathcal{F}(\mathbb{R}, \mathbb{R}) \to \mathcal{F}(\mathbb{R}, \mathbb{R})$ ($\mathcal{F}$ is the vector space of all real functions on $\mathbb{R}$) are defined by

$$\mathrm{T}f(x) \stackrel{\text{def}}{=} f(x+1) \qquad \text{(shift operator)},$$
$$\nabla f(x) \stackrel{\text{def}}{=} f(x) - f(x-1) \qquad \text{(backward difference operator)}.$$

Powers of these operators are understood as compositions and the zeroth power is the identity. For every $f, g \in \mathcal{F}(\mathbb{R},\mathbb{R})$ the operator T fulfills $\mathrm{T}(fg) = \mathrm{T}f \cdot \mathrm{T}g$ and moreover, T is a bijection with $\mathrm{T}^{-1}f(x) = f(x-1)$. If $p \in \mathcal{P}_k(\mathbb{R},\mathbb{R})$ (the vector space of all polynomials defined on $\mathbb{R}$ with degree less than or equal to k) is a polynomial with $\deg(p) = \nu$, then $\nabla^{\nu+1}p(x) = 0$ for every $x \in \mathbb{R}$. This leads to the following simple equation needed later on.

Lemma 2.1. *Let $p \in \mathcal{P}_k(\mathbb{R},\mathbb{R})$ and $n \in \mathbb{N}$; then negative n-th powers of* T *can be expressed in terms of ∇ by*

$$\mathrm{T}^{-n}p = \sum_{\nu=0}^{\min\{n,k\}} \binom{n}{\nu}(-1)^\nu \nabla^\nu p.$$

Proof. For any $f \in \mathcal{F}(\mathbb{R},\mathbb{R})$ the relation $\mathrm{T}^{-n}f = \sum_{\nu=0}^{n} \binom{n}{\nu}(-1)^\nu\nabla^\nu f$ can be proved easily by induction, using $\mathrm{T}^{-(n+1)}f = \mathrm{T}^{-n}f - \mathrm{T}^{-n}(\nabla f)$. Since $\nabla^\nu p$ is the zero function for $\nu > \deg(p)$, the assertion follows. □

In the sequel we look at polynomials orthogonal on the set $M \stackrel{\text{def}}{=} \{0,\dots,m\}$ (which will be our time window under consideration) with respect to an inner product $(\cdot|\cdot) : \mathcal{P}(\mathbb{R},\mathbb{R}) \times \mathcal{P}(\mathbb{R},\mathbb{R}) \to \mathbb{R}$ defined by

$$(p|\,q) \stackrel{\text{def}}{=} \sum_{j=0}^{m} p(j)q(j)w_j,$$

where $w_j > 0$ for $j \in M$. In general, polynomials orthogonal with respect to such discrete inner products can be computed by several methods, see e.g. [2], [3], or [6].

In the simplest case, if all the weights are equal, the polynomials orthogonal with respect to $(p|\,q) = \sum_{j=0}^{m} p(j)q(j)$ are the well-known **Discrete Chebyshev Polynomials** (or Gram Polynomials), which correspond to the class of Hahn Polynomials. Here the coefficients of the three-term recurrence relation are explicitly known, see e.g. [6] or [9]. The monic Discrete Chebyshev Polynomials fulfill

$$q_{i+1}(x) = \left(x - \frac{m}{2}\right) q_i(x) - \delta_i\, q_{i-1}(x)$$

for $i = 1,\dots,n-1$, where $q_0(x) = 1$, $q_1(x) = x - \frac{m}{2}$ and $\delta_i \stackrel{\text{def}}{=} \dfrac{i^2\left((m+1)^2 - i^2\right)}{4\,(4\,i^2 - 1)}$ with $i \geq 1$. Setting $\delta_0 \stackrel{\text{def}}{=} m+1$, the squared norms of these polynomials induced by the inner product $(\cdot|\cdot)$ are given by

$$\|q_i\|^2 = \prod_{k=0}^{i} \delta_k = \frac{(i\,!)^4}{(2\,i)\,!\,(2\,i+1)\,!} \cdot \prod_{k=-i}^{i} (m+1+k).$$

The least-squares approximation with a polynomial of degree $n \leq m$ for the data $y_0, \ldots, y_m$ measured at $0, \ldots, m$, i.e., the polynomial q minimizing $\sum_{j=0}^{m} (q(j)-y_j)^2$, is given by the orthogonal expansion

$$q = \sum_{i=0}^{n} \frac{\sum_{j=0}^{m} y_j q_i(j)}{\|q_i\|^2} q_i$$

in Discrete Chebyshev Polynomials. As an additional preparatory step we have to investigate how a set of orthogonal polynomials shifted along the real axis can be represented in the basis of the original orthogonal polynomials.

Lemma 2.2. *Let $\{p_0, \ldots, p_n\}$ be a set of not necessarily monic orthogonal polynomials fulfilling a three-term recurrence relation*

$$p_0(x) = a_{-1}, \tag{1}$$
$$p_{k+1}(x) = (a_k x - b_k)\, p_k(x) - c_k\, p_{k-1}(x) \text{ for } 0 \leq k \leq n-1,$$

where $p_{-1} \equiv 0$. Given $r \in \mathbb{Z}$, $i \in \{0, \ldots, n\}$ and $l \in \{0, \ldots, i\}$, the coefficients $\gamma_l^{(r,i)}$ are defined by the representation of a polynomial shifted r times in the basis of the orthogonal polynomials:

$$\mathrm{T}^r p_i \stackrel{\text{def}}{=} \sum_{l=0}^{i} \gamma_l^{(r,i)} p_l. \tag{2}$$

With the settings $\gamma_{-1}^{(r,i)} \stackrel{\text{def}}{=} 0$ for $i \in \{0, \ldots, n\}$ and $\gamma_j^{(r,i)} \stackrel{\text{def}}{=} 0$ for $j > i \geq -1$, the following recurrence relation holds: $\gamma_0^{(r,0)} = 1$ and for $i = 0, \ldots, n-1$ and $l = 0, \ldots, i+1$ the coefficients $\gamma_l^{(r,i+1)}$ can be obtained by

$$\gamma_l^{(r,i+1)} = \frac{a_i}{a_{l-1}} \gamma_{l-1}^{(r,i)} + c_{l+1} \frac{a_i}{a_{l+1}} \gamma_{l+1}^{(r,i)} + \left(\frac{a_i b_l}{a_l} + a_i r - b_i \right) \gamma_l^{(r,i)} - c_i \gamma_l^{(r,i-1)}. \tag{3}$$

If the orthogonal polynomials are monic, the recurrence relation for the coefficients $\gamma_l^{(r,i+1)}$ simplifies significantly to

$$\gamma_l^{(r,i+1)} = \gamma_{l-1}^{(r,i)} + c_{l+1} \gamma_{l+1}^{(r,i)} + (r - b_i + b_l)\, \gamma_l^{(r,i)} - c_i \gamma_l^{(r,i-1)}. \tag{4}$$

Proof. There are several ways to prove these recurrence relations. A direct proof is possible by multiple application of the three-term recurrence relation to the expression $\mathrm{T}^r p_{i+1} = \sum_{l=0}^{i+1} \gamma_l^{(r,i+1)} p_l$. Another way of deriving equation (4) is by using mixed moments in the orthogonal expansion $\mathrm{T}^r p_i = \sum_{l=0}^{i} \frac{(p_l|\,\mathrm{T}^r p_i)}{\|p_l\|^2} p_l$ and taking advantage of the fact that $\{T^r p_0, \dots, T^r p_n\}$ fulfill a three-term recurrence relation with the coefficients $b_k - r$ and c_k. Using a matrix formulation, the change of polynomial basis can be expressed as a Sylvester equation which can be explicitly solved, owing to the structure of the matrices involved. □

Based on these recurrence relations, it is easy to derive direct formulas for small polynomial degrees. After these preparations we can derive fast iterative approximation formulas in moving time windows.

3. Modifying Least-Squares Approximations

In real-time applications, processing a stream of continuously arriving data, it is often necessary to provide a polynomial approximation on a fixed number of the most recently measured data after each receipt of one (or more) new value(s). In particular, the orthogonal expansion coefficients of such a polynomial contain signal trend information, which can be used for trend correction or for trend extrapolation in conjunction with threshold crossing methods. In this section we will restrict ourselves to the simple case of equally weighted least-squares approximations to equidistant discrete data.

3.1. The problem and its solution

If we use a time window with fixed length for looking at the measuring data $y_0, \dots, y_t, y_{t+1}, \dots, y_{t+m}, y_{t+m+1}, \dots$, we can fix it within a coordinate system at $0, \dots, m$ and use the monic Discrete Chebyshev Polynomials for every approximation. Whenever a certain section $y_s, \dots, y_{s+m}$ passes by, we want to have a polynomial approximating this part of the signal. With the definition $\alpha_{i,s} \stackrel{\text{def}}{=} \sum_{j=0}^{m} y_{s+j}\, q_i(j)$ for $i = 0, \dots, n$ with $n \le m$ we get the approximations

$$p_t(x) = \sum_{i=0}^{n} \frac{\alpha_{i,t}}{\|q_i\|^2} q_i(x), \qquad p_{t+r}(x) = \sum_{i=0}^{n} \frac{\alpha_{i,t+r}}{\|q_i\|^2} q_i(x)$$

for the data $y_t, \dots, y_{t+m}$ and $y_{t+r}, \dots, y_{t+r+m}$, respectively. Since the polynomial basis used and the corresponding squared norms are independent of the data, they can be determined once and for all, and we have to solve the following **problem:** develop a fast and stable algorithm suitable for real-time applications, which updates $\alpha_{i,t}$ to $\alpha_{i,t+r}$ for $i = 0, \dots, n$ with as few floating-point operations as possible.

With the notation introduced in Lemma 2.2 applied to the monic Discrete Chebyshev Polynomials, the equation $T^{-r} q_i - q_i = -\sum_{l=0}^{i-1} \gamma_l^{(r,i)} T^{-r} q_l$ follows by simple manipulations. This result leads to the following update formula.

Theorem 3.1. (Iterative approximation updates) *For $i = 0, \dots, n$ the orthogonal expansion coefficients $\alpha_{i,t+r}$ can be computed from $\alpha_{i,t}$ by*

$$\alpha_{i,t+r} = \alpha_{i,t} + \sum_{j=m+1}^{m+r} y_{t+j} q_i(j) - \sum_{j=0}^{r-1} y_{t+j} q_i(j) - \sum_{l=0}^{i-1} \gamma_l^{(r,i)} \alpha_{l,t+r}, \tag{5}$$

where the empty sum $\sum_{l=0}^{-1} \gamma_l^{(r,0)} \alpha_{l,t+r} \stackrel{\text{def}}{=} 0.$

Proof. For $i = 0, \dots, n$ the following is valid:

$$\begin{aligned}
\alpha_{i,t+r} &= \sum_{j=0}^{m} y_{t+r+j} q_i(j) = \sum_{j=r}^{m+r} y_{t+j} \left(\mathrm{T}^{-r} q_i(j) + q_i(j) - q_i(j)\right) \\
&= \sum_{j=r}^{m+r} y_{t+j} q_i(j) + \sum_{j=r}^{m+r} y_{t+j} \left(- \sum_{l=0}^{i-1} \gamma_l^{(r,i)} \mathrm{T}^{-r} q_l(j) \right) \\
&= \alpha_{i,t} + \sum_{j=m+1}^{m+r} y_{t+j} q_i(j) - \sum_{j=0}^{r-1} y_{t+j} q_i(j) - \sum_{l=0}^{i-1} \gamma_l^{(r,i)} \alpha_{l,t+r}.
\end{aligned}$$

□

With these formulas it is possible to compute $\alpha_{i,t+r}$ iteratively, starting from $i = 0$ and going up to $i = n$, with a fixed number of arithmetic operations independent of m. If $r = 1$, the update formula simplifies to

$$\alpha_{i,t+1} = \alpha_{i,t} + y_{t+m+1}\, q_i(m+1) - y_t\, q_i(0) - \sum_{l=0}^{i-1} \gamma_l^{(1,i)} \alpha_{l,t+1}, \tag{6}$$

allowing an update of the approximating polynomial after each receipt of a new measured value. The main advantage of this important case ($r = 1$) are the computational costs: if every data-independent coefficient ($\gamma_l^{(r,i)}$, $q_i(j)$, $\|q_i\|^2$) is computed and stored as a constant in advance, one single update requires for a polynomial degree $n \leq 5$, which is the usual case in practice, not more than $5n+2$ multiplications and $5n - 2$ additions (disregarding trivial multiplications by 1.0, e.g. with $q_0(j)$). This can be seen from Table 1, where $am : \mathbb{N}^2 \to \mathbb{N}^2$ is a map showing the number of multiplications and additions needed for the computation of $\alpha_{i,t+r}$ for a given parameter constellation $(n, r) \in \mathbb{N}^2$. These results were obtained by counting the additions and multiplications needed in equations (5) and (6), respectively. In the general case, $\frac{n^2+5n+4}{2}$ multiplications and $\frac{n^2+5n-4}{2}$ additions are necessary when $r = 1$. If $n \leq 5$, the estimation $n^2 \leq 5n$ leads to the claimed $5n + 2$ multiplications and $5n - 2$ additions for small polynomial degrees used in practice.

Since the number of arithmetic operations required is not influenced by the approximation length m, the algorithm is especially suitable for large m. With increasing degree n, the advantage of calculating one update with $r = k > 1$

TABLE 1. Complexity of iterative approximation updates

	$r > 1$		$r = 1$	
n	$am(n,r)$	$\sum_{i=0}^{n} am(i,r)$	$am(n,r)$	$\sum_{i=0}^{n} am(i,r)$
0	$(2r, 0)$		$(2, 0)$	
1	$(2r+1, 2r+1)$	$(4r+1, 2r+1)$	$(3, 2)$	$(5, 2)$
2	$(2r+2, 2r+2)$	$(6r+3, 4r+3)$	$(4, 3)$	$(9, 5)$
$n \geq 3$	$(2r+n, 2r+n)$	$\left(\frac{(n+4r)(n+1)}{2}, \frac{(n+4r+1)n}{2}\right)$	$(n+2, n+2)$	$\left(\frac{n^2+5n+4}{2}, \frac{n^2+5n-4}{2}\right)$

grows compared to carrying out k updates with $r = 1$; the choice between these possibilities, however, is application-dependent.

Another advantage is that the iterative updates can be performed without square roots or divisions; for an implementation of these algorithms on DSPs this fact is an important prerequisite, because these operations are normally not available as a single instruction on such specific processors. Furthermore, starting with the signal data initially set to 0, the algorithm does not require an exceptional initial computation of $\alpha_{i,t}$, thus keeping the implementation simple and regular. Finally, equation (5) allows the efficient computation of a new approximation after a window shift $r > 1$ with *one* modification applied to the orthogonal expansion coefficients. This is not possible with other methods.

3.2. Numerical results

Many approximation updates are necessary if high sampling rates are used; this requires a careful analysis of possible rounding errors due to the iterative scheme of the algorithm. Errors introduced at the computation of $\alpha_{i,t}$ may be propagated to orthogonal coefficients of higher degree. These numerical problems (which occur especially if the equations (5) and (6) are implemented on specific processors (DSPs) having fewer bits available than usual) can be avoided by implementing an analogue of the Kahan-Babuška summation, which is generally a tool for floating-point additions of relatively small values to already large sums [8]. On Digital Signal Processors like TMS320C31 (Texas Instruments, 32-bit floating-point arithmetic), the accuracy may also be increased by implementing 40-bit addition algorithms.

Comparing the presented iterative approximation updates with the up- and downdating methods developed in [2] (in the special case of equally weighted equidistant data) shows that the latter loose accuracy faster; this is caused by the well-known instability of downdating (the condition depends on the data) and by rounding errors coming from the change of the orthogonal polynomial basis and of the Fourier coefficients in every step.

For applications it is important how accurately the approximating polynomial is computed: we therefore calculate the relative difference

$$\text{error}^2 \stackrel{\text{def}}{=} \sum_{k=0}^{m} \frac{\left(\sum\limits_{i=0}^{n} \frac{\sum\limits_{j=0}^{m} y_{t+j} q_i(j)}{\|q_i\|^2} q_i(k) - \sum\limits_{i=0}^{n} \frac{\alpha_{i,t}}{\|q_i\|^2} q_i(k) \right)^2}{y_{t+k}^2},$$

with $\alpha_{i,t}$ computed iteratively (starting at $t = 0$) for several t and low degrees n and plot the number of correct decimal digits ($-\log|\text{error}|$). Since this quantity mirrors the accuracy of the orthogonal expansion coefficients, we dispense with plotting these values. The following results were obtained by implementing the methods compared in C using the datatype double without applying any additional improvement of accuracy like Kahan-Babuška summation. Fig. 1 shows on

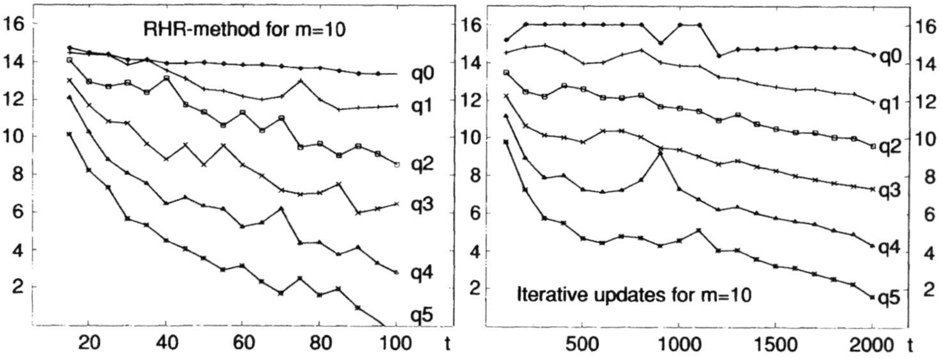

FIGURE 1. Accuracy of the approximating polynomial ($m = 10$)

the left the decay of accuracy for approximation length $m = 10$ of the RHR-method described in [2], where the relative difference is taken between a reference solution computed with RHR-updates only and an iterative solution (the underlying signal data were cutting force values measured in a CNC-lathe). At $t = 100$ the approximating polynomials of degrees 4 and 5 do not guarantee sufficient relative accuracy anymore. On the right side of Fig. 1 the number of correct decimal digits of the iterative approximation update method (with $r = 1$) is represented: although the accuracy decreases significantly with increasing t, the decay is much slower (after 100 updates the approximating polynomial of degree 5 still has 10 correct decimal digits).

With increasing approximation length m, both methods become more stable: the equal weights together with large m result in better condition numbers for the hyper-rotations of the RHR-downdate method. The stability of the iterative algorithm improves, because the orthogonal expansion coefficients $\alpha_{i,t}$ become larger with m and rounding errors introduced are reduced by divisions with relatively large squared norms $\|q_i\|^2$. Fig. 2 shows an example of the RHR-method with

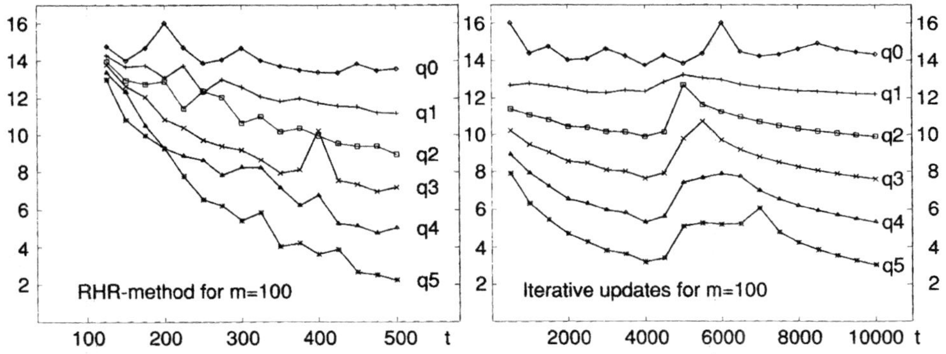

FIGURE 2. Accuracy of the approximating polynomial ($m = 100$)

$m = 100$ on the left side and next to it the accuracy of the iterative algorithm for the same parameters. Again, the accuracy of the RHR-method deteriorates faster; the increase of accuracy at around $t = 5,000$ in the right side of Fig. 2 is data-dependent, because there is a rapid ascent within the signal leading to better conditions for the computation of $\alpha_{i,t}$. The condition numbers for all input data of the algorithm (including $\gamma_l^{(r,i)}$ and $q_i(j)$) are small as long as $\alpha_{i,t}$ is not very close to 0, which is already avoided as much as possible with the choice of *monic* Discrete Chebyshev Polynomials resulting in relatively large $\alpha_{i,t}$ and $\|q_i\|^2$.

Statistical tests applying the method to 25 different randomly generated signals showed the same behavior: the mean relative difference error increases with the polynomial degree n in magnitude as already described; sufficient (a term, which is certainly application-dependent) accuracy can be reached for $n = 2$ or $n = 3$, even after 200,000 updates depending on m. The increasing accuracy with ascending m, already mentioned, is described in Fig. 3, where the mean correct number of digits is plotted over m after 50,000 updates on the left, and after 100,000 updates on the right. We remark that for $r > 1$ similar results were obtained, and the errors are of the same magnitude.

Although the condition of the problem is still data-dependent as in other modification algorithms too, better numerical results can be achieved with equations (5) and (6). The improvement of robustness is due to the combination of updating and downdating into one procedure. Furthermore, the use of a fixed polynomial basis avoids the (probably ill-conditioned) calculation of a new polynomial basis.

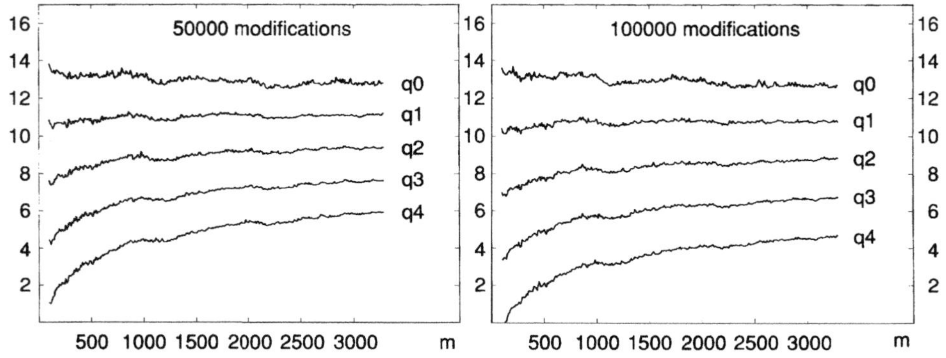

FIGURE 3. Accuracy depending on the approximation length m

4. Update Formulas for Other Weights

In the proof of Theorem 3.1 the discrete weights did not play any role, because they were all equal and set to 1. In this section we want to analyze whether other discrete weights allow analogous approximation update formulas. In order to be able to apply the shift and difference operators introduced in §2, we will consider *continuous* weight functions. These functions will be evaluated at equidistant points $j \in \mathbb{Z}$ thus again leading to discrete weights.

Just like the orthogonal polynomials are consistent with the shift operation, owing to the evident fact that shifted orthogonal polynomials can be represented in the original orthogonal polynomial basis (compare Lemma 2.2), the weight functions used have to be compatible with the shift operation as well. A closer look shows that this can be achieved if a weight function $\rho : \mathbb{R} \to \mathbb{R}$ fulfills $\nabla\rho = c \cdot \rho$ (c being a constant) or $\nabla^k \rho = 0$ for some k.

4.1. Geometric weights

The property $\nabla \exp(x) = (1 - e^{-1}) \exp(x)$ of the exponential function motivates the following definition.

Definition 4.1. (Geometric weights) *Let $w > 0$ with $w \neq 1$; then the function $\rho : \mathbb{R} \to \mathbb{R}$ with $\rho(x) = w^x$ defines discrete weights $\rho(j)$ for $j \in \mathbb{Z}$, which we call geometric weights, because they form a geometric sequence $1, w, \ldots, w^m$ for $j \in M$.*

The monic polynomials $\{p_0, \ldots, p_n\}$ orthogonal with respect to the inner product $(p|\,q)_w \overset{\text{def}}{=} \sum_{j=0}^{m} p(j)q(j)\rho(j)$ fulfill a three-term recurrence relation, and with $\tilde{\alpha}_{i,t} \overset{\text{def}}{=} \sum_{j=0}^{m} y_{t+j} p_i(j)\rho(j)$ the least-squares polynomial $\tilde{p}_t(x) = \sum_{i=0}^{n} \frac{\tilde{\alpha}_{i,t}}{\|p_i\|_w^2} p_i(x)$ solves the problem of finding a polynomial which minimizes $\sum_{j=0}^{m} (p(j) - y_{t+j})^2 \rho(j)$.

Suppose $\mathrm{T}^r p_i = \sum_{l=0}^{i} \tilde{\gamma}_l^{(r,i)} p_l$ for $i \in \{0, \ldots, n\}$; then the following update formula for approximations in moving time windows can be derived.

Corollary 4.2. (Approximation updates for geometric weights) *Given $\tilde{\alpha}_{i,t}$, the coefficients $\tilde{\alpha}_{i,t+r}$ can be obtained going from $i = 0$ up to $i = n$ by*

$$\tilde{\alpha}_{i,t+r} = w^{-r}\left[\tilde{\alpha}_{i,t} + \sum_{j=m+1}^{m+r} y_{t+j} p_i(j)\rho(j) - \sum_{j=0}^{r-1} y_{t+j} p_i(j)\rho(j)\right] - \sum_{l=0}^{i-1} \tilde{\gamma}_l^{(r,i)} \tilde{\alpha}_{l,t+r}.$$

Proof. In analogy to the proof of Theorem 3.1, the assertion follows from

$$\begin{aligned}
\tilde{\alpha}_{i,t+r} &= \sum_{j=0}^{m} y_{t+r+j} p_i(j)\rho(j) \\
&= \sum_{j=r}^{m+r} y_{t+j} \left[(p_i\,(\mathrm{T}^{-r}\rho))(j) + (\mathrm{T}^{-r}(p_i\rho))(j) - (p_i\,(\mathrm{T}^{-r}\rho))(j)\right] \\
&= w^{-r}\left[\tilde{\alpha}_{i,t} + \sum_{j=m+1}^{m+r} y_{t+j} p_i(j)\rho(j) - \sum_{j=0}^{r-1} y_{t+j} p_i(j)\rho(j)\right] \\
&\qquad + \sum_{j=r}^{m+r} y_{t+j} (\mathrm{T}^{-r}\rho)(j)\left[\left(\mathrm{T}^{-r} p_i - p_i\right)(j)\right] \\
&= w^{-r}\left[\tilde{\alpha}_{i,t} + \sum_{j=m+1}^{m+r} y_{t+j} p_i(j)\rho(j) - \sum_{j=0}^{r-1} y_{t+j} p_i(j)\rho(j)\right] - \sum_{l=0}^{i-1} \tilde{\gamma}_l^{(r,i)} \tilde{\alpha}_{l,t+r}.
\end{aligned}$$

□

If all the data-independent constants are computed in advance, the algorithm for geometric weights offers the same advantages as the one for equal weights. Furthermore, the computational costs differ only by one additional multiplication from the costs for equal weights. The statements about numerical behavior made above are valid in this case as well.

4.2. Polynomial weights

The application of the difference operator to a polynomial p with $\deg(p) = k > 0$ gives again a polynomial with $\deg(\nabla p) = k - 1$. For polynomial weights, this property eventually leads to an update algorithm for least-squares approximations with multiple levels.

Definition 4.3. (Discrete polynomial weights) *Let $\sigma : \mathbb{R} \to \mathbb{R}$ be a polynomial with $\deg(\sigma) = \kappa \geq 0$; if $\sigma(j) > 0$ is fulfilled for $j \in M$, we call $\{\sigma(0), \ldots, \sigma(m)\}$ discrete polynomial weights and σ an admissible polynomial weight function.*

The required positivity guarantees the existence of the discrete inner product $(p|q)_\sigma \stackrel{\text{def}}{=} \sum_{j=0}^{m} p(j)q(j)\sigma(j)$. Besides the usual methods, monic polynomials $\{\check{p}_0, \ldots, \check{p}_n\}$ orthogonal with respect to $(\cdot|\cdot)_\sigma$ can be determined with specific algorithms (compare [4] and [5]). Applying the shift operator to $\{\check{p}_0, \ldots, \check{p}_n\}$ defines again constants $\check{\gamma}_l^{(r,i)}$ with $\mathrm{T}^r \check{p}_i \stackrel{\text{def}}{=} \sum_{l=0}^{i} \check{\gamma}_l^{(r,i)} \check{p}_l$, and the least-squares solution is given by $\sum_{i=0}^{n} \frac{\sum_{j=0}^{m} y_j \check{p}_i(j)\sigma(j)}{\|\check{p}_i\|_\sigma^2} \check{p}_i$. The observation $\deg(\nabla^\nu \sigma) = \kappa - \nu$ leads to an extension of the orthogonal expansion coefficients $\alpha_{i,t}$ used previously.

Definition 4.4. *Let σ be an admissible polynomial weight function of degree κ; the real coefficients $\alpha_{i,t}^{(\kappa,\nu)}$ for $\nu \in \{0, \ldots, \kappa\}$ and $i \in \{0, \ldots, n\}$ are defined by*

$$\alpha_{i,t}^{(\kappa,\nu)} \stackrel{\text{def}}{=} \sum_{j=0}^{m} y_{t+j} \check{p}_i(j) \left(\nabla^{(\kappa-\nu)} \sigma\right)(j).$$

In the case $\nu = \kappa$ the values $\alpha_{i,t}^{(\kappa,\kappa)}$ are exactly the coefficients of the orthogonal expansion of the sought approximation polynomial. Theorem 4.5 now provides an efficient formula for updating these coefficients from one time window to the next.

Theorem 4.5. (Iterative approximation updates for polynomial weights) *The values $\alpha_{i,t+r}^{(\kappa,\nu)}$ can be computed in ascending order going from $i = 0$ up to $i = n$ for $\nu \in \{0, \ldots, \kappa\}$ by*

$$\begin{aligned}\alpha_{i,t+r}^{(\kappa,\nu)} &= \sum_{k=0}^{\min\{\nu,r\}} \binom{r}{k} (-1)^k \alpha_{i,t}^{(\kappa,\nu-k)} - \sum_{l=0}^{i-1} \check{\gamma}_l^{(r,i)} \alpha_{l,t+r}^{(\kappa,\nu)} \\ &+ \sum_{j=m+1}^{m+r} y_{t+j} \check{p}_i(j) \left(\mathrm{T}^{-r} \circ \nabla^{(\kappa-\nu)} \sigma\right)(j) - \sum_{j=0}^{r-1} y_{t+j} \check{p}_i(j) \left(\mathrm{T}^{-r} \circ \nabla^{(\kappa-\nu)} \sigma\right)(j).\end{aligned}$$

Proof. The representation of negative powers of the shift operator T as a sum of powers of the backward difference operator ∇, which was proven in Lemma 2.1, applied to the function $\nabla^{(\kappa-\nu)}\sigma$, i.e., $\mathrm{T}^{-r} \circ \nabla^{(\kappa-\nu)} \sigma = \sum_{k=0}^{\min\{\nu,r\}} \binom{r}{k} (-1)^k \nabla^{(\kappa-\nu+k)} \sigma$, allows analogous transformations as above:

$$\alpha_{i,t+r}^{(\kappa,\nu)} = \sum_{j=r}^{m+r} y_{t+j} \mathrm{T}^{-r} \check{p}_i(j) \left(\mathrm{T}^{-r} \circ \nabla^{(\kappa-\nu)} \sigma\right)(j)$$

$$
\begin{aligned}
&= \sum_{j=r}^{m+r} y_{t+j} \Big\{ \mathrm{T}^{-r}\check{p}_i(j) \left(\mathrm{T}^{-r}\circ\nabla^{(\kappa-\nu)}\sigma\right)(j) \\
&\quad - \check{p}_i(j) \underbrace{\Big[\left(\mathrm{T}^{-r}\circ\nabla^{(\kappa-\nu)}\sigma\right)(j) - \Big(\sum_{k=0}^{\min\{\nu,r\}} \binom{r}{k} (-1)^k \nabla^{(\kappa-\nu+k)}\sigma \Big)(j) \Big]}_{=\,0 \text{ according to Lemma 2.1}} \Big\} \\
&= \sum_{j=r}^{m+r} y_{t+j} \left(\mathrm{T}^{-r}\circ\nabla^{(\kappa-\nu)}\sigma\right)(j) \Big\{ \mathrm{T}^{-r}\check{p}_i(j) - \check{p}_i(j) \Big\} \\
&\quad + \sum_{j=r}^{m+r} y_{t+j}\check{p}_i(j) \Big[\Big(\sum_{k=0}^{\min\{\nu,r\}} \binom{r}{k} (-1)^k \nabla^{(\kappa-(\nu-k))}\sigma \Big)(j) \Big] \\
&= -\sum_{l=0}^{i-1} \check{\gamma}_l^{(r,i)} \alpha_{l,t+r}^{(\kappa,\nu)} + \sum_{j=0}^{m} y_{t+j}\check{p}_i(j) \Big[\Big(\sum_{k=0}^{\min\{\nu,r\}} \binom{r}{k} (-1)^k \nabla^{(\kappa-(\nu-k))}\sigma \Big)(j) \Big] \\
&\quad + \sum_{j=m+1}^{m+r} y_{t+j}\check{p}_i(j) \left(\mathrm{T}^{-r}\circ\nabla^{(\kappa-\nu)}\sigma\right)(j) - \sum_{j=0}^{r-1} y_{t+j}\check{p}_i(j) \left(\mathrm{T}^{-r}\circ\nabla^{(\kappa-\nu)}\sigma\right)(j) \\
&= \sum_{k=0}^{\min\{\nu,r\}} \binom{r}{k} (-1)^k \alpha_{i,t}^{(\kappa,\nu-k)} - \sum_{l=0}^{i-1} \check{\gamma}_l^{(r,i)} \alpha_{l,t+r}^{(\kappa,\nu)} \\
&\quad + \sum_{j=m+1}^{m+r} y_{t+j}\check{p}_i(j) \left(\mathrm{T}^{-r}\circ\nabla^{(\kappa-\nu)}\sigma\right)(j) - \sum_{j=0}^{r-1} y_{t+j}\check{p}_i(j) \left(\mathrm{T}^{-r}\circ\nabla^{(\kappa-\nu)}\sigma\right)(j).
\end{aligned}
$$

□

In the case $\kappa = 0$ the result obtained is equivalent to the one in Theorem 3.1. The algorithm resulting from Theorem 4.5 is an iterative update method with $\kappa+1$ levels. If the parameters n, m, r, and κ are fixed and the data-independent constants $\check{\gamma}_l^{(r,i)}$ and $\check{p}_i(j)\left(\mathrm{T}^{-r}\circ\nabla^{(\kappa-\nu)}\sigma\right)(j)$ are computed in advance, then the computation order $\alpha_{0,t}^{(\kappa,\kappa)}, \alpha_{0,t}^{(\kappa,\kappa-1)}, \dots, \alpha_{0,t}^{(\kappa,0)}, \alpha_{1,t}^{(\kappa,\kappa)}, \dots, \alpha_{n,t}^{(\kappa,0)}$ leads to an efficient *in place* implementation. Changing the order by swapping the loops over $i = 0\dots n$ and $\nu = \kappa\dots 0$ provides the desired coefficients $\alpha_{i,t}^{(\kappa,\kappa)}$ already after the first execution of the (then inner) i-loop. Therefore, a natural possibility of a parallel execution of the algorithm in a multiprocessor environment is readily available.

The computational costs increase linearly with κ based on the complexity numbers established in §3.2; therefore, the savings in floating-point operations are worthwhile if a small degree κ for the polynomial weight function and a large approximation length m is used. For degrees $n \leq 2$ of the approximation polynomial the rounding errors of this multilevel algorithm do not exclude satisfactory numerical results, even after modifications in the magnitude of 10^5.

5. Applications and Concluding Remarks

Typical applications of the algorithms derived are threshold crossing methods under real-time conditions. A very important example is a time-critical aspect of monitoring a CNC-lathe, namely the collision detection based on three (orthogonal) measured force signals. The earlier a collision is detected, the less is the economical damage. For this purpose we implemented the method described in §3.1 on a DSP using a polynomial approximating the last 64 (or 128) values at a time with a short extrapolation into the future. The reaction time for detecting a threshold crossing in the case of a collision could be significantly reduced without a worsening of the false alarm behavior. To achieve more accuracy, we used a 40-bit addition algorithm requiring about 50 processor cycles for one modification from $\alpha_{i,t}$ to $\alpha_{i,t+1}$, $0 \leq i \leq 2$; therefore the sampling rate 10 kHz of each of the signals can easily be handled. The utilization of geometric weights or discrete polynomial weights (of degree 1) amplifying the influence of the most recent measuring data improved the results further.

In addition, the approximating polynomial was used to eliminate an ascent or descent in the signal as a preprocessing step for frequency analysis. These long-term trends, interpreted as periodical by an FFT, falsify the spectrum in the low-frequency area. Therefore, the trend correction was a by-product of the collision detection.

Tracing objects in image sequences is another field of application of the method presented. By considering each coordinate separately, it is possible to estimate accurately the center (or other significant points) of an object in a subsequent image by extrapolation. This allows the determination of a region of interest in which the traced object has to be searched for. Two- (or more-) dimensional versions of the methods presented can be obtained easily by a tensor product approach.

Summing up, a valuable algorithm for the processing of equidistant signal data in a real-time environment was developed; the modification mechanism for polynomial least-squares approximations is especially suitable for DSP-implementations, providing sufficient accuracy for long monitoring intervals in a variety of applications.

References

[1] Å. Björck, *Numerical methods for least squares problems,* SIAM Publications, Philadelphia, 1996.

[2] S. Elhay, G.H. Golub, and J. Kautsky, *Updating and downdating of orthogonal polynomials with data fitting applications,* SIAM J. Matrix Anal. Appl., **12** (1991), 327–353.

[3] H.-J. Fischer, *On generating orthogonal polynomials for discrete measures,* Z. Anal. Anwendungen, **17** (1998), 183–205.

[4] D. Galant, *An implementation of Christoffel's theorem in the theory of orthogonal polynomials,* Math. Comp., **25** (1971), 111–113.

[5] W. Gautschi, *An algorithmic implementation of the generalized Christoffel theorem,* in: G. Hämmerlin, Ed., *Numerical integration*, Internat. Ser. Numer. Math. **57**, Birkhäuser, Basel, 1982, 89–106.

[6] W. Gautschi, *On generating orthogonal polynomials,* SIAM J. Sci. Statist. Comput., **3** (1982), 289–317.

[7] G.H. Golub and C.F. Van Loan, *Matrix computations,* 3d ed., Johns Hopkins University Press, Baltimore, 1996.

[8] G. Maess, *Vorlesungen über numerische Mathematik I*, Birkhäuser, Basel, 1985.

[9] A.F. Nikiforov, S.K. Suslov, and V.B. Uvarov, *Classical orthogonal polynomials of a discrete variable*, Springer, Berlin, 1991.

Erich Fuchs
Fakultät für Mathematik und Informatik
Universität Passau
Innstr. 33
D-94032 Passau, Germany
E-mail address: `fuchs@fmi.uni-passau.de`

International Series of Numerical Mathematics
Vol. 131, © 1999 Birkhäuser Verlag Basel/Switzerland

Quadrature Rules Based on s-Orthogonal Polynomials for Evaluating Integrals with Strong Singularities

Laura Gori and Elisabetta Santi

Abstract. Integrals defined by Cauchy principal values or by Hadamard finite parts are involved in many boundary integral equations; in this paper we present a method based on the use of Gauss-Turán quadrature formulas, that is, formulas with multiple nodes, for approximating these kinds of singular or hypersingular integrals. We show that the rules here proposed are particularly effective when the s-orthogonal polynomials involved in the construction of the method are invariant with respect to s. Finally, we introduce another type of integration rules, also with multiple nodes, for numerically evaluating hypersingular integrals.

1. Introduction

Integrals with strong singularities are involved in several singular boundary integral equations; such integrals are defined in the Cauchy principal value (CPV) sense or as the Hadamard finite part (fp), and their numerical evaluation is of interest in many problems of applied mathematics. In the recent literature, for both these types of integrals, some quadrature methods have been given, a few of them based on Gauss type rules [8,9,10 and references therein, 12, 14, 16, 18]; for CPV integrals also Gauss-Turán type formulas, based on multiple nodes, have been proposed [7]. Each of these rules has an exactness degree depending on the number, or on the number and the multiplicity, of the nodes. To achieve a higher precision, a new calculation of all the nodes is generally required. This is not only expensive computationally, but can also produce the undesirable effect of one of the new nodes being too close to (or coinciding with) the singularity, giving rise to severe numerical cancellations. This is true, in particular, for Gaussian quadrature rules, where the precision can be increased by varying the number of the nodes. For Gauss-Turán [3] type formulas, on the other hand, whose nodes have multiplicity $2s+1$ (s a nonnegative integer) and coincide with the zeros of some s-orthogonal polynomials, the exactness degree can be increased by varying s. In this context, it is of particular interest to have at disposal s-orthogonal polynomials whose zeros are independent of s. This is achievable when the results in [6] are exploited. Indeed, in [6] a large class of weight functions is provided for which the corresponding s-orthogonal polynomials are invariant with respect to s.

This paper is organized as follows; in §2, some notations and a few results from [6] are recalled; in §3, we construct a class of quadrature rules of Gauss-Turán type for approximating fp integrals, that is integrals of the form

$$\int_a^b k(x,t)f(x)w(x)dx, \qquad t \in (a,b), \tag{1.1}$$

where the kernel $k(x,t)$ has a pole of multiplicity $p+1$, $p \geq 0$, at $t = x$, f is smooth, and w is a weight function, possibly containing endpoint singularities. When $p = 0$, one gets methods for evaluating CPV integrals. Section 4 analyzes the behavior of the error term of these rules and considers in detail the case when the weight function w belongs to the set of functions introduced in [6]. Section 5 contains some remarks and a few numerical examples; in §6, another type of multiple-nodes integration rules for singular integrals is presented.

2. Preliminaries

Given an interval A, finite or infinite, on the real line, a weight function w nonnegative on A having all the moments m_i, $i \in \mathbb{N}$, finite (in particular $m_0 > 0$), and a nonnegative integer s, the monic polynomials of the sequence $\{P_{ms}\}_{m\in\mathbb{N}}$, $P_{ms} \in \mathbb{P}_m$ (the set of polynomials of degree m), are said to be *s-orthogonal* in A with respect to w if the following conditions are fulfilled:

$$\int_A x^k [P_{ms}(x)]^{2s+1} w(x)dx = 0, \qquad k = 0,1,\dots,m-1.$$

For instance, when $w(x) = (1-x^2)^{-1/2}$, $x \in [-1,1]$, the system $\{P_{ms}\}_{m\in\mathbb{N}}$ of the corresponding s-orthogonal polynomials is independent of s, and coincides with the set of (monic) Chebyshev polynomials of the first kind $\{T_m\}_{m\in\mathbb{N}}$.

Every P_{ms} has m zeros $\{x_i\}_{i=1}^m$ which are real and simple, and in the interior of A. With these zeros are associated the Gauss-Turán quadrature rules [5, 3 and references therein], also termed *hypergaussian* rules, having nodes of multiplicity $2s+1$ at $\{x_i\}_{i=1}^m$. Namely, for a given function F having the required derivatives at least at the nodes, one has

$$\int_A F(x)w(x)dx = \sum_{i=1}^{m}\sum_{h=0}^{2s} A_{hi}F^{(h)}(x_i) + r_{ms}(F) = G_{ms}(F) + r_{ms}(F) \tag{2.1}$$

with

$$r_{ms}(F) = 0 \quad \text{for} \quad F \in \mathbb{P}_\nu, \qquad \nu = (2s+2)m-1, \tag{2.2}$$

$$A_{hi} = \int_A U_{hi}(x)w(x)dx, \tag{2.3}$$

where U_{hi}, $h = 0, 1, \ldots, 2s$, $i = 1, 2, \ldots, m$, are the fundamental Lagrange-Hermite polynomials relative to the given nodes. Moreover, if $F \in C^{\nu+1}(A)$, the following expression for the error holds [17]:

$$r_{ms}(F) = \frac{F^{(\nu+1)}(\xi)}{(\nu+1)!} \int_A [P_{ms}(x)]^{2s+2} w(x) dx, \qquad \xi \in (-1, 1). \tag{2.4}$$

An efficient algorithm for evaluating the zeros of s-orthogonal polynomials and the coefficients (or weights) A_{hi} of the associated formulas (2.1) is given in [4].

In the following, we shall limit ourselves to finite intervals, which we take, without loss of generality, to be $[-1, 1]$.

Since the zeros of P_{ms} and $P_{m+1,s}$ interlace [15], if m is increased with the aim of increasing the exactness of (2.1), a new calculation of all the nodes, the weights, and the functional values in (2.1) is required. But for rules (2.1), unlike rules with simple nodes, it is possible to achieve higher precision also by increasing s (cf. (2.2)), and this may result in low computational cost in all those cases when w is in the class W introduced in [6]. We mention a large class W of weight functions which can be considered as a generalization of the Chebyshev weight of the first kind; they are characterized by the property that, for a given m, the corresponding s-orthogonal polynomials P_{ms} of degree m do not vary with s, and, moreover, coincide with T_m. These weights can be described as follows: for a given positive m a weight function, say $w_m(x)$, $x \in [-1, 1]$, is in W if the following relation holds:

$$\frac{w_m(x)}{w_\infty(x)} = \sum_{\ell=0}^{\infty} \rho_\ell T_{2m\ell}(x),$$

where w_∞ denotes the Chebyshev weight function of the first kind and $\{\rho_\ell\}$ is a suitable sequence of real numbers.

A subset of W is provided by the following weight functions:

$$w_{m,\mu}(x) = |V_{m-1}(x)/m|^{2\mu+1}(1 - x^2)^\mu, \tag{2.5}$$

where $\mu > -1$ is a parameter and V_j, $j \in \mathbb{N}$, are the Chebyshev polynomials of the second kind; in particular $\mu = -1/2$ gives again the Chebyshev polynomials of the first kind. The weight functions in (2.5) belong to the set of *generalized Jacobi smooth weights*, which are of the form

$$\psi(x) \prod_{j=1}^{n} |x - \tau_j|^{\alpha_j},$$

where τ_j are in $[-1, 1]$, $\alpha_j > -1$, $j = 1, 2, \ldots, n$, and ψ is a positive Dini type function defined on $[-1, 1]$. The corresponding orthogonal polynomials have been thoroughly investigated by P. Nevai and appear in several applications [11].

If the weight is in W, the rules (2.1) become

$$\int_{-1}^{1} f(x) w(x)\, dx = \sum_{i=1}^{m} \sum_{h=0}^{2s} A_{hi}^{(m)} f^{(h)}\left(\cos\frac{2i-1}{2m}\pi\right), \quad f \in \mathbb{P}_\nu.$$

Moreover, the coefficients $A_{hi}^{(m)}$ have explicit expressions. In [6], these formulas have been used for the approximation of Fourier coefficients.

Finally, we point out that, in the sequel, we shall use the following definition of finite part integral in the sense of Hadamard,

$$\begin{aligned}\oint_a^b \frac{f(x)}{(x-t)^{p+1}} w(x)dx = \int_a^b & \frac{f(x) - \sum_{k=0}^{p} f^{(k)}(t)(x-t)^k/k!}{(x-t)^{p+1}} w(x)dx \\ & + \sum_{k=0}^{p} \frac{f^{(k)}(t)}{k!} \oint_a^b \frac{w(x)}{(x-t)^{p+1-k}} dx\,,\end{aligned} \tag{2.6}$$

where $p \geq 0$ is an integer, $t \in (a,b)$, w a weight function possibly containing integrable endpoint singularities, and f is in $C^p[a,b]$, with $f^{(p)}$ Hölder continuous. This definition makes sense because the assumptions on f assure that the last integrals on the right of (2.6) exist in the sense of CPV integrals (cf. also Remark 2.4 in [10]).

3. Gauss-Turán Quadrature Rules for Finite Part Integrals

As we mentioned, we assume without loss of generality $[a,b] = [-1,1]$. Applying the rule (2.1) to the ordinary integral in (2.6) yields the following quadrature rule with multiple nodes for a Hadamard finite part integral,

$$\begin{aligned} I_p(wf;t) &:= \oint_{-1}^{1} \frac{f(x)}{(x-t)^{p+1}} w(x)dx \\ &= \sum_{i=1}^{m} \sum_{h=0}^{2s} B_{hi}^{(p)}(t) f^{(h)}(x_i) + \sum_{k=0}^{p} C_{p-k}(t) \frac{f^{(k)}(t)}{k!} + e_{ms}^{(p)}(f;t) \\ &=: Q_{ms}^{(p)}(f;t) + e_{ms}^{(p)}(f;t)\,, \end{aligned} \tag{3.1}$$

where $\{x_i\}_{i=1}^m$ are the zeros of the mth polynomial s-orthogonal in $[-1,1]$ with respect to w; the function f is assumed to have the required derivatives at least at the nodes, t is assumed to be different from any x_j; and there results $e_{ms}^{(p)}(f;t) = 0$ when $f \in \mathbb{P}_N$, $N = (2s+2)m + p$. Furthermore, by (2.1) and (2.3), we have

$$\begin{cases} B_{hi}^{(p)}(t) = \sum_{k=h}^{2s} (-1)^{k-h} \binom{k}{h} \frac{(k-h)!}{(x_i - t)^{p+1+k-h}} A_{ki}\,, \\ C_{p-k}(t) = I_{p-k}(w;t) - G_{ms}[(x-t)^{k-p-1}]\,. \end{cases} \tag{3.2}$$

The knowledge of $I_0(w;t)$ is required, which, differentiating with respect to t, then yields I_{p-k}, $k = 0, 1, \dots, p-1$.

We point out that (3.1) for $s = 0$ is the Gauss type quadrature rule given in [8].

When one of the nodes, say x_j, coincides with t, and f is assumed to have derivatives up to the $(p+2s+1)$st, a formula similar to (3.1) can be derived. To do so, introduce the following function,

$$g(x,t) = \frac{f(x) - \sum_{k=0}^{p} f^{(k)}(t)(x-t)^k/k!}{(x-t)^{p+1}}\,; \tag{3.3}$$

by the above assumptions on f one has

$$g(x,t) = \frac{1}{p!(x-t)^{p+1}} \int_t^x f^{(p+1)}(u)(x-u)^p du\,. \tag{3.4}$$

Changing the variable u by $u = t + (x-t)v$ and differentiating (3.4) h times with respect to x yields

$$\left[\frac{\partial^h g(x,t)}{\partial x^h}\right]_{x=x_j} = \frac{1}{p!}\int_0^1 f^{(p+1+h)}(t+(x_j-t)v)(1-v)^p v^h dv\,.$$

Then, for $x_j \to t$, one gets

$$\left[\frac{\partial^h g(x,t)}{\partial x^h}\right]_{x=t} = f^{(p+1+h)}(t)\frac{h!}{(p+1+h)!}\,. \tag{3.5}$$

Consequently, the quadrature sum $Q_{ms}^{(p)}(f;t)$ of (3.1) becomes $\overline{Q}_{ms}^{(p)}(f;t)$, where

$$\overline{Q}_{ms}^{(p)}(f;t) = \sum_{\substack{i=1\\ i\neq j}}^{m}\sum_{h=0}^{2s} A_{hi}\left[\frac{\partial^h g(x,t)}{\partial x^h}\right]_{x=x_i} + \sum_{h=0}^{2s} A_{hi}\frac{h!}{(p+1+h)!} f^{(p+1+h)}(t)$$

$$+\sum_{k=0}^{p}\frac{f^{(k)}(t)}{k!} I_{p-k}(w;t) \quad (x_j = t)\,.$$

4. Analysis of the Error Term $e_{ms}^{(p)}(f;t)$

We recall that it is possible to associate with any system of s-orthogonal polynomials and with the corresponding Gauss-Turán quadrature rule (2.1) a *characteristic function* $K_{ms}(z;w)$ [13] given by

$$K_{ms}(z;w) = \frac{\displaystyle\int_{-1}^{1}\frac{[P_{ms}(\sigma)]^{2s+1}}{z-\sigma}d\sigma}{[P_{ms}(z)]^{2s+1}}\,, \qquad z \notin [-1,1]\,, \tag{4.1}$$

which plays a fundamental role in the representation of the error of (2.1) by means of a contour integral. In fact, when F is analytic in a domain D containing $[-1,1]$ and C is a curve in D surrounding $[-1,1]$, then the following expression holds [2]:

$$r_{ms}(F) = \frac{1}{2\pi i}\oint_C K_{ms}(z;w)F(z)dz\,. \tag{4.2}$$

We recall that, in [2], the authors observe that the estimates of the error term of the type (4.2) are good even for small values of m and less conservative than other estimates given in terms of high-order derivatives of the integrand.

Suppose now that the function f is single-valued holomorphic in a domain $D \supset [-1,1]$ and Γ a contour contained in D and surrounding $[-1,1]$. Then the following representation of the error in (3.1) can be given.

Theorem 4.1. *One has*

$$e_{ms}^{(p)}(f;t) = \frac{1}{2\pi i}\oint_\Gamma K_{ms}(z;w)\frac{f(z)}{(z-t)^{p+1}}dz\,. \tag{4.3}$$

Proof. The proof follows by applying the residue theorem, with respect to z, to the function

$$\frac{f(z)[P_{ms}(\sigma)]^{2s+1}}{(z-t)^{p+1}(z-\sigma)[P_{ms}(z)]^{2s+1}}$$

and then integrating with respect to the variable σ on the interval $[-1,1]$. □

For $p=0$, a result from [7] is recovered; in the same paper it was proved that, when w is even in $[-1,1]$ and $z \in \mathbb{C} = \{z : |z| = \rho,\ \rho > 1\}$, the characteristic function satisfies the relation

$$\max_{z\in C}|K_{ms}(z;w)| = K_{ms}(\rho;w)\,, \tag{4.4}$$

and thus from (4.3) and (4.4) the estimate below of $|\ e_{ms}^{(p)}(f;t)\ |$ is easily derived.

Proposition 4.2. *There results*

$$|e_{ms}^{(p)}(f;t)| \le \frac{\rho}{(\rho-|t|)^{p+1}}K_{ms}(\rho;w)\max_{z\in C}|f(z)|. \tag{4.5}$$

In the case when f is a real-valued smooth function, another expression of $e_{ms}^{(p)}(f;t)$ is stated in the following theorem.

Theorem 4.3. *Let* $f \in C^{N+1}[-1,1]$, $N=(2s+2)m+p$; *then*

$$e_{ms}^{(p)}(f;t) = \frac{f^{(N+1)}(\xi)}{(N+1)!}\int_{-1}^{1}[P_{ms}(x)]^{2s+2}w(x)dx, \qquad \xi\in(-1,1)\,. \tag{4.6}$$

Proof. Consider the function $g(x,t)$ defined by (3.3); from the way (3.1) was constructed, and by comparing (3.1) with (2.1), one sees that

$$e_{ms}^{(p)}(f;t) = r_{ms}(g)\,.$$

On the other hand, (2.4) yields

$$r_{ms}(g) = \frac{\left[\dfrac{\partial^{\nu+1}g(x,t)}{\partial x^{\nu+1}}\right]_{x=\xi}}{(\nu+1)!}\int_{-1}^{1}[P_{ms}(x)]^{2s+2}w(x)dx, \quad \xi\in(-1,1)\,.$$

From this relation and (3.5), it is easily seen that the expression (4.6) can also be written as

$$e_{ms}^{(p)}(f;t) = \gamma_{ms}^{(p)}f^{(N+1)}(\xi), \quad \xi\in(-1,1), \tag{4.7}$$

where

$$\gamma_{ms}^{(p)} := \frac{1}{(N+1)!}\int_{-1}^{1}[P_{ms}(x)]^{2s+2}w(x)dx\,. \tag{4.8}$$

□

5. Some Remarks and Numerical Examples

It has already been observed that in the proposed rules having precision $N = (2s+2)m+p$, the increase of N can be realized by passing from m to $m+1$. This costs $(2s+2)(m+1)$ new functional evaluations. In contrast, if the increase of N is achieved passing from s to $s+1$, and moreover, one has $P_{ms} = P_{m,s+1}$, as in the case of weight functions in W, only $2m$ new functional evaluations are required. Moreover, the evaluation of higher derivatives can be performed, in several cases, by exploiting suitable recurrence relations between successive derivatives of the function under consideration (cf., for instance, [1]). Also, for the number of coefficients in the quadrature sums, analogous remarks hold. Thus, in conclusion, the integration formulas given here turn out to be particularly effective for this special class of weights. Furthermore, since the s-orthogonal polynomials corresponding to the weight functions in W coincide with the Chebyshev polynomials of the first kind, it is easy, in many cases, to evaluate in closed form the constant factors appearing in both the estimates (4.5) and (4.7) of the remainder $e_{ms}^{(p)}(f;t)$. From (4.8) we have

$$\gamma_{ms}^{(p)} = \frac{1}{[(2s+2)m+p+1]!}\int_{-1}^{1}[T_m(x)]^{2s+2}w(x)dx, \quad w\in W\,. \tag{5.1}$$

So, for instance, if we consider w of the type (2.5), for $m = 2$ we have $w = w_{2,\mu}(x) = |x|^{2\mu+1}(1-x^2)^{\mu}$ and

$$\gamma_{2s}^{(p)} = \frac{B(s+3/2,\mu+1)}{(4s+p+5)!2^{2\mu+2s+3}}\,,$$

where B denotes the Beta function.

Analogously, for $m = 3$, $w = w_{3,\mu}(x) = |4x^2-1|^{2\mu+1}(1-x^2)^{\mu}$, there results, for 2μ a nonnegative integer,

$$\gamma_{3s}^{(p)} = \frac{2(2\mu)!!(2s+1)!!}{(6s+p+7)!3^{2\mu+1}4^{2s+2}(2s+2\mu+3)!!}\,.$$

Just to give an idea of the magnitude of these constants, we report a few of them in Table 1.

As for the constants $K_{ms}(\rho;w)$ (cf. (4.4) and (4.5)), they have been calculated for different values of ρ and s when $w = w_{3,\mu}$; the results are displayed in Table 2.

In Tables 3 and 4 we report the values of $Q_{2s}^{(2)}(f;0)$ for $f(x) = e^{3x}$, $w = w_{2,\mu}$, respectively for $\mu = -1/2$ and $\mu = 0$, which give approximations with 2 nodes of

TABLE 1

m	μ	s	p	$\gamma_{ms}^{(p)}$
2	$-1/2$	3	0	$0.12\mathrm{E}-13$
3	0	2	1	$0.48\mathrm{E}-23$
3	$1/2$	1	1	$0.31\mathrm{E}-15$

TABLE 2

m	μ	s	ρ	$K_{ms}(\rho; w)$
3	0	1	2	$0.69\mathrm{E}-06$
3	$-1/2$	2	3	$0.37\mathrm{E}-12$
3	$-1/2$	1	5	$0.44\mathrm{E}-11$

multiplicities $2s+1$, $s = 0, 1, \ldots, 4$, of the following integrals

$$J_1 = \fint_{-1}^{1} \frac{e^{3x}}{x^3} \frac{dx}{\sqrt{1-x^2}}; \quad J_2 = \fint_{-1}^{1} \frac{e^{3x}}{x^3} \mid x \mid dx\,.$$

TABLE 3

s	$Q_{2s}^{(2)}(e^{3x}; 0);\ w(x) = (1-x^2)^{-1/2}$
0	17.681033689316350
1	17.886131626436580
2	17.886508674077390
3	17.886508862951530
4	17.886508862986840

TABLE 4

s	$Q_{2s}^{(2)}(e^{3x}; 0);\ w(x) = \lvert x\rvert$
0	6.294714487924726
1	6.338237621938917
2	6.338301631534444
3	6.338301659018102
4	6.338301659022664

TABLE 5

	$Q_{3s}^{(1)}(e^{3x};0.25)$	$Q_{3s}^{(1)}(e^{3x};0.99)$
$s=0$	32.573213409809650	150.637198327232100
$s=1$	32.596221000418060	150.677176284030600
$s=2$	32.596221318644190	150.677176723816300
$s=3$	32.596221318644680	150.677176723821500

In Table 5 we report the values of $Q_{3s}^{(1)}(f;0.25)$ and $Q_{3s}^{(1)}(f;0.99)$ for $f(x) = e^{3x}$, $w = w_{3,-1/2}$, which give approximations with 3 nodes of multiplicities $2s+1$, $s = 0,1,\dots,3$, of the following integrals

$$J_3 = \oint_{-1}^{1} \frac{e^{3x}}{(x-0.25)^2} \frac{dx}{\sqrt{1-x^2}}; \quad J_4 = \oint_{-1}^{1} \frac{e^{3x}}{(x-0.99)^2} \frac{dx}{\sqrt{1-x^2}}.$$

6. Another Multiple-Nodes Quadrature Rule for Finite Part Integrals

We observe that (3.1) can be obtained also by replacing f in the left-hand side by its Lagrange-Hermite interpolatory polynomial relative to the nodes $\{x_i\}_{i=1}^m$ of multiplicity $2s+1$ and t of multiplicity p and then integrating.

Another type of quadrature rules with multiple nodes can be derived by using the Lagrange-Hermite interpolatory polynomial relative only to the nodes $\{x_i\}_{i=1}^m$ of multiplicity $2s+1$ and then integrating. The procedure then amounts to evaluating the finite part integrals of the type

$$I_p(wU_{hi};t) = \oint_{-1}^{1} \frac{U_{hi}(x)}{(x-t)^{p+1}} w(x)dx,$$

where U_{hi} are the fundamental polynomials of Lagrange-Hermite already mentioned in §2. Indeed, denoting by H the Lagrange-Hermite interpolatory polynomial in question, one has

$$H(x) = \sum_{i=1}^{m} \sum_{h=0}^{2s} U_{hi}(x) f^{(h)}(x_i).$$

Taking into account the definition (2.6), we have

$$I_p(wU_{hi};t) = \int_{-1}^{1} \frac{U_{hi}(x) - \sum_{\ell=0}^{p} U_{hi}^{(\ell)}(t)(x-t)^\ell/\ell!}{(x-t)^{p+1}} w(x)dx$$
$$+ \sum_{\ell=0}^{p} \frac{U_{hi}^{(\ell)}(t)}{\ell!} \oint_{-1}^{1} \frac{w(x)}{(x-t)^{p+1-\ell}} dx.$$

In view of the degree of the polynomials in the first integral on the right and the exactness degree of the Gauss-Turán quadrature, one gets after some algebra the quadrature sum below:

$$\widetilde{Q}^{(p)}_{ms}(f,t) = \sum_{i=1}^{m}\sum_{h=0}^{2s} B^{(p)}_{hi}(t) f^{(h)}(x_i) + \frac{1}{p!}\left\{\frac{d^p}{dt^p}[H(t)I_0(w;t)] - \sum_{i=1}^{m}\sum_{h=0}^{2s} A_{hi}\left[\frac{d^h}{dx^h}\left(\frac{d^p}{dt^p}\frac{H(t)}{x-t}\right)\right]_{x=x_i}\right\},$$

and so

$$I_p(wf;t) = \widetilde{Q}^{(p)}_{ms}(f;t) + E^{(p)}_{ms}(f;t)\,. \tag{6.1}$$

Since $Q^{(p)}_{ms}(f,t)$ in (3.1) can also be written, more explicitly, in the form

$$Q^{(p)}_{ms}(f;t) = \sum_{i=1}^{m}\sum_{h=0}^{2s} B^{(p)}_{hi}(t) f^{(h)}(x_i) + \frac{1}{p!}\left\{\frac{d^p}{dt^p}[f(t)I_0(w;t)] - \sum_{i=1}^{m}\sum_{h=0}^{2s} A_{hi}\left[\frac{d^h}{dx^h}\left(\frac{d^p}{dt^p}\frac{f(t)}{x-t}\right)\right]_{x=x_i}\right\}, \tag{1}$$

the following relation between the error terms in (3.1) and (6.1) can be established:

$$E^{(p)}_{ms}(f;t) = e^{(p)}_{ms}(f;t) + \frac{1}{p!}\frac{d^p}{dt^p}\left\{[f(t)-H(t)]\left[I_0(w;t) - \sum_{i=1}^{m}\sum_{h=0}^{2s} A_{hi}\left[\frac{d^h}{dx^h}\frac{1}{x-t}\right]_{x=x_i}\right]\right\}.$$

References

[1] R.A. Cicenia, *Numerical integration formulae involving derivatives,* J. Inst. Math. Appl., **18** (1976), 79–85.

[2] J.D. Donaldson and D. Elliott, *A unified approach to quadrature rules with asymptotic estimates of their remainders,* SIAM J. Numer. Anal., **9** (1972), 573–602.

[3] W. Gautschi, *A survey of Gauss-Christoffel quadrature formulae,* in: P.L. Butzer and F. Fehér, Eds., *E.B. Christoffel,* Birkhäuser, Basel, 1981, 72–147.

[4] W. Gautschi and G.V. Milovanović, *S-orthogonality and construction of Gauss-Turán-type quadrature formulae,* J. Comput. Appl. Math., **86** (1997), 205–218.

[5] A. Ghizzetti and A. Ossicini, *Quadrature formulae,* Birkhäuser, Basel/Stuttgart, 1970.

[6] L. Gori and C.A. Micchelli, *On weight functions which admit explicit Gauss-Turán quadrature formulas,* Math. Comp., **65** (1996), 1567–1581.

[7] L. Gori and E. Santi, *On the evaluation of Hilbert transforms by means of a particular class of Turán quadrature rules,* Numer. Algorithms, **10** (1995), 27–39.

[8] N.I. Ioakimidis, *On the uniform convergence of Gaussian quadrature rules for Cauchy principal value integrals and their derivatives,* Math. Comp., **44** (1985), 191–198.

[9] H.R. Kutt, *The numerical evaluation of principal value integrals by finite-part integration,* Numer. Math., **24** (1975), 205–210.

[10] G. Monegato, *Numerical evaluation of hypersingular integrals,* J. Comput. Appl. Math., **50** (1994), 9–31.

[11] P. Nevai, *Orthogonal polynomials,* Mem. Amer. Math. Soc., **18** (1979), no. 213.

[12] A. Ossicini, *Alcune formule di quadratura per il calcolo della parte finita e del valore principale di integrali divergenti,* Rend. Mat. (6), **2** (1969), 385–401.

[13] A. Ossicini and F. Rosati, *Funzioni caratteristiche nelle formule di quadratura gaussiane con nodi multipli,* Boll. Un. Mat. Ital. (4), **11** (1975), 224–237.

[14] D.F. Paget, *The numerical evaluation of Hadamard finite-part integrals,* Numer. Math., **36** (1981), 447–453.

[15] A. Pinkus and Z. Ziegler, *Interlacing properties of the zeros of the error functions in best L^p-approximations,* J. Approx. Theory, **27** (1979), 1–18.

[16] H.V. Smith and D.B. Hunter, *The numerical evaluation of a class of divergent integrals,* in: H. Brass and G. Hämmerlin, Eds., *Numerical integration III*, Birkhäuser, Basel, 1988, 274–284.

[17] D.D. Stancu, *Sur quelques formules générales de quadrature du type Gauss-Christoffel,* Mathematica (Cluj), **1** (**24**) (1959), 167–182.

[18] G. Tsamasphyros and P.S. Theocaris, *On the convergence of some quadrature rules for Cauchy principal-value and finite-part integrals,* Computing, **31** (1983), 105–114.

Laura Gori
Dipartimento di Metodi e Modelli Matematici
per le Scienze Applicate
Università di Roma "La Sapienza"
I-00161 Roma, Italy
E-mail address: gori@dmmm.uniroma1.it

Elisabetta Santi
Dipartimento di Energetica
Università di L'Aquila
I-67100 L'Aquila, Italy
E-mail address: esanti@dsiaq1.ing.univaq.it

International Series of Numerical Mathematics
Vol. 131, © 1999 Birkhäuser Verlag Basel/Switzerland

Gegenbauer Weight Functions Admitting L_2 Duffin and Schaeffer Type Inequalities

David B. Hunter and Geno Nikolov[1]

Abstract. For a given set Ω of $n+1$ distinct points in $[-1,1]$ and algebraic polynomial Q_n of degree n, sufficient conditions are found ensuring that for every other polynomial f of degree at most n the inequality $|f| \le |Q_n|$ on Ω induces inequalities between some weighted L_2 norms of the derivatives of f and Q_n.

1. Introduction

Denote by π_n the set of all algebraic polynomials of degree less than or equal to n. The classical Markov inequality asserts that the Chebyshev polynomial of the first kind

$$T_n(x) := \cos n \arccos x$$

has the greatest uniform norm in $[-1,1]$ of its kth derivative ($k = 1, 2, \dots, n$) among all polynomials $p \in \pi_n$ satisfying on the interval $[-1,1]$

$$|p(x)| \le 1. \tag{1}$$

A remarkable extension of this result was found by Duffin and Schaeffer [4], who showed that this extremal property of T_n persists even in the larger class of polynomials satisfying the inequality (1) at the points of local extrema of T_n, i.e.,

$$|p(x)| \le |T_n(x)| \text{ at the zeros of } (x^2-1)T_n'(x).$$

Bojanov and Nikolov [2] recently proved that the ultraspherical polynomials $P_n^{(\lambda)}$ enjoy the same extremal property. Recall that the ultraspherical polynomial $P_n^{(\lambda)}$ is the nth orthogonal polynomial in $[-1,1]$ with respect to the Gegenbauer weight function

$$w_\lambda(x) := (1-x^2)^{\lambda-1/2} \qquad (\lambda > -1/2). \tag{2}$$

The result obtained by Bojanov and Nikolov reads as follows.

---

[1]The second author was supported by a grant of the Royal Society and by the Bulgarian Ministry of Education, Science, and Technologies, under Grant MM-414/94.

Theorem A. ([2]) *Let* $t_j := t_j(P_n^{(\lambda)})$ $(j = 0, \dots, n)$ *be the extremal points of* $P_n^{(\lambda)}$ *in* $[-1, 1]$ (i.e., the zeros of $(1-t^2)P_{n-1}^{(\lambda+1)}(t)$). *Let* $f \in \pi_n$ *satisfy*

$$|f(t_j)| \le |P_n^{(\lambda)}(t_j)|, \ j = 0, \dots, n.$$

Then the inequality

$$\|f^{(k)}\| \le \left\| \frac{d^k}{dx^k} P_n^{(\lambda)} \right\|$$

holds for all $k \in \{1, \dots, n\}$, *if* $\lambda \ge 0$, *and for all* $k \in \{2, \dots, n\}$, *if* $-1/2 \le \lambda < 0$. (Here $\|\cdot\|$ denotes the uniform norm in the interval $[-1, 1]$.)

In 1986 Bojanov proposed a method for deriving inequalities of Duffin and Schaeffer type in some weighted L_2 norms. Briefly, Bojanov's idea is the following: According to some observations of V. Markov, inequalities between the absolute values of two polynomials of degree n at $n+1$ interpolation points induce inequalities between the absolute values of their kth derivatives on some set $I_{n,k}$ of points (see Lemma 2.2 below). The L_2 norms of these derivatives can be evaluated with the help of a Gaussian quadrature formula. Should the nodes of this quadrature formula belong to the set $I_{n,k}$, we immediately obtain an L_2 inequality of Duffin and Schaeffer type. Bojanov obtained the following analogue of Theorem A.

Theorem B. ([1]) *Let* $-1 = t_0 < t_1 < \cdots < t_n = 1$ *be the extremal points of* $P_n^{(\lambda)}$ $(\lambda \in (-1/2, 1/2])$. *If* $p \in \pi_n$ *satisfies*

$$|p(t_j)| \le |P_n^{(\lambda)}(t_j)| \ \text{for}\ j = 0, 1, \dots, n,$$

then

$$\int_{-1}^{1} w_{\lambda+k-1}(t)|p^{(k)}(t)|^2\, dt \le \int_{-1}^{1} w_{\lambda+k-1}(t) \left| \frac{d^k}{dt^k} P_n^{(\lambda)}(t) \right|^2 dt \tag{3}$$

for $k = 1, \dots, n$. *Equality is possible if and only if* $p(x) = \alpha P_n^{(\lambda)}(x)$ *with* $|\alpha| = 1$.

Using a similar approach, it was recently proved by Guessab and Rahman [5] that under the assumptions of Theorem B

$$\int_{-1}^{1} w_{\lambda+k-2}(t)|p^{(k)}(t)|^2\, dt \le \int_{-1}^{1} w_{\lambda+k-2}(t) \left| \frac{d^k}{dt^k} P_n^{(\lambda)}(t) \right|^2 dt \ \text{for all}\ k \ge 2. \tag{4}$$

For some results of a similar nature see [11], [3], [7].

The inequalities (3)–(4) are in the spirit of the Duffin and Schaeffer extension of Markov's inequality, namely, the set Ω of the "checking points" consists of the points of local extrema of the majorant $P_n^{(\lambda)}$. It seems not to have been noticed before that more general results can be obtained if this requirement is dropped, i.e., if the set Ω is not necessarily formed by the critical points of the majorant. In fact, the following theorem holds true.

Theorem 1.1. *Let* $Q_n \in \pi_n$ *have distinct zeros which interlace with the zeros* $\{t_j\}_{j=0}^n$ *of* $(1-t^2)\frac{d}{dt}P_n^{(\lambda)}(t)$ $(-1/2 < \lambda \le 1/2)$. *If* $p \in \pi_n$ *satisfies*

$$|p(t_j)| \le |Q_n(t_j)| \text{ for } j = 0, 1, \dots, n,$$

then

$$\int_{-1}^{1} w_{\lambda+k-1}(t)|p^{(k)}(t)|^2\,dt \le \int_{-1}^{1} w_{\lambda+k-1}(t)|Q_n^{(k)}(t)|^2\,dt \qquad (k = 1, \dots, n) \quad (5)$$

and

$$\int_{-1}^{1} w_{\lambda+k-2}(t)|p^{(k)}(t)|^2\,dt \le \int_{-1}^{1} w_{\lambda+k-2}(t)|Q_n^{(k)}(t)|^2\,dt \qquad (k = 2, \dots, n). \quad (6)$$

Equality in (5)–(6) *is possible if and only if* $p(x) = \alpha Q_n(x)$ *with* $|\alpha| = 1$.

The inequalities (3) and (4) follow from (5) and (6), respectively, if we choose $Q_n = P_n^{(\lambda)}$. More generally, the conditions of Theorem 1.1 are fulfilled for $Q_n = P_n^{(\mu)}$, $\mu \in [\lambda, \lambda+1]$.

There are three objects involved: the majorant Q_n, the set of comparison points $\Omega = \{t_j\}_{j=0}^n$ and the weight functions. We observe that in (5)–(6) the weight functions and the set of checking points Ω are closely interlinked, while there is a relative freedom in the choice of the majorant. In §4 we show that, for Ω consisting of the zeros of $(1-t^2)P_n'(t)$, P_n being the nth Legendre polynomial, we may also vary the weight functions (Theorem 4.1). Section 2 contains some auxiliary results. The proof of Theorem 1.1 is given in §3 (it is not essentially different from the proofs of (3) and (4) given in [1] and [5]). In §5 we show that for the highest-order derivatives ($k = n-1$ and $k = n-2$) there is a great variety of Gegenbauer weight functions which admit L_2 inequalities.

2. Auxiliary Results

We start with a simple lemma which is due to V. Markov (for a proof, see [8, Lemma 2.7.1]).

Lemma 2.1. *Let* p *and* q *be two algebraic polynomials having only real and distinct zeros which interlace. Then the zeros of* p' *and* q' *also interlace.*

Lemma 2.1 is the basic ingredient of the proof of the next lemma, which is also due essentially to V. Markov (a proof can be found, e.g., in [9, Lemma 2]).

Lemma 2.2. *Let* $\omega(t) = c\prod_{j=0}^n (t - t_j)$ $(c \ne 0)$ *with* $-1 = t_0 < t_1 < \cdots < t_n = 1$. *Let* $Q_n \in \pi_n$ *be a fixed polynomial which has* n *distinct real zeros interlacing with the zeros of* ω. *If* $f \in \pi_n$ *satisfies the inequalities*

$$|f(t_j)| \le |Q_n(t_j)| \quad \text{for } j = 0, \dots, n,$$

then for every $k \in \{1, \dots, n\}$ *there exists a set* $I_{n,k} = I_{n,k}(\omega)$ *such that*

$$|f^{(k)}(x)| \le |Q_n^{(k)}(x)| \quad \text{for all } x \in I_{n,k}.$$

The set $I_{n,k}$ is given by

$$I_{n,k} = [-1, \alpha_1^k] \cup [\beta_1^k, \alpha_2^k] \cup \cdots \cup [\beta_{n-k-1}^k, \alpha_{n-k}^k] \cup [\beta_{n-k}^k, 1],$$

where $\{\alpha_j^k\}_1^{n-k}$ and $\{\beta_j^k\}_1^{n-k}$ are the ordered zeros of $\omega_n^{(k)}$ and $\omega_0^{(k)}$, respectively, with $\omega_j(x) = \omega(x)/(x - t_j)$.

Moreover, equality occurs at an interior point of $I_{n,k}$ if and only if $f(x) \equiv \alpha Q_n(x)$ with $|\alpha| = 1$.

Remark 2.3. *We shall call the $n - k + 1$ intervals forming $I_{n,k}$ the Chebyshev intervals. In the case $k = n$ we have $I_{n,n} \equiv [-1, 1]$. An important characterization of $I_{n,k}(\omega)$ is the following: $x^* \in I_{n,k}(\omega)$ if and only if $\omega_0^{(k)}(x^*)\omega_n^{(k)}(x^*) \geq 0$.*

We formulate as a lemma a variant of Sturm's comparison theorem (see [10, Theorem 1.82.1]).

Lemma 2.4. *Let $f(x)$ and $F(x)$ be functions continuous in $x_0 < x < X_0$ with $f(x) \leq F(x)$. Let the functions $y(x)$ and $Y(x)$, both not identically zero, satisfy the differential equations*

$$y'' + f(x)y = 0, \quad Y'' + F(x)Y = 0,$$

respectively. Let x' and x'', $x' < x''$, be two consecutive zeros of $y(x)$. Then the function $Y(x)$ has at least one variation of sign in the interval $x' < x < x''$ provided $f(x) \not\equiv F(x)$ in $[x', x'']$.

Finally, we list in a lemma some well-known facts about ultraspherical polynomials (see [10, §4.7]).

Lemma 2.5. (i) $y = P_n^{(\lambda)}$ *satisfies the differential equation*

$$(1 - x^2)y'' - (2\lambda + 1)xy' + n(n + 2\lambda)y = 0 \;;$$

(ii) $u = (1 - x^2)^{\lambda/2+1/4} P_n^{(\lambda)}$ *satisfies the differential equation*

$$\frac{d^2u}{dx^2} + \left\{ \frac{(n+\lambda)^2}{1-x^2} + \frac{1/2 + \lambda - \lambda^2 + x^2/4}{(1-x^2)^2} \right\} u = 0 \;;$$

$$\text{(iii)}\ P_n^{(\lambda)}(x) = \sum_{m=0}^{[n/2]} (-1)^m \frac{\Gamma(n - m + \lambda)}{\Gamma(\lambda)\Gamma(m+1)\Gamma(n - 2m + 1)} (2x)^{n-2m} \ (\lambda \neq 0).$$

3. Proof of Theorem 1.1

We begin by showing that the zeros of $\frac{d^{k-1}}{dx^{k-1}} P_n^{(\lambda)} = \text{const} \cdot P_{n-k+1}^{(\lambda+k-1)}$ belong to the set $I_{n,k}(\omega)$, where $\omega(x) = (1 - x^2)\frac{d}{dx} P_n^{(\lambda)}(x)$. We prove this first in the case $k = 1$ by showing that if $P_n^{(\lambda)}(x_*) = 0$, then $\omega_0'(x_*)\omega_n'(x_*) > 0$. Indeed, using Lemma 2.5(i), we obtain

$$\omega_0'(x_*) = -\left[1 - (2\lambda + 1)\frac{x_*}{1 + x_*}\right] \frac{d}{dx} P_n^{(\lambda)}(x)\bigg|_{x=x_*},$$

$$\omega_n'(x_*) = -\left[1 + (2\lambda+1)\frac{x_*}{1-x_*}\right] \frac{d}{dx} P_n^{(\lambda)}(x)\Big|_{x=x_*}.$$

By assumption, $-1/2 < \lambda \leq 1/2$; therefore the terms in the square brackets are both positive, and hence $x_* \in I_{n,1}(\omega)$. Moreover, if x_* and y_* are two consecutive zeros of $P_n^{(\lambda)}$, then

$$\omega_0'(x_*)\omega_0'(y_*) < 0, \quad \omega_n'(x_*)\omega_n'(y_*) < 0,$$

which means that the zeros of $P_n^{(\lambda)}$ interlace with the zeros of both ω_0' and ω_n'. By Lemma 2.1 we conclude that the zeros of $\frac{d^{k-1}}{dx^{k-1}} P_n^{(\lambda)}$ interlace with the zeros of both $\omega_0^{(k)}$ and $\omega_n^{(k)}$ for $k = 1, \dots, n-1$. Hence, the zeros of $\frac{d^{k-1}}{dx^{k-1}} P_n^{(\lambda)}$ belong to $I_{n,k}(\omega)$. More precisely, since the interlacing is strict, the zeros of $\frac{d^{k-1}}{dx^{k-1}} P_n^{(\lambda)}$ are interior points for $I_{n,k}(\omega)$.

Denote the zeros of $\frac{d^{k-1}}{dx^{k-1}} P_n^{(\lambda)}$ by $\{z_j\}_{j=1}^{n-k+1}$. If $p \in \pi_n$ satisfies the assumptions of Theorem 1.1, then, by Lemma 2.2,

$$|p^{(k)}(z_j)| \leq |Q_n^{(k)}(z_j)|, \qquad j = 1, \dots, n-k+1, \tag{7}$$

and all the inequalities are strict unless $p = \alpha Q_n$ for some α, $|\alpha| = 1$. The inequality (5) follows from the Gaussian quadrature formula

$$\int_{-1}^{1} w_{\lambda+k-1}(t) f(t)\, dt = \sum_{j=1}^{n-k+1} w_j^G f(z_j) + R^G(f). \tag{8}$$

Since $R^G(f) = 0$ for $f \in \pi_{2n-2k+1}$, we have

$$\begin{aligned} \int_{-1}^{1} w_{\lambda+k-1}(t) |p^{(k)}(t)|^2 dt &= \sum_{j=1}^{n-k+1} w_j^G |p^{(k)}(z_j)|^2 \\ \leq \sum_{j=1}^{n-k+1} w_j^G |Q_n^{(k)}(z_j)|^2 &= \int_{-1}^{1} w_{\lambda+k-1}(t) |Q_n^{(k)}(t)|^2 dt, \end{aligned} \tag{9}$$

and equality is possible if and only if $p = \alpha Q_n$ with $|\alpha| = 1$.

For the proof of (6), we apply the generalized Gauss-Lobatto quadrature formula

$$\int_{-1}^{1} w_\lambda(t) f(t)\, dt = \sum_{\nu=0}^{k-2} a_\nu [f^{(\nu)}(-1) + (-1)^\nu f^{(\nu)}(1)] + \sum_{j=1}^{n-k+1} w_j^{\mathrm{Lo}} f(z_j) + R^{\mathrm{Lo}}(f), \tag{10}$$

which uses the same interior nodes and all of whose weights $\{a_\nu\}_{\nu=0}^{k-2}$, $\{w_j^{\mathrm{Lo}}\}_{j=1}^{n-k+1}$ are positive. Moreover, $R^{\mathrm{Lo}}(f) = 0$ for all $f \in \pi_{2n-1}$. We apply this quadrature formula to $f(x) = (1-x^2)^{k-2}|p^{(k)}(x)|^2$ to obtain

$$\int_{-1}^{1} w_{\lambda+k-2}(t)|p^{(k)}(t)|^2 dt = \int_{-1}^{1} w_\lambda(t)(1-t^2)^{k-2}|p^{(k)}(t)|^2 dt$$

$$= 2^{k-2}(k-2)!a_{k-2}[|p^{(k)}(-1)|^2 + |p^{(k)}(1)|^2] + \sum_{j=1}^{n-k+1} w_j^{\mathrm{Lo}}(1-z_j^2)^{k-2}|p^{(k)}(z_j)|^2$$

$$\le 2^{k-2}(k-2)!a_{k-2}[|Q_n^{(k)}(-1)|^2 + |Q_n^{(k)}(1)|^2] + \sum_{j=1}^{n-k+1} w_j^{\mathrm{Lo}}(1-z_j^2)^{k-2}|Q_n^{(k)}(z_j)|^2$$

$$= \int_{-1}^{1} w_{\lambda+k-2}(t)|Q_n^{(k)}(t)|^2 dt. \tag{11}$$

Again, equality in (11) is possible if and only if $p = \alpha Q_n$ with $|\alpha| = 1$.

4. Varying the Weight for Fixed Comparison Points

We showed in the preceding section that for $-1/2 < \lambda \le 1/2$ the zeros of $P_n^{(\lambda)}$ are interior points for the Chebyshev set $I_{n,1}(\omega)$ formed by $\omega(t) := (1-t^2)\frac{d}{dt}P_n^{(\lambda)}(t)$. Specifically, each of the $n-k+1$ Chebyshev intervals forming $I_{n,k}(\omega)$ contains in its interior exactly one zero of $\frac{d^{k-1}}{dt^{k-1}}P_n^{(\lambda)}(t)$. By continuity arguments, there exists a small $\epsilon = \epsilon(n) > 0$ such that for every $\mu \in (\lambda-\epsilon, \lambda+\epsilon)$ the zeros of $\frac{d^{k-1}}{dt^{k-1}}P_n^{(\mu)}(t)$ also possess this property.

Here, we illustrate this by considering the special case $\lambda = 1/2$. The reason for choosing probably the worst case (remember that $\lambda = 1/2$ is the limit value for which the above mentioned phenomenon occurs) is that the calculations are somewhat simpler.

We prove the following theorem.

Theorem 4.1. *Let $Q_n \in \pi_n$ have n distinct zeros which interlace with the zeros $\{t_j\}_{j=0}^n$ of $(1-t^2)P_n'(t)$. If $p \in \pi_n$ satisfies*

$$|p(t_j)| \le |Q_n(t_j)| \quad \textit{for } j = 0, 1, \dots, n,$$

then

$$\int_{-1}^{1} w_{\lambda+k-1}(t)|p^{(k)}(t)|^2 dt \le \int_{-1}^{1} w_{\lambda+k-1}(t)|Q_n^{(k)}(t)|^2 dt \qquad (k = 1, \dots, n) \tag{12}$$

for every $\lambda \in [\frac{1}{2} - \frac{2}{2n+1}, \frac{1}{2}]$, if n is odd, and for every $\lambda \in [\frac{1}{2} - \frac{2}{2n+5}, \frac{1}{2}]$, if n is even . Moreover, equality in (12) *is possible if and only if $p = \alpha Q_n$ with $|\alpha| = 1$.*

Proof. Set $\omega(t) := (1-t^2)P_n'(t)$. All we need to show is that, for the values of λ postulated in the theorem, the zeros $\{z_j\}_{j=1}^{n-k+1}$ of $\frac{d^{k-1}}{dt^{k-1}}P_n^{(\lambda)}(t)$ belong to $I_{n,k}(\omega)$.

Then the inequalities (7) will follow, and the proof of Theorem 4.1 is completed by application of the Gaussian quadrature formula (8).

The assertion that $\{t_j\}_{j=1}^{n-k+1}$ belong to $I_{n,k}(\omega)$ will be proved if we can show that the zeros of $P_n^{(\lambda)}$ interlace with the zeros of both ω_0' and ω_n' (here $\omega_0(t) = (1-t)P_n'(t)$ and $\omega_n(t) = (1+t)P_n'(t)$). Indeed, this establishes the assertion for $k = 1$, and then for $k > 1$ it follows from Lemma 2.1.

By symmetry, it suffices to show that the zeros of $P_n^{(\lambda)}$ and ω_0' interlace. We already know from the preceding section that the zeros $\{\beta_j\}_{j=1}^{n-1}$ of ω_0' interlace with the zeros of P_n. Let $\{\tau_j(\lambda)\}_{j=1}^{n}$ be the zeros of $P_n^{(\lambda)}$, $-1 < \tau_1(\lambda) < \cdots < \tau_n(\lambda) < 1$. It is well known that the zeros of $P_n^{(\lambda)}$ depend monotonically on λ; specifically, the positive τ's decrease and the negative τ's increase as λ increases. Let us mention two consequences of this fact:

(a) For $\lambda \in (-1/2, 1/2]$, the $\{\beta_j\}_{j=1}^{n-1}$ belong to the interval $(\tau_1(\lambda), \tau_n(\lambda))$;

(b) If $\{\beta_j\}_{j=1}^{n-1}$ and $\{\tau_j(\lambda)\}_{j=1}^{n}$ interlace for $\lambda = a$ and $\lambda = b$ $(-1/2 < a < b)$, then the interlacing property holds for every $\lambda \in [a, b]$. The assertion remains also valid if only the nonnegative (nonpositive) β's and τ's are considered.

By Lemma 2.5(ii), $Y = (1-x^2)^{\lambda/2+1/4} P_n^{(\lambda)}(x)$ satisfies the differential equation

$$\frac{d^2Y}{dx^2} + F(x)Y = 0, \quad F(x) = \frac{(n+\lambda)^2}{1-x^2} + \frac{1/2+\lambda-\lambda^2+x^2/4}{(1-x^2)^2}.$$

Using Lemma 2.5(i), we find that the function

$$y(x) = (1-x)^{1/2}(1+x)^{3/2}\omega_0'(x)$$

satisfies the differential equation

$$\frac{d^2y}{dx^2} + f(x)y = 0, \quad f(x) = \frac{n(n+1)}{1-x^2} + \frac{2x-1}{(1-x^2)^2}.$$

In order to apply Lemma 2.4, we need to compare the functions f and F. We have

$$(1-x^2)^2(f(x) - F(x)) = (2\lambda-1)[n + \tfrac{1}{4}(2\lambda+1)]x^2 + 2x - (2\lambda-1)n - \lambda - \tfrac{3}{2}$$
$$=: g(\lambda; x).$$

Consider the quadratic equation

$$g(\lambda; x) = 0, \tag{13}$$

where the parameter λ is in $(-1/2, 1/2)$. We have

$$g(\lambda; -1) = (\lambda + \frac{3}{2})(\lambda - \frac{5}{2}) < 0, \quad g(\lambda; 1) = (\lambda - \frac{1}{2})^2 > 0;$$

therefore (13) has a unique root $x_* = x_*(\lambda)$ in $(-1, 1)$, and this is the smaller root of this equation. Differentiation in (13) yields

$$\frac{dx_*}{d\lambda} = -\left(\frac{\partial g}{\partial \lambda} \Big/ \frac{\partial g}{\partial x}\right)\Bigg|_{x=x_*} = -[(2n+2\lambda)x_*^2 - 2n - 1]\left(\frac{\partial g}{\partial x}\Bigg|_{x=x_*}\right)^{-1} > 0;$$

therefore x_* is a strictly monotone increasing function of λ. We consider separately the cases of odd and even n.

Case A: $n = 2m+1$. We choose $\lambda = \lambda_0 = \frac{1}{2} - \frac{2}{2n+1}$, so that $x_*(\lambda_0) = 0$. With this choice

$$f(x)\begin{cases} \geq F(x) & \text{for } 0 \leq x \leq 1, \\ < F(x) & \text{for } x < 0. \end{cases} \tag{14}$$

By virtue of (14) and Lemma 2.4 each interval $(\tau_j(\lambda_0), \tau_{j+1}(\lambda_0))$ $(j = m+1, \dots, 2m)$ contains at least one zero of ω_0', and since ω_0' has exactly m positive zeros, there holds

$$0 = \tau_{m+1}(\lambda_0) < \beta_{m+1} < \cdots < \tau_{2m}(\lambda_0) < \beta_{2m} < \tau_{2m+1}(\lambda_0).$$

For the same reason, each interval (β_j, β_{j+1}) $(j = 1, \dots, m-1)$ contains at least one zero of $P_n^{(\lambda_0)}$. Since, by (a), $\tau_1(\lambda_0) < \beta_1$, we get

$$\tau_1(\lambda_0) < \beta_1 < \tau_2(\lambda_0) < \cdots < \tau_m(\lambda_0) < \beta_m (< 0).$$

Thus, the zeros of ω_0' and $P_n^{(\lambda_0)}$ interlace, and according to (b) so do the zeros of ω_0' and $P_n^{(\lambda)}$ when $\lambda_0 \leq \lambda \leq 1/2$. This proves Theorem 4.1 in Case A.

Case B: $n = 2m$. In this case, ω_0' has m positive and $m-1$ negative zeros. As in Case A we find that

$$(0 <)\beta_m < \tau_{m+1}(\lambda) < \cdots < \beta_{2m-1} < \tau_{2m}(\lambda) \text{ for every } \lambda \in [\lambda_0, 1/2] \tag{15}$$

(for the first inequality we have used $\beta_m < \tau_{m+1}(1/2) \leq \tau_{m+1}(\lambda)$). Unfortunately, these arguments can not yield interlacing of the negative zeros of ω_0' and $P_n^{(\lambda)}$ for every $\lambda \geq \lambda_0$. Suppose that $\lambda = \lambda_1$ $(\lambda_0 < \lambda_1 < 1/2)$ is chosen such that

$$f(x) < F(x) \text{ for every } x \leq \beta_m. \tag{16}$$

By Lemma 2.4 each interval (β_j, β_{j+1}) $(j = 1, \dots, m-1)$ contains at least one zero of $P_n^{(\lambda_1)}$, and since, by (a), $\tau_1(\lambda_1) < \tau_1(1/2) < \beta_1$, there must hold

$$\tau_1(\lambda_1) < \beta_1 < \tau_2(\lambda_1) < \cdots < \tau_m(\lambda_1) < \beta_m.$$

It follows from (b) that

$$\tau_1(\lambda) < \beta_1 < \tau_2(\lambda) < \cdots < \tau_m(\lambda) < \beta_m \text{ for every } \lambda \in [\lambda_1, 1/2]. \tag{17}$$

From (15) and (17) the interlacing property holds for every $\lambda \in [\lambda_1, 1/2]$. The proof of Case B will be complete if we show that (16) holds for $\lambda = \lambda_1 := 1/2 - 2/(2n+5)$.

The inequality (16) is fulfilled exactly when $x_*(\lambda) > \beta_m$, with $x_*(\lambda)$ being the zero of (13) in $(-1, 1)$. We know that $\beta_m < \tau_{m+1}(1/2)$, the smallest positive zero of the Legendre polynomial P_n, and an upper bound for $\tau_{m+1}(1/2)$ is given by (see [10, Eq. 6.6.5])

$$\tilde{x} = \sin \frac{\pi}{2n+1}.$$

Therefore, (16) will certainly hold if $x_*(\lambda) \geq \tilde{x}$, or, equivalently, if the inverse function $\lambda(x)$ satisfies

$$\lambda(x) \geq \lambda(\tilde{x}). \tag{18}$$

It is more convenient to work with $\mu = \lambda - 1/2$, rewriting (13) as

$$x^2\mu^2 - (2n+1)(1-x^2)\mu + 2x - 2 = 0.$$

Equivalently to (18), the negative root (with respect to μ) of this equation must satisfy $\mu \geq \mu(\tilde{x})$, i.e.,

$$\mu \geq \frac{-4}{(2n+1)(1+\tilde{x}) + \sqrt{(2n+1)^2(1+\tilde{x})^2 + 8\tilde{x}^2/(1-\tilde{x})}}. \tag{19}$$

Making use of the fact that the sequence $\{(2n+1)\sin\frac{\pi}{2n+1}\}$ increases monotonically, after some manipulation we find that $\mu \geq \frac{-2}{2n+5}$ is a sufficient condition for (19) to hold. Performing this reasoning backward, we conclude that (16) holds for $\lambda = \lambda_1$. The proof of Theorem 4.1 is complete. □

5. The Case of Highest Derivatives

In the preceding section we were able to vary the parameter λ for a fixed set of comparison points Ω only by very little, so that the zeros of $P_n^{(\lambda)}$ still belong to the Chebyshev set $I_{n,1}(\omega)$. The situation is not so bad if we require that the zeros of $\frac{d^{k-1}}{dx^{k-1}}P_n^{(\lambda)}(x)$ belong to $I_{n,k}(\omega)$ only for $k = n-2, n-1, n$. In fact, as was mentioned in Remark 2.3, for $k = n$, the Chebyshev set $I_{n,n}(\omega)$ consists of the whole interval $[-1,1]$, and we have the following result.

Proposition 5.1. *Let* $\Omega = \{t_j\}_{j=0}^n$, $-1 = t_0 < t_1 < \cdots < t_n = 1$, *and let* $Q_n \in \pi_n$ *have* n *distinct zeros which interlace with* $\{t_j\}_{j=0}^n$. *If* $p \in \pi_n$ *satisfies*

$$|p(t_j)| \leq |Q_n(t_j)| \quad \text{for } j = 0,1,\ldots,n,$$

then

$$\int_{-1}^{1} w(t)|p^{(n)}(t)|^2 dt \leq \int_{-1}^{1} w(t)|Q_n^{(n)}(t)|^2 dt \tag{20}$$

for every weight function $w(t)$. *Equality in* (20) *is possible if and only if* $p = \alpha Q_n$ *with* $|\alpha| = 1$.

An easy calculation reveals that if the set Ω is symmetric with respect to the origin, then the Chebyshev set is given by

$$I_{n,n-1} = \left[-1, -\frac{1}{n}\right] \cup \left[\frac{1}{n}, 1\right].$$

Since the nodes of the two-point Gaussian quadrature formula with respect to the Gegenbauer weight $w_\lambda(t)$ are $z_{1,2} = \pm(2\lambda+2)^{-1/2}$, these nodes belong to $I_{n,n-1}$ if and only if $\lambda \in (-1/2, n^2/2 - 1]$. Thus we obtain

Proposition 5.2. *Let* $\Omega = \{t_j\}_{j=0}^n$, $-1 = t_0 < t_1 < \cdots < t_n = 1$ *be a symmetric set with respect to the origin, and let* $Q_n \in \pi_n$ *have* n *distinct zeros which interlace with* $\{t_j\}_{j=0}^n$. *If* $p \in \pi_n$ *satisfies*

$$|p(t_j)| \le |Q_n(t_j)| \quad \textit{for } j = 0, 1, \ldots, n,$$

then

$$\int_{-1}^{1} w_\lambda(t)|p^{(n-1)}(t)|^2 dt \le \int_{-1}^{1} w_\lambda(t)|Q_n^{(n-1)}(t)|^2 dt \tag{21}$$

for every $\lambda \in (-\frac{1}{2}, \frac{n^2}{2} - 1]$. *Equality in* (21) *is possible if and only if* $p = \alpha Q_n$ *with* $|\alpha| = 1$.

A similar result holds in the case $k = n - 2$ $(n \ge 3)$.

Proposition 5.3. *Let* $\Omega = \{t_j\}_{j=0}^n$ *consist of the zeros of* $(1 - t^2)\frac{d}{dt}P_n^{(\lambda)}(t)$. *Let* $Q_n \in \pi_n$ *have* n *distinct zeros which interlace with* $\{t_j\}_{j=0}^n$. *If* $p \in \pi_n$ *satisfies*

$$|p(t_j)| \le |Q_n(t_j)| \quad \textit{for } j = 0, 1, \ldots, n,$$

then

$$\int_{-1}^{1} w_\mu(t)|p^{(n-2)}(t)|^2 dt \le \int_{-1}^{1} w_\mu(t)|Q_n^{(n-2)}(t)|^2 dt \tag{22}$$

for every $\mu \in (-\frac{1}{2}, \frac{3n^2}{2(\sigma(n,\lambda)+1)^2} - 2] \cup [\frac{3n^2}{2(\sigma(n,\lambda)-1)^2} - 2, \infty)$, *where*

$$\sigma(n, \lambda) = \sqrt{\frac{n^2 + 2\lambda - 2}{2n + 2\lambda - 2}}.$$

Equality in (22) *is possible if and only if* $p = \alpha Q_n$ *with* $|\alpha| = 1$.

Proof. Using Lemma 2.5(iii), we find that the Chebyshev set $I_{n,n-2}(\omega)$ is

$$I_{n,n-2}(\omega) = \left[-1, \tfrac{-1-\sigma(n,\lambda)}{n}\right] \cup \left[\tfrac{1-\sigma(n,\lambda)}{n}, \tfrac{-1+\sigma(n,\lambda)}{n}\right] \cup \left[\tfrac{1+\sigma(n,\lambda)}{n}, 1\right].$$

The nodes of the three-point Gaussian quadrature formula associated with the weight function $w_\mu(t)$ are

$$z_1 = -\sqrt{\frac{3}{2(\mu+2)}}, \; z_2 = 0, \; z_3 = \sqrt{\frac{3}{2(\mu+2)}},$$

and these nodes belong to $I_{n,n-2}(\omega)$ if

$$z_3 \le n^{-1}(\sigma(n,\lambda) - 1) \text{ or } z_3 \ge n^{-1}(\sigma(n,\lambda) + 1).$$

This is the case exactly when μ belongs to the intervals postulated in the proposition. □

References

[1] B. Bojanov, *An inequality of Duffin and Schaeffer type,* East J. Approx., **1** (1995), 37–46.

[2] B. Bojanov and G. Nikolov, *Duffin and Schaeffer type inequality for ultraspherical polynomials*, J. Approx. Theory, **84** (1996), 129–138.

[3] D.K. Dimitrov, *Markov inequalities for weight functions of Chebyshev type*, J. Approx. Theory, **83** (1995), 175–181.

[4] R.J. Duffin and A.C. Schaeffer, *A refinement of an inequality of the brothers Markoff*, Trans. Amer. Math. Soc., **50** (1941), 517–528.

[5] A. Guessab and Q.I. Rahman, *Quadrature formulae and polynomial inequalities*, J. Approx. Theory, **90** (1997), 255–282.

[6] V.A. Markov, *On functions least deviated from zero in given interval*, St. Petersburg, 1892 (in Russian). German translation: *Über Polynome, die in einem gegebenen Intervalle möglichst wenig von Null abweichen*, Math. Ann., **77** (1916), 213–258.

[7] G. Nikolov, *Weighted L_2 inequalities of Duffin and Schaeffer type via quadrature formulae*, submitted for publication.

[8] T.J. Rivlin, *The Chebyshev polynomials*, Pure and Applied Mathematics, Wiley, New York, 1974.

[9] A.Yu. Shadrin, *Interpolation with Lagrange polynomials. A simple proof of Markov inequality and some of its generalizations*, Approx. Theory Appl., **8** (1992), 51–61.

[10] G. Szegö, *Orthogonal Polynomials*, Amer. Math. Soc. Colloquium Publ. **23**, 4th ed., Amer. Math. Soc., Providence, RI, 1975.

[11] A.K. Varma, *On some extremal properties of algebraic polynomials*, J. Approx. Theory, **69** (1992), 48–54.

David B. Hunter
Department of Computing
University of Bradford
Bradford BD7 1DP, England
E-mail address: `d.b.hunter@bradford.ac.uk`

Geno Nikolov
Department of Mathematics
University of Sofia
BG-1126 Sofia, Bulgaria
E-mail address: `geno@fmi.uni-sofia.bg`

International Series of Numerical Mathematics
Vol. 131, © 1999 Birkhäuser Verlag Basel/Switzerland

Questions Related to Gaussian Quadrature Formulas and Two-Term Recursions

Dirk P. Laurie

Abstract. There are theoretical and practical reasons for preferring coupled two-term recursion to three-term recursion when working with orthogonal polynomials, especially in problems related to Gaussian quadrature.

1. Three Terms Good, Two Terms Better

It has been known for a long time that there are two ways of generating orthogonal polynomials: by a single three-term recursion or by two coupled two-term recursions. The choice between the two has been a subject of debate among practitioners of conjugate-gradient (two-term) and Lanczos (three-term) iterative methods for linear equations, but in the more traditional applications of orthogonal polynomials, the three-term formulas have reigned unchallenged. It is the opinion of this author that the time has come to reassess the relative merits of the two approaches.

There are theoretical as well as practical reasons for preferring two-term recursion. From the theoretical point of view, the classical orthogonal polynomials on finite and semi-infinite intervals have simpler coefficients when written as two-term recursions — in fact, having seen the two-term coefficients, one cannot escape a feeling that they are more fundamental than the three-term ones. We suggest that it may be a useful option, in cases where recursion coefficients are investigated, to attempt finding the two-term coefficients first. Questions dealing with the location of zeros near the left point of the interval of orthogonality are answered by a glance at the two-term coefficients but require quite a bit of work when starting from the three-term coefficients.

From the practical point of view, it has been argued by Fernando and Parlett [5] that the 'differential' form of Rutishauser's *qd* algorithm offers high floating-point accuracy in the calculation of the singular values and vectors of a bidiagonal matrix. This implies that Gaussian quadrature formulas can be computed with high relative accuracy from two-term coefficients. For the converse problem, the present author [13] has recently exhibited an algorithm that recovers the two-term coefficients with high relative accuracy from the weights and gaps (i.e. differences between nodes) of the quadrature formula. This result is of particular interest since the problem of recovering the three-term coefficients from the weights and nodes can be ill posed [10] in the sense of perturbations in a vector norm.

The purpose of this paper is to convince the reader that two-term recursion should be tried in nearly all cases and used in most. There is therefore little that is really new here (the two-term analogue of the Sack-Donovan algorithm is the most substantial new item). Rather, the bulk of the paper consists of known but not always well-known results, mustered to support the main thesis. I hope that this will give the reader a warm sense of familiarity rather than boredom.

2. Summary of (Mostly) Known Results

This section makes no attempt at completeness or great originality. Rather, it is a collection in one place and using a common notation, of some useful facts from various sources, in particular [1, 2, 4, 5, 10]. I omit proofs, but have tried when possible to expound the results in a way that suggests a proof outline. In some cases, especially when dealing with two-term recursion coefficients, I have not been able to find the precise result spelled out in the literature, but the proofs (omitted) are mostly one-liners.

2.1. The fundamental identity of Gaussian quadrature

Let w be a nondecreasing function with at least n points of increase on an interval (a, b). Define the inner product $(\cdot,\cdot)_w$ by the Stieltjes integral

$$(p, q) = \int_a^b p(t)q(t)\,\mathrm{d}w(t)$$

and let

$$\mu_0 = \int_a^b \mathrm{d}w(t).$$

A fundamental tool in the study of orthogonal polynomials and Gaussian quadrature is the generating function r defined by

$$r(w; x) = \int_a^b \frac{1}{x-t}\,\mathrm{d}w(t). \tag{1}$$

We are not interested here in analytical questions about the existence of the integral defining r; all results are derived in the setting of formal power series.

The n-point Gaussian quadrature formula

$$G_n f := \sum_{j=1}^{n} \gamma_j f(\xi_j) = \int_a^b f(t)\,\mathrm{d}w(t) - R_n f$$

has the property that $R_n f = 0$ when f is a polynomial of degree $2n-1$. One can view a quadrature formula as tantamount to replacing the weight $\mathrm{d}w$ by a discrete measure

$$\mathrm{d}w_n(t) = \sum_{j=1}^{n} \gamma_j \delta(t-\xi_j)\,\mathrm{d}t,$$

where δ is the Dirac delta: using G_n to approximate an integral with respect to dw does the same as replacing dw by dw_n and integrating formally. Do so in (1) to obtain the partial fraction

$$r_n(w;x) = \sum_{j=1}^{n} \frac{\gamma_j}{x - \xi_j}.$$

Since r_n is an $(n-1,n)$ rational function which has a power series agreeing with that of r up to degree $2n-1$, it equals the $(n-1,n)$ formal Padé approximation to r. When r_n is represented as a Jacobi continued fraction, we obtain the fundamental identity

$$\sum_{j=1}^{n} \frac{\gamma_j}{x-\xi_j} = \frac{\mu_0}{x - a_0 -} \frac{b_1}{x - a_1 -} \cdots \frac{b_{n-1}}{x - a_{n-1}}$$

that relates a Gaussian formula to the three-term recursion coefficients of the orthogonal polynomials.

2.2. Three-term recursion formulas

The monic polynomials p_k orthogonal with respect to $(\cdot,\cdot)_w$ are given by

$$\begin{aligned} p_{-1}(x) &= 0, \\ p_0(x) &= 1, \\ p_{k+1}(x) &= (x - a_k)p_k(x) - b_k p_{k-1}(x), \end{aligned}$$

where $b_0 := \mu_0$.

The *Darboux-Stieltjes formulas* for the three-term coefficients are

$$b_k = \frac{(p_k, p_k)}{(p_{k-1}, p_{k-1})}, \tag{2}$$

$$a_k = \frac{(xp_k, p_k)}{(p_k, p_k)}, \tag{3}$$

where by an abuse of notation xp_k is the polynomial such that $(xp_k)(x) = xp_k(x)$.

The *Stieltjes algorithm* for generating orthogonal polynomials, given a method of evaluating the inner product, boils down to the use of the above formulas to calculate in turn $a_0, p_1, b_1, a_1, p_2, \ldots, p_n$.

Gautschi [6, 7] proposes that a suitable numerical quadrature be used for the evaluation of the required inner products. This process is known as the *discrete Stieltjes algorithm*.

2.3. The matrix connection

The connection between three-term recursion and the Jacobi matrix is clearest when one considers the unsymmetric matrix

$$J = \mathsf{tridiag}\begin{pmatrix} & 1 & & 1 & & \bullet & & 1 & \\ a_0 & & a_1 & & \bullet & & \bullet & & a_{n-1} \\ & b_1 & & b_2 & & \bullet & & b_{n-1} & \end{pmatrix}.$$

It is easy to see that when ξ is a zero of p_n, the vector $\mathbf{y} = [p_0(\xi)\ p_1(\xi)\ \ldots\ p_{n-1}(\xi)]^\mathsf{T}$ satisfies $J\mathbf{y} = \xi\mathbf{y}$.

The symmetric matrix

$$T = \mathsf{tridiag}\begin{pmatrix} & \sqrt{b_1} & & \sqrt{b_2} & & \bullet & & \sqrt{b_{n-1}} & \\ a_0 & & a_1 & & \bullet & & \bullet & & a_{n-1} \\ & \sqrt{b_1} & & \sqrt{b_2} & & \bullet & & \sqrt{b_{n-1}} & \end{pmatrix},$$

is diagonally similar to J. The Golub-Welsch algorithm [9] for calculating the Gaussian formula from the recursion coefficients rests on the observation that the nodes and weights can be obtained by solving the eigenproblem for T as follows: let U be an orthogonal matrix containing eigenvectors of T; then

$$U^\mathsf{T} T U = \mathrm{diag}(\xi_j), \tag{4}$$

$$w_j = \mu_0 u_{1,j}^2. \tag{5}$$

The implementation of the algorithm in [9] avoids forming more than the first row of U by using a suitably restricted version of the QR algorithm. A similar modification can be made to the QL algorithm [13]. Both these algorithms offer excellent numerical stability when roundoff errors are measured in a vector norm.

2.4. Two-term recursion

When T is positive definite, then it has the Cholesky factorization $T = LL^\mathsf{T}$ where

$$L = \mathsf{bidiag}\begin{pmatrix} \sqrt{q_1} & & \sqrt{q_2} & & \bullet & & \bullet & & \sqrt{q_n} \\ & \sqrt{e_1} & & \sqrt{e_{n-2}} & & \bullet & & \sqrt{e_{n-1}} & \end{pmatrix}$$

and $e_0 = 0$ by definition. An equivalent factorization is $J = EQ$ where

$$E = \mathsf{bidiag}\begin{pmatrix} 1 & & 1 & & \bullet & & \bullet & & 1 \\ & e_1 & & e_{n-2} & & \bullet & & e_{n-1} & \end{pmatrix};$$

$$Q = \mathsf{bidiag}\begin{pmatrix} & 1 & & 1 & & \bullet & & 1 & \\ q_1 & & q_2 & & \bullet & & \bullet & & q_n \end{pmatrix}.$$

Algorithms involving two-term recursion are easiest to follow in the square-root-free setting of the unsymmetric matrix J, but it should be kept in mind that there exists an equivalent formulation in terms of symmetric positive definite matrices.

The coefficients e_k, q_k are related to a_k, b_k by

$$b_k = q_k e_k,\ k = 1, 2, \ldots, \tag{6}$$

$$a_k = q_{k+1} + e_k,\ k = 0, 1, \ldots. \tag{7}$$

The two-term recursion is given in terms of the entries in L by

$$u_0(x) = 1, \quad p_0(x) = 1,$$
$$u_{k+1}(x) = x p_k(x) - e_k u_k(x),$$
$$p_{k+1}(x) = u_{k+1}(x) - q_{k+1} p_k(x).$$

TABLE 1. Two- and three-term recursion coefficients for the classical orthogonal polynomials. In the Jacobi case, $\kappa = 2k + \alpha + \beta$.

	Jacobi over $(0,2)$	Laguerre over $(0,\infty)$
a_k	$1 + \dfrac{\beta^2 - \alpha^2}{\kappa(\kappa+2)}$	$2k+1+\alpha$
b_k	$\dfrac{4k(k+\alpha)(k+\beta)(k+\alpha+\beta)}{\kappa^2(\kappa+1)(\kappa-1)}$	$k(k+\alpha)$
e_k	$\dfrac{2k(k+\alpha)}{\kappa(\kappa+1)}$	k
q_k	$\dfrac{2(k+\beta)(k+\alpha+\beta)}{\kappa(\kappa-1)}$	$k+\alpha$

The fundamental identity now involves a Stieltjes continued fraction

$$r_n(w;x) = \sum_{j=1}^{n} \frac{\gamma_j}{x-\xi_j} = \frac{\mu_0}{x-}\,\frac{q_1}{1-}\,\frac{e_1}{x-}\,\frac{q_2}{1-}\cdots\frac{e_{n-1}}{x-}\,\frac{q_n}{1}.$$

The two-term analogue to the Darboux-Stieltjes formulas is

$$q_l = \frac{(u_l, p_{l-1})}{(p_{l-1}, p_{l-1})}, \tag{8}$$

$$e_l = \frac{(p_l, p_l)}{(u_l, p_{l-1})}, \tag{9}$$

and the Stieltjes algorithm now involves computing $u_1, q_1, p_1, e_1, u_2, q_2, p_2, e_2, \ldots$ in this order.

The Cholesky factorization exists when T is positive definite, which is equivalent to requiring that all its eigenvalues ξ_j are positive. In the usual case of positive w this is always true when the left endpoint $a \geq 0$. When working with two-term recursion, it is therefore natural to transform the interval of integration affinely so that $a = 0$.

The Golub-Welsch algorithm can now be viewed in terms of the singular value decomposition $L = U\Sigma V^{\mathsf{T}}$, where U is the same as in (4), and $\Sigma = \operatorname{diag}(\sigma_j)$. This gives $\xi_j = \sigma_j^2$. The implementation of the algorithm in [5] involves the 'differential' form of the *qd* algorithm, which offers excellent componentwise floating-point accuracy.

2.5. Three terms versus two

In Table 1 we give the three- and two-term recurrence coefficients for the Jacobi polynomials over $[0,2]$ and the generalized Laguerre polynomials over $[0,\infty]$. Note that the coefficients b_k are $(4,4)$ rational functions of k in the Jacobi case, and quadratic polynomials in the Laguerre case, whereas the coefficients q_k and e_k are $(2,2)$ and linear respectively. This leads us to believe that it is more natural to think of the equation $b_k = q_k e_k$ as a definition of b_k in terms of basic quantities than vice versa.

We conclude with three remarks connected with Gaussian and related quadrature formulas.

1. The n-point Radau formula containing 0 as a prescribed node can be obtained by the Golub-Welsch algorithm when the three-term recursion coefficients have been modified by

$$a_{n-1} \leftarrow a_{n-1} + p_n(0)/p_{n-1}(0).$$

The corresponding modification of the two-term coefficients is much simpler:

$$q_n \leftarrow 0.$$

2. The $(n+1)$-point anti-Gaussian formula [11] is likewise found by modifying the three-term coefficients:

$$b_n \leftarrow 2b_n.$$

The corresponding modification of the two-term coefficients is hardly more complicated:

$$q_{n+1} \leftarrow q_{n+1} - e_n, \quad e_n \leftarrow 2e_n.$$

3. The criterion whether the left endpoint in an anti-Gaussian quadrature formula is internal is stated in terms of three-term coefficients as

$$p_{n+1}(0) \geq b_n p_{n-1}(0).$$

A good deal of work was expended in [11] to examine this question in the case of the Jacobi polynomials. In terms of two-term coefficients the condition to be tested simplifies to

$$q_{n+1} \geq e_n.$$

The result so laboriously found in [11] could have been obtained much more easily by substituting the expressions from Table 1 into the above inequality.

3. Two Sets of Orthogonal Polynomials

In certain applications it is useful to consider together polynomials orthogonal with respect to two weight functions. The situation arises in the generation of orthogonal polynomials from modified moments [6, 16] and in the calculation of Gauss-Kronrod quadrature formulas [12].

Together with the polynomials p_k with respect to w, we consider polynomials π_k orthogonal with respect to a different weight function ω that is also such that $\int_a^b \mathrm{d}\omega(t) = \mu_0$. The use of the same interval for both weight functions imposes no loss of generality (the functions can be defined to be 0 outside their smallest support), but it is in any case a bad idea to use different intervals: it is known [3] that the condition number of the map from the one polynomial basis to the other rises exponentially if the supports of the weight functions do not coincide.

Let the three-term recursion for these polynomials be given by

$$\begin{aligned}\pi_{-1}(x) &= 0,\\ \pi_0(x) &= 1,\\ \pi_{k+1}(x) &= (x-\alpha_k)\pi_k(x) - \beta_k\pi_{k-1}(x),\end{aligned}$$

where $\beta_0 = \mu_0$. The inner product $(\cdot,\cdot)_\omega$ is defined as

$$(p,q)_\omega = \int_a^b p(t)q(t)\,\mathrm{d}\omega(t).$$

It is possible to read off the degree to which the two inner products agree by counting the number of terms to which the sequences $a_0, b_1, a_1, b_2, a_2, \ldots$ and $\alpha_0, \beta_1, \alpha_1, \beta_2, \alpha_2, \ldots$ agree (remember $b_0 = \beta_0$ because both equal μ_0). This fact is exploited in [12], where Gauss-Kronrod formulas are calculated by the Golub-Welsch algorithm applied to the Jacobi matrix of order $2n+1$, after modifying the last n entries of the sequence $\ldots, b_{2n-2}, a_{2n-2}, b_{2n-1}, a_{2n-1}, b_{2n}, a_{2n}$.

The key to many computational questions in this situation is to use *mixed moments*

$$\sigma_{k,l} = (\pi_k, p_l)_\omega.$$

These satisfy the five-term recursion relation

$$\begin{pmatrix} & -\beta_k & \\ b_l & a_l-\alpha_k & 1\\ & -1 & \end{pmatrix} \cdot \begin{pmatrix} & \sigma_{k-1,l} & \\ \sigma_{k,l-1} & \sigma_{k,l} & \sigma_{k,l+1}\\ & \sigma_{k+1,l} & \end{pmatrix} = 0.$$

The orthogonality of the polynomials π_k implies that $\sigma_{k,l} = 0$ for $k > l$. The mixed moments $\sigma_{k,l}$ can therefore be arranged in an upper triangular matrix

$$\begin{bmatrix} \sigma_{0,0} & \sigma_{0,1} & \sigma_{0,2} & \cdots & \sigma_{0,m}\\ 0 & \sigma_{1,1} & \sigma_{1,2} & \cdots & \sigma_{1,m}\\ 0 & 0 & \sigma_{2,2} & \cdots & \sigma_{2,m}\\ \vdots & \vdots & \vdots & \ddots & \vdots\\ 0 & 0 & 0 & \cdots & \sigma_{m,m} \end{bmatrix}.$$

A survey of how the elements of this matrix are generated in various applications appears in [11]. Its transpose appears in [8], where it is used to represent some interesting relations in matrix form.

For two-term recursion, let the second set of polynomials be given by

$$\begin{aligned}\upsilon_0(x) &= 1, \quad \pi_0(x) = 1,\\ \upsilon_{k+1}(x) &= x\pi_k(x) - \eta_k\upsilon_k(x),\\ \pi_{k+1}(x) &= \upsilon_{k+1}(x) - \psi_{k+1}\pi_k(x),\end{aligned}$$

where $\eta_0 = 0$. As before, it is possible to read off the degree to which the two inner products agree: the sequences involved are now $q_1, e_1, q_2, e_2, q_3, \dots$ and $\psi_1, \eta_1, \psi_2, \eta_2, \psi_3, \dots$.

Here we need two kinds of mixed moments, namely

$$\nu_{k,l} = (v_k, p_l)_\omega, \tag{10}$$

$$\rho_{k,l} = (\pi_k, u_l)_\omega. \tag{11}$$

These are related by two four-term recursion formulas or *rhombus rules* as follows:

$$\begin{pmatrix} & -\psi_k & \\ q_l & & 1 \\ & -1 & \end{pmatrix} \cdot \begin{pmatrix} & \rho_{k-1,l} & \\ \nu_{k,l-1} & & \nu_{k,l} \\ & \rho_{k,l} & \end{pmatrix} = 0;$$

$$\begin{pmatrix} & -\eta_k & \\ e_l & & 1 \\ & -1 & \end{pmatrix} \cdot \begin{pmatrix} & \nu_{k,l} & \\ \rho_{k,l} & & \rho_{k,l+l} \\ & \nu_{k+1,l} & \end{pmatrix} = 0.$$

Instead of a matrix, it is convenient to arrange the mixed moments in a 'rhombus' array

$$\begin{array}{ccccccccc}
 & \nu_{0,0} & & \nu_{0,1} & & \nu_{0,2} & & \cdots \\
\rho_{0,0} & & \rho_{0,1} & & \rho_{0,2} & & \cdots & \\
 & \nu_{1,0} & & \nu_{1,1} & & \nu_{1,2} & & \cdots \\
\rho_{1,0} & & \rho_{1,1} & & \rho_{1,2} & & \cdots & \\
 & \nu_{2,0} & & \nu_{2,1} & & \nu_{2,2} & & \cdots \\
\rho_{2,0} & & \rho_{2,1} & & \rho_{2,2} & & \ddots & \\
 & \vdots & & \vdots & & \ddots & & \\
\vdots & & \vdots & & \vdots & & &
\end{array}$$

The orthogonality of the polynomials π_k implies that $\rho_{k,l} = 0$ for $k > l$. Since v_k is a linear combination of π_k and π_{k-1} it also follows that $\nu_{k,l} = 0$ for $k > l+1$. The array is therefore zero below the diagonal $(\rho_{0,0}, \nu_{1,0}, \rho_{1,1}, \nu_{2,1}, \dots)$.

The analogue of the Sack-Donovan algorithm [16] for generating recursion coefficients from mixed moments is easy in terms of the rhombus array. The mixed moments are given in the first row $(\nu_{0,0}, \nu_{0,1}, \nu_{0,2}, \cdots, \nu_{0,2n-1})$ and the array is generated downwards by the rhombus rules, using the zero entries to define the recursion coefficients $\psi_1, \eta_1, \psi_2, \eta_2, \dots$ just before they are needed. The whole process is given in the Matlab code in Fig. 1.

4. Gaussian Formulas From Two-Term Coefficients and Vice Versa

In Rutishauser's monograph [14] it is suggested that one should first calculate the nodes of a Gaussian formula (using, of course, Rutishauser's *qd* algorithm) and then obtain the weights from two-term coefficients by the scheme in Fig. 2. The

```
function Snew=sms(mu,Sold)
% SMS Two-term recurrence coefficients from modified moments.
%     Snew=SMS(MU,Sold) returns the coefficients Snew
%     for the weight function defined by the modified moments MU,
%     i.e the integrals of the basis polynomials generated
%     by the coefficients in Sold.  If Sold is omitted, the
%     coefficients appropriate for the Chebyshev polynomials
%     shifted to the interval [0,2] are used.

% DPL  1997-09-28

n=floor(length(mu)/2);
if nargin<2, Sold=jacobi(2*n,-0.5);
  else [two,m]=size(Sold); n=min(floor(m/2),n);
  end

l=1:2*n; u=zeros(1,2*n); v=u; u(:)=mu(1:2*n);
e=zeros(1,n); q=e;
psi=Sold(2,l); eta=Sold(1,l); eta(1)=0;
v(1)=u(1); l(1)=[];
v(l)=u(l)+psi(l-1).*u(l-1);

for k=2:n+1,
  l=k-1:2*n-k+1;
  u(l)=eta(l).*v(l)+v(l+1)-e(k-1)*u(l);
  q(k-1)=u(k-1)/v(k-1);
  if k>n, break; end
  l(1)=[];
  v(l)=u(l)+psi(l-1).*u(l-1)-q(k-1)*v(l);
  e(k)=v(k)/u(k-1);
  end
e(1)=mu(1);
Snew=[e; q];
```

FIGURE 1. Matlab code for generating two-term recursion coefficients from modified moments. The arrays Sold and Snew have first row $[\mu_0, e_1, e_2, \ldots, e_{n-1}]$ and second row $[q_1, q_2, \ldots, q_n]$.

'rhombus rules' for defining the scheme are:

(12) **Additive rule:** West + South = North + East + *shift*;

(13) **Multiplicative rule:** West × South = North × East.

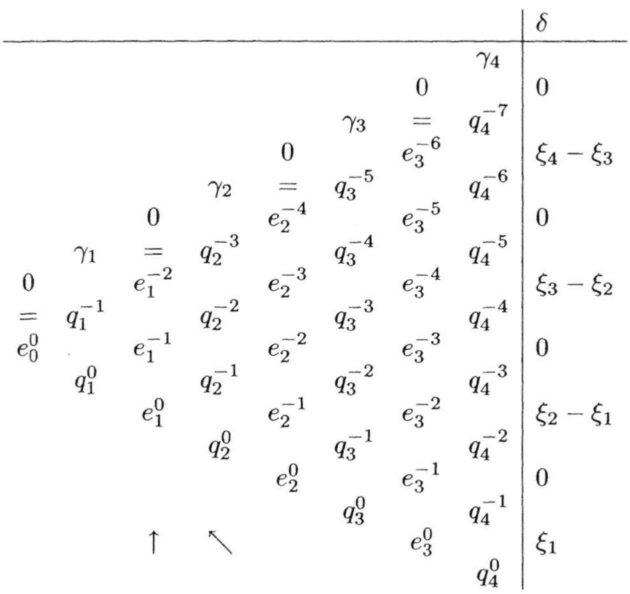

FIGURE 2. Rutishauser's algorithm for the weights of a Gaussian quadrature formula. Rhombuses centred on q and e columns respectively satisfy the additive and multiplicative rhombus rules, except the rhombuses marked with '$=$' which satisfy the relation $e_k^{-2k} = q_{k+1}^{-2k-1}$. The shift δ is constant along diagonals running southeast to northwest, which are built up bottom to top.

The additive rule applies when the North and South elements are q's, and the multiplicative rule when they are e's.

Rutishauser later [15] showed that the qd algorithm for finding eigenvalues of $J = EQ$ is stable when all e_k, q_k are positive (which of course simply means that L is real). Fernando and Parlett [5] pointed out that the 'differential' form (known to Rutishauser but neglected by him because it requires n more multiplications per step than the simple rhombus rules) of the qd algorithm gives componentwise floating-point backward stability.

The implementation of the scheme in Fig. 2 via the differential form of the qd algorithm is therefore a candidate when looking for an accurate way of calculating a quadrature formula from the two-term coefficients. Apart from Rutishauser's own work, this possibility seems not to have attracted much attention.

An intriguing feature of Fig. 2 is that it is fully reversible; in other words, the weights and inter-node gaps could be taken as data, and the NW-SE diagonals generated top to bottom, each diagonal being computed left to right. This algorithm is discussed in [13], where it is shown that it contains no subtractions, so that the algorithm possesses componentwise floating-point *forward* stability, and the two-term coefficients thus found have high relative accuracy. Of course, if only the nodes are initially available, the calculation of the inter-node gaps introduces cancellation which may be serious in the case of closely spaced nodes.

This procedure offers an alternative to Gautschi's discrete Stieltjes algorithm. However, it may be a problem to find a discretization of the measure in which the inter-node gaps are known to high relative accuracy.

5. Conclusion

I am not claiming that two-term recursion is better in all cases. Two exceptions, caused by the fact that two-term recursions work best on an interval of orthogonality with left endpoint 0, are: polynomials orthogonal with respect to a weight function supported on the whole real line; even weight functions, for which three-term coefficients display the symmetry in an obvious way as $a_k = 0$.

What I am claiming is that there are enough advantages to two-term recursion to make it a useful addition to our toolbox for computations involving orthogonal polynomials. Moreover, in cases where analytic expressions for the coefficients are sought, it may well be simpler to obtain these for the two-term coefficients.

Thanks

An anonymous referee drew my attention to [8]. And of course, it is impossible to thank the organizers and sponsors of the Oberwolfach meetings enough for providing the most stimulating and enjoyable conferences anywhere in the world.

References

[1] M. Abramowitz and I.A. Stegun, Eds., *Handbook of mathematical functions*, National Bureau of Standards, Appl. Math. Ser. **55**, Washington, D.C., 1964.

[2] N.I. Akhiezer, *The classical moment problem and some related questions in analysis*, translated by N. Kemmer, Hafner, New York, 1965.

[3] B. Beckermann, *On the numerical condition of polynomial bases: estimates for the condition number of Vandermonde, Krylov and Hankel matrices*, Habilitation thesis, Universität Hannover, 1995.

[4] E.W. Cheney, *Introduction to approximation theory*, Chelsea, New York, 1982.

[5] K.V. Fernando and B.N. Parlett, *Accurate singular values and differential qd algorithms*, Numer. Math., **67** (1994), 191–229.

[6] Walter Gautschi, *On generating orthogonal polynomials*, SIAM J. Sci. Statist. Comput., **3** (1982), 289–317.

[7] Walter Gautschi, *Algorithm 726: ORTHPOL — a package of routines for generating orthogonal polynomials and Gauss-type quadrature rules*, ACM Trans. Math. Software, **20** (1994), 21–62.

[8] G.H. Golub and M.H. Gutknecht, *Modified moments for indefinite weight functions*, Numer. Math., **57** (1990), 607–624.

[9] G.H. Golub and J.H. Welsch, *Calculation of Gauss quadrature rules*, Math. Comp., **23** (1969), 221–230, loose microfiche suppl. A1–A10.

[10] W.B. Gragg and W.J. Harrod, *The numerically stable reconstruction of Jacobi matrices from spectral data*, Numer. Math., **44** (1984), 317–335.

[11] D.P. Laurie, *Anti-Gaussian quadrature formulas*, Math. Comp., **65** (1996), 739–747.

[12] D.P. Laurie, *Calculation of Gauss-Kronrod quadrature rules*, Math. Comp., **66** (1997), 1133–1145.

[13] D.P. Laurie, *Accurate recovery of recursion coefficients from Gaussian quadrature formulas*, J. Comput. Appl. Math. (1998), submitted to special issue on integration.

[14] H. Rutishauser, *Der Quotienten-Differenzen-Algorithmus*, Mitt. Inst. Angew. Math. ETH Zurich, Nr. 7, Birkhäuser, Basel, 1957.

[15] H. Rutishauser, *Stabile Sonderfälle des Quotienten-Differenzen-Algorithmus*, Numer. Math., **5** (1963), 95–112.

[16] R.A. Sack and A.F. Donovan, *An algorithm for Gaussian quadrature given modified moments*, Numer. Math., **18** (1971/72), 465–478.

Dirk P. Laurie
Department of Mathematics and Computer Science
Potchefstroom University for Christian Higher Education
P.O. Box 1174
Vanderbijlpark, South Africa
E-mail address: `dlaurie@na-net.ornl.gov`

International Series of Numerical Mathematics
Vol. 131,

Construction and Computation of a New Set of Orthogonal Polynomials

Shikang Li

Abstract. In this paper we will discuss how to construct and compute a new set of orthogonal polynomials from an existing one. For a given pair of positive integers (n, r) and a given positive measure $d\sigma(t)$, we will construct a set of orthogonal polynomials corresponding to the modified measure $d\hat{\sigma}(t) = (\pi_n(t))^{2r}\, d\sigma(t)$. For $r = 2$ and the first-kind Chebyshev measure we are able to find explicit formulas for the recurrence coefficients of the new set of polynomials. A conjecture is made on those coefficients for any positive integer r and the first-kind Chebyshev measure. For $r = 2$ and arbitrary measures, a computational method is proposed. Other results are also stated.

1. Introduction

For a given positive measure $d\sigma$ on $\mathbb{R}$ having finite moments of all orders, there is a set of orthogonal polynomials $\{\pi_m\}$ that is associated with the measure. It is well known that they satisfy a three-term recurrence relation

$$\pi_{k+1}(t) = (t - \alpha_k)\pi_k(t) - \beta_k \pi_{k-1}(t), \;\; k = 0, 1, 2, \dots, \tag{1.1}$$

where

$$\pi_0(t) = 1, \;\; \pi_{-1}(t) = 0.$$

For a given pair of positive integers (n, r), we wish to introduce and construct a new set of orthogonal polynomials $\{\hat{\pi}_{k,n}(t)\}$ that corresponds to the modified measure

$$d\hat{\sigma}(t) = (\pi_n(t))^{2r} d\sigma(t). \tag{1.2}$$

The case $r = 1$ was studied by Gautschi and the author [3]. The existence of this set of orthogonal polynomials follows from the assumptions made for $d\sigma$ and the fact that the same assumptions are true for the modified measure. The main problem that we wish to consider is the following:

Given the recursion coefficients α_k, β_k for $\{\pi_m\}$, find the recursion coefficients $\hat{\alpha}_k$ and $\hat{\beta}_k$ for the modified measure $d\hat{\sigma}$ which depends both on n and r.

The case $r = 1$ is motivated by research related to constrained polynomial least squares approximation [1]. Other modifications like Stieltjes polynomials are motivated by extension of quadrature formulas. More general modifications of the measure $d\sigma$ are studied by Nevai [8]. The modification we propose here can provide a source of additional interpolation points —the zeros of $\hat{\pi}_{n+1,n}$— in the process of extending Lagrange interpolation at the zeros of π_n.

In §2, we consider the first-kind Chebyshev measure and will derive explicit formulas for the desired coefficients by elementary methods for the case $r = 2$. Explicit formulas for the orthogonal polynomials $\hat{\pi}_{k,1}$ when $r = 2$ are obtained. In §3 we will study the relation between the zeros of π_n and those of $\hat{\pi}_{n+1,n}$. In the same section we will demonstrate that the set of polynomials $\{\hat{\pi}_{n+1,n}\}$ forms an orthogonal polynomial sequence relative to a measure of Bernstein-Szegö type. In §4 we present an algorithm based on successive QR decompositions with shifts to compute the new recursion coefficients.

2. Recurrence Coefficients for the First-Kind Chebyshev Measure

Generally, one will not be able to derive analytic expressions for the coefficients $\hat{\alpha}_j$, $\hat{\beta}_j$ in the three-term recurrence relation satisfied by the new set of orthogonal polynomials $\{\hat{\pi}_{k,n}\}$. Because of the special properties enjoyed by Chebyshev polynomials of the first kind, we are able to obtain explicit formulae for the three-term recurrence coefficients. Let $d\sigma(t) = (1-t^2)^{-1/2}dt$ on the interval $[-1,1]$. In this case, $\pi_n(t) = 2^{-(n-1)}T_n(t)$, which is the monic Chebyshev polynomial of the first kind, and $T_n(t) = \cos n\theta$ for $t = \cos\theta$. The modified measure is given by

$$d\hat{\sigma}(t) = (\pi_n(t))^{2r}(1-t^2)^{-\frac{1}{2}}dt$$

for given positive integers n and r.

Lemma 2.1. *For any positive integer s, we have the following expansion formula:*

$$(1+T_{2n}(t))^s = \frac{1}{2^{s-1}}\left\{\sum_{k=1}^{s}\binom{2s}{s-k}T_{2kn}(t) + \frac{1}{2}\binom{2s}{s}\right\}. \tag{2.1}$$

One can find a proof of this lemma in [7].

Theorem 2.2. *For any positive integers n and r, there holds*

$$\hat{\pi}_{k,n}(t) = \hat{T}_k(t),\ 0 \le k \le n,$$

where $\hat{T}_k(t)$ is the nth-degree monic Chebyshev polynomial.

Proof. It is sufficient to show

$$\int_{-1}^{1} T_k(t)T_r(t)T_n^{2r}(t)(1-t^2)^{-\frac{1}{2}}\,dt = 0,\ 0 \le r < k \le n. \tag{2.2}$$

We use (2.1) to write

$$T_n^{2r}(t) = \left(\frac{1+T_{2n}(t)}{2}\right)^r = \frac{1}{2^{2r-1}}\left\{\sum_{k=1}^{r}\binom{2r}{r-k}T_{2kn}(t) + \frac{1}{2}\binom{2r}{r}\right\}$$

and then use $T_kT_r \in \mathbb{P}_{2k-1} \subseteq \mathbb{P}_{2n-1}$ together with orthogonality to conclude that

$$\int_{-1}^{1} T_k(t)T_r(t)T_n^{2r}(t)(1-t^2)^{-\frac{1}{2}}\,dt = \int_{-1}^{1} \frac{1}{2^{2r}}\binom{2r}{r}T_k(t)T_r(t)(1-t^2)^{-\frac{1}{2}}\,dt = 0.$$

□

Theorem 2.3. *If $r = 2$ and $n = 1$, then*

$$\hat{\pi}_{k,1}(t) = \begin{cases} \sum_{j=0}^{k/2}(-1)^j 4^{-j}\frac{(k+1-j)(j+1)}{k+1}\hat{T}_{k-2j}(t) & \text{if } k(\text{even}) \geq 4, \\ \sum_{j=0}^{(k-1)/2}(-1)^j 4^{-j}\frac{(j+1)(k-j+1)(k-2j)}{k(k+1)}\hat{T}_{k-2j}(t) & \text{if } k(\text{odd}) \geq 1 \end{cases} \tag{2.3}$$

and $\hat{\pi}_{2,1}(t) = \hat{T}_2(t) - \frac{1}{3}$. Moreover, the recurrence relation for the polynomials $\hat{\pi}_{k,1}$ is given by

$$\hat{\pi}_{k+1,1}(t) = t\hat{\pi}_{k,1}(t) - \hat{\beta}_k\hat{\pi}_{k-1,1}(t),\ k = 0, 1, 2, \dots,$$

$$\hat{\pi}_{0,1}(t) = 1,\ \hat{\pi}_{-1,1}(t) = 0,$$

where $\hat{\beta}_0 = \int_{-1}^{1}(T_1(t))^2(1-t^2)^{-1/2}\,dt = \frac{\pi}{2}$ and

$$\hat{\beta}_k = \begin{cases} \frac{1}{4}\frac{k(k-1)}{(k+1)(k+2)}, & k(\text{even}) \geq 2, \\ \frac{1}{4}\frac{(k+3)(k+4)}{(k+1)(k+2)}, & k(\text{odd}) \geq 1. \end{cases} \tag{2.4}$$

The proof of this theorem is very similar to the one given for Theorem 3.2 in [2].

Theorem 2.4. *For $r = 2$ and any $n \geq 2$, the polynomials $\hat{\pi}_{k,n}$ satisfy the three-term recurrence relation*

$$\hat{\pi}_{k+1,n}(t) = t\hat{\pi}_{k,n}(t) - \hat{\beta}_{k,n}\hat{\pi}_{k-1,n}(t),\ k = 0, 1, 2, \dots,$$

$$\hat{\pi}_{0,n}(t) = 1,\ \hat{\pi}_{-1,n}(t) = 0,$$

where

$$\hat{\beta}_{k,n} = \begin{cases} \frac{3\pi}{2^{4n-1}} & \text{if } k = 0, \\ \frac{1}{4}\left(1 - \frac{2(-1)^{\frac{k}{n}}}{\frac{k}{n}+2}\right) & \text{if } k = 0 \bmod n\ (k \neq 0), \\ \frac{1}{4}\left(1 + \frac{2(-1)^{\frac{k-1}{n}}}{\frac{k-1}{n}+2}\right) & \text{if } k = 1 \bmod n, \\ \frac{1}{4} & \text{otherwise.} \end{cases} \tag{2.5}$$

Below we sketch the proof of this theorem.

Proof. The polynomial $\hat{\pi}_{k+1,n}(t)$ can be written as

$$\hat{\pi}_{k+1,n}(t) = \frac{1}{2^k}(T_{k+1}(t) + a_{k-1}T_{k-1}(t) + a_{k-3}T_{k-3}(t) + \cdots). \tag{2.6}$$

In order to satisfy the required orthogonality, the coefficients $a_{k-1}, a_{k-3}, \dots$ must be chosen so as to have, for $0 \le r < k$,

$$\int_{-1}^{1} (T_{k+1}(t) + a_{k-1}T_{k-1}(t) + \cdots)(6T_r(t) + 4T_{|r-2n|}(t) \\ + T_{|r-4n|}(t) + 4T_{r+2n}(t) + T_{r+4n}(t))(1-t^2)^{-\frac{1}{2}}dt = 0. \tag{2.6}$$

To find the coefficients in the recurrence relation, it suffices to find a_{k-1} for $k \ge 1$. To do this, we solve a linear system of equations for $a_{k-1}, a_{|k-1-2n|}, \dots, a_{|k-1-2sn|}$, where s is defined by

$$s = \begin{cases} \frac{k}{n} - 1 \text{ if } k = 0 \bmod n, \\ \lfloor \frac{k}{2n} \rfloor \text{ if } k = 1 \bmod n, \\ \lfloor \frac{k}{n} \rfloor \text{ otherwise.} \end{cases} \tag{2.8}$$

The system is obtained by taking $r = k-1, |k-1-2n|, \dots, |k-1-2sn|$ in (2.7). Solving it yields $a_{k-1} = 0$ if $k \ne 0 \bmod n$ and

$$a_{k-1} = \frac{2(-1)^{\frac{k}{n}}}{\frac{k}{n}+2} \text{ if } k = 0 \bmod n. \tag{2.9}$$

It thus follows that

$$\hat{\pi}_{k+1,n} = \hat{T}_{k+1} + \{\hat{T}_{k-3}, \hat{T}_{k-5}, \dots\}, \; k \ne 0 \bmod n, \tag{2.10}$$

where $\{\dots\}$ denotes a linear combination of the entries enclosed by the braces, and

$$\hat{\pi}_{k+1,n} = \hat{T}_{k+1} + \frac{1}{4}\frac{2(-1)^{\frac{k}{n}}}{\frac{k}{n}+2}\hat{T}_{k-1} + \{\hat{T}_{k-3}, \hat{T}_{k-5}, \dots\}, \; k = 0 \bmod n. \tag{2.11}$$

The rest of the proof is exactly the same as in [2, pp. 12–13]. □

Conjecture 2.5. *The formula* (2.5) *in Theorem* 2.4 *holds also for* $r \ge 2$ *if the number "2" is replaced by "r" throughout.*

3. Interlacing Property and Orthogonality

In some applications, for example to extended interpolation, one wants to interpolate a function at the zeros τ_i of $\pi_n(\,\cdot\,; d\sigma)$ and the zeros $\hat{\tau}_j$ of $\hat{\pi}_{n+1,n}(\,\cdot\,)$. A question of particular interest then is whether the zeros τ_i and $\hat{\tau}_j$ interlace, i.e., whether, if ordered decreasingly and contained in $(-1,1)$, they satisfy

$$-1 < \hat{\tau}_{n+1} < \tau_n < \hat{\tau}_n < \cdots < \hat{\tau}_2 < \tau_1 < \hat{\tau}_1 < 1. \tag{3.1}$$

We assume in this section that $r = 2$.

Theorem 3.1. *For the first-kind Chebyshev measure, and for any fixed* $n \ge 1$, *the zeros* τ_i *of* π_n *interlace with the zeros* $\hat{\tau}_j$ *of* $\hat{\pi}_{n+1,n}$.

Proof. In this case, π_n is the first-kind Chebyshev polynomial and its zeros are $\tau_i = \cos\theta_i,\ \theta_i = (2i-1)\pi/2n$. If $n \geq 2$,

$$\hat{\pi}_{n+1,n}(t) = t\hat{T}_n(t) - \frac{5}{12}\hat{T}_{n-1}(t), \tag{3.2}$$

as follows from Theorems 2.3 and 2.4. It can easily be shown that

$$\hat{\pi}_{n+1,n}(1) = \frac{1}{2^n 3}, \quad \hat{\pi}_{n+1,n}(-1) = \frac{(-1)^{n+1}}{2^n 3}, \tag{3.3}$$

and

$$\hat{\pi}_{n+1,n}(\tau_i) = \frac{5}{12}\frac{1}{2^{n-2}}(-1)^i \sin\theta_i, \ i = 1, 2, \dots, n. \tag{3.4}$$

Since $\hat{\pi}_{n+1,n}(t)$ takes alternating signs at $1, \tau_1, \tau_2, \dots, \tau_n, -1$, it follows that $\hat{\pi}_{n+1,n}$ has a zero on each of the intervals $(\tau_1, 1), (\tau_2, \tau_1), \dots, (-1, \tau_n)$. Hence the interlacing property holds for $n \geq 2$. For $n = 1$, one has $\pi_1(t) = t,\ \hat{\pi}_{2,1}(t) = t^2 - 5/6$ and the interlacing property is clearly true. □

The special sequence of induced polynomials $\{\hat{\pi}_{n,n-1}\}$ whose degrees exceed the degrees of the generating polynomials by one, are of some use in extended interpolation. It is interesting to find out whether they form an orthogonal sequence. Since $\hat{\pi}_{2,1}(t) = t^2 - 5/6,\ \hat{\pi}_{1,0}(t) = t,\ \hat{\pi}_{0,-1}(t) = 1$, it is easy to verify that $\hat{\pi}_{2,1}(t) = t\hat{\pi}_{1,0}(t) - \beta_1\hat{\pi}_{0,-1}(t)$ where $\beta_1 = 5/6$. Generally it follows from the identity

$$\hat{\pi}_{n+1,n}(t) = \frac{1}{2^{n-1}}(tT_n(t) - \frac{5}{6}T_{n-1}(t)) \tag{3.5}$$

that

$$\hat{\pi}_{n+1,n}(t) = t\hat{\pi}_{n,n-1}(t) - \beta_n\hat{\pi}_{n-1,n-2}(t), \tag{3.6}$$

where

$$\beta_n = \begin{cases} \frac{5}{6} & \text{if } n = 1, \\ \frac{11}{12} & \text{if } n = 2, \\ \frac{1}{4} & \text{otherwise}. \end{cases} \tag{3.7}$$

Since all the β-coefficients in the three-term recurrence relation are positive, we apply Favard's Theorem to conclude that there exists a positive symmetric measure with respect to which the polynomial sequence $\{\hat{\pi}_{n,n-1}\}$ is orthogonal. From [5, Eq. (3.8) and Prop. 2.1] it follows indeed that

$$\hat{\pi}_{n,n-1}(\,\cdot\,) = \pi_n(\,\cdot\,; ds), \ n \geq 0,$$

where

$$ds(t) = \frac{(1-t^2)^{-\frac{1}{2}}}{1 - \frac{24}{25}t^2}dt \ \text{ on } (-1, 1).$$

4. Computational Method

For other measures we are not able to derive explicit formulas for the three-term recurrence coefficients. Here we describe a computational method to generate the recurrence coefficients $\hat{\alpha}_{k,n}$ and $\hat{\beta}_{k,n}$ in the three-term recurrence relation satisfied by the desired orthogonal polynomials $\{\hat{\pi}_{k,n}\}$ from the recurrence coefficients α_k and β_k for the given orthogonal polynomials $\{\pi_m\}$. For simplicity, we assume $r = 2$.

Let $J = J(d\sigma)$ be the Jacobi matrix for the measure $d\sigma$, i.e., the infinite symmetric tridiagonal matrix

$$J = J(d\sigma) = \text{tri}(\alpha_0, \alpha_1, \alpha_2, \dots, \sqrt{\beta_1}, \sqrt{\beta_2}, \dots) \tag{4.1}$$

with the recursion coefficients $\alpha_k = \alpha_k(d\sigma)$ on the main diagonal and the $\sqrt{\beta_k} = \sqrt{\beta_k(d\sigma)}$ on the two side diagonals. In the same manner, let

$$\hat{J} = J(d\hat{\sigma}) = \text{tri}(\hat{\alpha}_0, \hat{\alpha}_1, \hat{\alpha}_2, \dots, \sqrt{\hat{\beta}_1}, \sqrt{\hat{\beta}_2}, \dots) \tag{4.2}$$

be the Jacobi matrix for the measure $d\hat{\sigma} = (\pi_n(t))^4 d\sigma(t)$. By J_m and $\hat{J}_m$ we denote the leading principal minor matrices of order m of $J(d\sigma)$ and $J(d\hat{\sigma})$, respectively.

The main idea of our method comes from Kautsky and Golub [6]. Given the Jacobi matrix $J_{m+2}(d\sigma)$ and a real τ, one can obtain the Jacobi matrix of order m, $\hat{J}_m((t-\tau)^4 d\sigma)$ by two steps of the well-known QR algorithm with shift τ: Let Q_1 (orthogonal) and R_1 (upper triangular with diag $R_1 \geq 0$) be the matrices of order $m+2$ such that $J_{m+2}(d\sigma) - \tau I_{m+2} = Q_1 R_1$; then $J_{m+1}((\cdot - \tau)^2 d\sigma) = (R_1 Q_1 + \tau I_{m+2})_{(m+1)\times(m+1)}$. Let Q_2 (orthogonal) and R_2 (upper triangular with diag $R_2 \geq 0$) be matrices of order $m+1$ such that $J_{m+1}((\cdot - \tau)^2 d\sigma) - \tau I_{m+1} = Q_2 R_2$; then $J_m((\cdot - \tau)^4 d\sigma) = (R_2 Q_2 + \tau I_{m+1})_{m\times m}$.

We define $d\hat{\sigma}(t; j) = \prod_{i=1}^{j} (t - \tau_i)^4 d\sigma(t)$, $j = 0, 1, \dots, n$, where τ_i are the zeros of $\pi_n(\,\cdot\,; d\sigma)$ and the empty product is defined to be 1. The following algorithm is used to obtain $\hat{J}_N = J_N(d\hat{\sigma})$:

Algorithm

for $j = 1, 2, \dots, n$ do
  let
    $J_{\text{old}} = J_{2n+N-j+2}(d\hat{\sigma}(\,\cdot\,; j-1))$
  do QR step with shift
    $J_{\text{old}} - \tau_j I = QR$
    $J_{\text{new}} = (RQ + \tau_j I)_{(2n+N-j+1)\times(2n+N-j+1)}$
  let
    $J_{\text{old}} = J_{2n+N-j+1}(d\hat{\sigma}(\,\cdot\,; j-1))$
  do QR step with shift τ_j
    $J_{\text{old}} - \tau_j I = QR$
    $J_{\text{new}} = (RQ + \tau_j I)_{(2n+N-j)\times(2n+N-j)}$
end

In principle, the method can be extended to arbitrary integers $r \geq 2$.

5. Open Problems

In this section we state some open problems related to the work in this paper.

Problem 5.1. *Let $L_{2n}(t)$ be the 2nth-degree polynomial interpolating $f(t)$ at the zeros of π_n and $\hat{\pi}_{n+1,n}$; do we have mean convergence for $\{L_{2n}(t)\}$ as $n \to \infty$, or only quadrature convergence* (cf. [4])*?*

Problem 5.2. *Find explicit formulae for the recurrence coefficients for the second-kind Chebyshev measure.*

Problem 5.3. *Does the interlacing property hold for other measures?*

Problem 5.4. *For what measure $d\sigma$ does the sequence $\{\hat{\pi}_{n,n-1}\}$ form an orthogonal sequence?*

Acknowledgment. The author wishes to thank Professor Walter Gautschi for his help in the preparation of the final version of the paper and his inspiration for taking up this work.

References

[1] A. Bellen, *Alcuni problemi aperti sulla convergenza in media dell'interpolazione Lagrangiana estesa*, Rend. Ist. Mat. Univ. Trieste, **20** (1988), Fasc. suppl., 1–9.

[2] W. Gautschi and S. Li, *A set of orthogonal polynomials induced by a given orthogonal polynomial*, Technical Report CSD-TR-92-075, Purdue University, October 1992.

[3] W. Gautschi and S. Li, *A set of orthogonal polynomials induced by a given orthogonal polynomial*, Aequationes Math., **46** (1993), 174–198.

[4] W.Gautschi and S. Li, *On quadrature convergence of extended Lagrange interpolation*, Math. Comp., **65** (1996), 1249–1256.

[5] W. Gautschi and S.E. Notaris, *Gauss-Kronrod quadrature formulae for weight functions of Bernstein-Szegö type*, J. Comput. Appl. Math., **25** (1989), 199–224.

[6] J. Kautsky and G.H. Golub, *On the calculation of Jacobi matrices*, Linear Algebra Appl., **52/53** (1983), 439–455.

[7] S. Li, *Kronrod extension of Turán formula*, Studia Sci. Math. Hungar., **29** (1994), 71–83.

[8] P. Nevai, *A new class of orthogonal polynomials*, Proc. Amer. Math. Soc., **91** (1984), 409–415.

Shikang Li
Department of Mathematics
Southeastern Louisiana University
Hammond, LA 70402, USA
E-mail address: `kli@selu.edu`

International Series of Numerical Mathematics
Vol. 131, © 1999 Birkhäuser Verlag Basel/Switzerland

Fourier Transforms of Orthogonal Polynomials of Singular Continuous Spectral Measures

Giorgio Mantica

Abstract. We discuss algorithms for the solution of the Schrödinger time-dependent equation, based on orthogonal polynomial decomposition of the exponential function. After reviewing the classical Chebyshev series approach and its iterated version, we show their inefficiency when applied to operators with singular continuous spectral measures. We then introduce new decompositions based on the spectral measure of the problem under consideration, which are especially suited to deal with this case. A fast version of these algorithms is also developed and shown to achieve the theoretical maximum performance.

1. Introduction

Let μ be a positive measure for which all moments exist and the moment problem is determined [1, 15]. The Fourier transforms of the sequence of associated orthogonal polynomials, $\{p_n(\mu;x)\}_{n\in\mathbb{N}}$, define the *generalized* Bessel functions $\mathbf{J}_n(\mu;t)$,

$$\mathbf{J}_n(\mu;t) := \int d\mu(x) e^{-itx} p_n(\mu;x). \tag{1}$$

Indeed, when μ is the orthogonality measure of Chebyshev polynomials, then $\mathbf{J}_n(\mu;t)$ are the usual, integer order, Bessel functions $J_n(t)$ (modulo a factor $(-i)^n$).

These functions are of physical significance, for they solve the evolution in $l_2(\mathbb{Z}_+)$ generated by Schrödinger's equation, when the Hamiltonian is the Jacobi matrix J_μ associated with μ, and the initial state is the vector e_0 (as usual, $\{e_n\}_{n\in\mathbb{Z}_+}$ is the canonical basis of $l_2(\mathbb{Z}_+)$). Letting $\psi(t) \in \mathbb{Z}_+$,

$$i\frac{d\psi}{dt} = J_\mu \psi, \quad \psi(0) = e_0, \tag{2}$$

we obtain in fact that [10]

$$(e_n, \psi(t)) = \mathbf{J}_n(\mu;t). \tag{3}$$

1991 *Mathematics Subject Classification.* 47B39, 65F50, 65L05, 81Q10. *PACS numbers.* 05.45.+b, 02.30.-f, 71.30.+h, 71.55.Jv.
Key words and phrases. Jacobi matrices, orthogonal polynomials, singular continuous measures, Fourier transform, Schrödinger equation, matrix exponential.

Here and in the following, round brackets indicate the scalar product in $l_2(\mathbb{Z}_+)$. The generalized Bessel functions are therefore the basis projections of the evolution of the linear wave equation (2).

Much theoretical work has been devoted to these functions when μ is a singular measure supported on a Cantor set. In particular, it has been shown (for $n = 0$ in [5, 8], for the general case in [7, 10]) that the Cesaro average of $|\mathbf{J}_n(\mu;t)|^2$ from time zero to T decays as $T^{-D_2(\mu)}$, where $D_2(\mu)$ is the *correlation dimension* of the measure μ. Typically, $D_2(\mu) < 1$, so that one observes a slower decay than in the usual case, $D_2(\mu) = 1$. Moreover, while in this second case one can speak of the *wave propagation* of an initial excitation over a *chain* labelled by the index n, in the singular case this propagation is *anomalous* and *intermittent* in many respects [7, 10, 11]. Consider for instance the quantities

$$\nu_\alpha(\mu;t) = \sum_{n=0}^{\infty} n^\alpha |\mathbf{J}_n(\mu;t)|^2,$$

which can be viewed as moments of order $\alpha \geq 0$ of the probability distribution of the wave $|\psi(t)|^2$ on the lattice. They are characterized by the asymptotic law [7]

$$\nu_\alpha(\mu;t) \simeq t^{\alpha\beta(\alpha)} \text{ as } t \to \infty,$$

which defines the function β of the moment index α. For singular measures, this function has nontrivial characteristics: for instance, when μ is the equilibrium measure of a Julia set (see [3, 2, 4] for definition and properties), it can be computed as $\beta(\alpha) = D_{1-\alpha}(\mu)$, where $D_q(\mu)$, $q \in \mathbb{R}$, are the *generalized* fractal dimensions of μ [10]. Further results on the relation between the measure μ and the function $\beta(\alpha)$ are contained in [10, 11].

In these investigations, two numerical problems of major concern arise: the computation of the Jacobi matrix associated with a singular measure, and the evaluation of the highly oscillatory integrals (1). The former has been discussed in [9] and [12]. This paper is devoted to the study of the latter, and presents unpublished material; we shall put the emphasis here on computational aspects of this problem, referring to [13] for further details and some of the proofs.

The integrals (1) can also be seen as the components of the infinite vector $\psi(t)$ in (3), and the formal solution of the Schrödinger equation,

$$\psi(t) = e^{-itJ_\mu}\psi(0), \tag{4}$$

introduces the familiar problem of evaluating the matrix exponential e^{-itJ_μ}, typically for large values of the time t. The techniques we shall present belong naturally to the family of methods developed to solve this problem [6, §§11.2,11.3], [14] and in particular to Chebyshev series expansions [16]. In this paper we introduce extensions of this method to treat conveniently the motion generated by operators possessing singular continuous spectral measures; for these latter, in fact, the classical procedure will be shown to be largely inefficient.

The paper is structured as follows: In §2 we discuss the application of the conventional Chebyshev series to the bounded Jacobi matrix of a singular continuous measure; while this algorithm is standard, our rederivation permits us to justify theoretically its inefficiency when applied to this case. In §3 we present a trivial modification of the algorithm (an iterative application of the same) which allows us to partially resolve the inefficiency. A further improvement is presented in §4: instead of the usual Chebyshev polynomials, we employ in the Chebyshev series the orthogonal polynomials of the Jacobi matrix itself. This idea requires a careful encoding, for the generalized Bessel functions, which are required, are precisely the solution of the evolution equation, that is, the unknowns of the problem. In §5 we put forward the idea of immediately re-employing the values of the Fourier transforms at time τ to compute those at time 2τ, obtaining in this way a *Fast* Self-Decomposition, Generalized Chebyshev Series. A few numerical examples follow in §6, and in the Conclusions, the applicability of these techniques to the general case of sparse Hamiltonian operators with localized initial states is briefly discussed.

2. Chebyshev Series for the Matrix Exponential

Let $S(\mu)$ be the support of μ and assume $\int d\mu(x) = 1$. Take σ to be a (yet to determine) measure whose support, $S(\sigma)$, includes $S(\mu)$. Let also $\{p_n(\sigma;x)\}$ be the set of orthonormal polynomials associated with σ, and define $\mathbf{J}_n(\sigma;t)$ to be the *generalized Bessel functions* associated with σ, as in Eq. (1). Suppose that these functions are explicitly known and readily computable. Then, $e^{-itx} = \sum_{n=0}^{\infty} \mathbf{J}_n(\sigma;t)p_n(\sigma;x)$. This sum obviously converges in the sense of $L_2(d\sigma)$; stronger convergence properties justify the formal expansion of the matrix exponential,

$$\psi(t) = e^{-itJ_\mu}\psi_0 = \sum_{n=0}^{\infty} \mathbf{J}_n(\sigma;t)p_n(\sigma;J_\mu)\psi_0, \tag{5}$$

which is the basis of the computational techniques developed in this work. Usually in the literature, σ is the orthogonality measure of Chebyshev polynomials, $\frac{d\sigma}{dx} = \frac{1}{\pi\sqrt{1-x^2}}$, and $\mathbf{J}_n(\sigma;t) = (-i)^n J_n(t)$. This choice gives rise to a common technique for solving Eq. (4) that we call, for convenience, the *Chebyshev series algorithm.*

Theorem 2.1. (Chebyshev Series) *Let σ be the Chebyshev measure, and let $S(\mu) \subset S(\sigma)$; then for any $t \in \mathbb{R}$ the series in*

$$\mathbf{J}_k(\mu;t) = \sum_{n=0}^{\infty} \mathbf{J}_n(\sigma;t)(e_k, p_n(\sigma;J_\mu)e_0)$$

converges uniformly in k.

Proof. Consider the scalar products occurring in Theorem 2.1; because of the spectral expansion of the Jacobi matrix, we have

$$h_{k,n} := (e_k, p_n(\sigma; J_\mu)e_0) = \int p_k(\mu; x)p_n(\sigma; x)\, d\mu(x). \tag{6}$$

Moreover, $|p_n(\sigma; x)| \leq 1$ on the spectrum of σ, and hence of μ, so that the norm of $p_n(\sigma; x)$ in $L_2(d\mu)$ is bounded:

$$\sum_{k=0}^{\infty} |h_{k,n}|^2 = \|p_n(\sigma; x)\|^2_{L_2(d\mu)} := \int |p_n(\sigma; x)|^2\, d\mu(x) \leq \int d\mu(x) = 1. \tag{7}$$

The first equality is nothing but Plancherel's Theorem. This implies $|h_{k,n}| \leq 1$ for all k, n. The sum in Theorem 2.1 then converges totally: $|\mathbf{J}_k(\mu; t)| \leq \sum_{n=0}^{\infty} |\mathbf{J}_n(\sigma; t)| < \infty$, and therefore absolutely, and uniformly in $k \in \mathbb{Z}_+$. □

A further property of the generalized Bessel functions $\mathbf{J}_k(\mu; t)$ can be obtained observing that the summation in Theorem 2.1 actually starts from $n = k$.

Theorem 2.2. *Let $S(\mu)$ be bounded; then at fixed time t, for any $\alpha > 0$, there exists a constant C_α so that the generalized Bessel functions $\mathbf{J}_k(\mu; t)$ decay at least exponentially in k:*

$$|\mathbf{J}_k(\mu; t)| \leq C_\alpha e^{-\alpha k}. \tag{8}$$

A simple proof of this result, which is probably not original, but instructive, follows from the inequality (7) [13].

These theorems justify the application of the classical *Chebyshev series algorithm* to our case:

cheby(e_0, t, ε):
$v_{-1} = 0,\ v_0 = e_0,\ w = J_0(t)v_0,\ n = 1$
while $\sum_{l=n}^{\infty} |J_l(t)| \geq \varepsilon$ **do**
  $v_n = 2J_\mu v_{n-1} - v_{n-2}, \quad w = w + (-i)^n J_n(t) v_n, \quad n = n + 1$
end

At the end of the calculation, the vector w contains the generalized Bessel functions $\mathbf{J}_n(\mu; t)$. The halting criterion is structured so that, by virtue of (7), the *absolute* error in the determination of each function is less than ε, in exact arithmetic. Actually, using the techniques of Theorems 2.1, 2.2, we can show that for any $\alpha > 0$ there exists a constant C_α such that the difference between $\mathbf{J}_k(\mu; t)$ and the M-truncation of the sum in Theorem 2.1 is less than $C_\alpha e^{-\alpha M}$.

Obviously, the algorithm computes only a *finite* number, $n_\varepsilon(t)$, of such functions. Because of the asymptotic properties (in the order n) of the usual Bessel functions $J_n(t)$, this number grows *linearly* with t, with a proportionality constant of the order of one, and is only mildly sensitive to ε, for ε sufficiently small. We can therefore estimate the number of floating-point operations required by this algorithm to be

$$\text{cost of } \mathbf{cheby}(e_0, t, \varepsilon) \simeq Qt^2 \text{ flops}, \tag{9}$$

where $Q = 15$ in a nonoptimized implementation. Is this performance satisfactory? When the spectral measure μ is absolutely continuous, and it has been rescaled so that $S(\mu) = S(\sigma)$, the answer is affirmative. However, when μ is singular continuous, a different situation arises. In fact, as a consequence of the properties of the *intermittency* function $\beta(\alpha)$, one observes [7] that the *effective dimension* $N(\mu; \varepsilon, t)$ of the vector $\mathbf{J}_n(\mu; t)$, defined via

$$|\mathbf{J}_n(\mu; t)| \leq \varepsilon \text{ for } n > N(\mu; \varepsilon, t), \tag{10}$$

grows only like a power of time smaller than one:

$$N(\mu; \varepsilon, t) \simeq A(\mu, \varepsilon) t^{\omega(\mu)}. \tag{11}$$

The dependence of $A(\mu, \varepsilon)$ on ε is mild for small ε, while that on μ is determined almost entirely by the *spectral radius* of μ; since we take the latter to be unity, this dependence is removed. On the other hand, the role played by the fine structure of μ in determining $\omega(\mu)$ is crucial: since typically $\omega(\mu) < 1$, we can say that *quantum intermittency* slows down the propagation of excitations. For simplicity of notation, the dependence of ω on μ will be left implicit in the following.

In the singular case, the Chebyshev series is vastly inefficient, more so the larger the value of the time t, because it contains an increasing number of practically zero components: the sums $\sum_{n=k}^{n_\varepsilon(t)} h_{k,n} \mathbf{J}_n(\sigma; t)$ give negligible results for all $k > N(\mu; \varepsilon, t)$. What complicates the situation is the fact that the estimate $|h_{k,n}| \leq 1$, which allows us to estimate $n_\varepsilon(t)$, cannot be significantly improved. In fact, the coefficients $h_{k,n}$, for fixed k, are the Fourier transforms of a distribution; setting $x = \cos\theta$ in Eq. (6) gives $h_{k,n} = \int d\tilde{\mu}(\theta) p_k(\mu; \arccos\theta) \cos(n\theta)$, which — for singular measures — cannot be expected to decay to assure a better truncation. Further details on this topic are presented in [13]. We now propose a first solution of this problem.

3. The Iterated Chebyshev Series for Jacobi Matrices of Singular Measures

A partial remedy to the situation presented in the previous section is to decompose the interval $[0, t]$ into a convenient number, M, of subintervals of length $\tau = \frac{t}{M}$, and to apply iteratively the conventional Chebyshev algorithm, **cheby**(v, t, ε), where v is the starting vector, t the time span, and ε the error parameter in the termination rule:

chebit(e_0, t, M, ε):
$v_0 = e_0$, $\tau = \frac{t}{M}$
for $j := 1$ to M **do**
  $v_j =$ **cheby**$(v_{j-1}, \tau, \varepsilon)$
end

Each step in this algorithm computes the general summation (5); its truncation error analysis is similar to that of the original Chebyshev series [13]. We now estimate its complexity in a form suitable for further analysis. Each iterated Chebyshev step proceeding from s to $s+\tau$ (i.e., Eq. (5) with $t=\tau$) involves $n(\tau)$ Chebyshev polynomial functions in the Jacobi matrix acting upon a vector of $N(s)$ nonzero components; thus,

$$\text{cost of iterated Chebyshev step} = Qn(\tau)[n(\tau)+2N(s)] + qQN(s) \text{ flops}, \quad (12)$$

where, as before $Q \simeq 15$, and $q \simeq \frac{1}{3}$. At the jth iteration we have

$$s_j := j\tau, \ \ n(\tau) \simeq \tau, \ \ N(s_j) \simeq j^\omega \tau^\omega, \quad (13)$$

and therefore, the cost of this algorithm over the interval $[0,t]$ is

$$\text{cost of } \mathbf{chebit}(e_0,t,M,\varepsilon) \simeq Q[\frac{t^2}{M} + D_\omega t^{\omega+1} + \frac{q}{2} D_\omega t^\omega M] \text{ flops}, \quad (14)$$

where $D_\omega := \frac{2}{\omega+1}$. Keeping $\tau = \frac{t}{M}$ fixed when t is increased, we obtain

$$\text{cost of } \mathbf{chebit}(e_0,t,t/\tau,\varepsilon) \simeq Q[t\tau + t^{\omega+1} D_\omega(1+\frac{q}{2\tau})] \text{ flops}. \quad (15)$$

The leading coefficient grows as $t^{\omega+1}$: we notice a smaller power-law dependence on time than in the original Chebyshev algorithm, Eq. (9). This result is easy to understand: at each iteration of the basic Chebyshev series, we are computing a *fixed* overhead of negligible Bessel functions, of cardinality at most τ, which accounts for the linear contribution in time. On top of that, we have to compute at each iteration a number of nonnegligible generalized Bessel functions, which increases as t^ω. To further improve the situation, we need to get away from using conventional Bessel functions and Chebyshev polynomials.

4. Self-Decomposition Generalized Chebyshev Series

Let again $\tau = \frac{t}{M}$, with M a suitable integer. With the aid of the standard Chebyshev algorithm (or any other means) we can compute the functions $\mathbf{J}_n(\mu;\tau)$ for $n = 0, \ldots, N(\mu;\varepsilon,\tau)$. Then, we write the formula (5) with $\sigma = \mu$, i.e., we assume that the integrating measure σ *coincides* with the dynamical measure μ, and we iterate as we did in the previous section:

self-deco(e_0,t,M,ε):
$v_0 = e_0$, $\tau = \frac{t}{M}$
$v_1 =$ **cheby** (v_0,τ,ε)
$\mathbf{J}_n(\mu;\tau) = (e_n, v_1)$
compute $N(\mu;\varepsilon,\tau)$
for $j := 1$ to $M-1$ **do**
$\quad v_{j+1} = \sum_{n=0}^{N(\mu;\varepsilon,\tau)} \mathbf{J}_n(\mu;\tau) p_n(\mu;J_\mu) v_j$,
end

The error analysis of this algorithm is presented in [13]. Here we focus on its cost. In the notations of Eq. (12) the first time step of the algorithm uses the classical Chebyshev series and is characterized by $N(0)=1$, $n(\tau)\simeq\tau$, which gives a cost of $Q\tau^2$, as expected from Eq. (9). However, the successive steps require only $n(\tau) \simeq \tau^\omega$. Keeping τ fixed as before and letting t vary, we get

$$\text{cost of } \mathbf{self\text{-}deco}(e_0, t, t/\tau, \varepsilon) \simeq Q[\tau^2 + \tau^{2\omega-1} t + t^{\omega+1} D_\omega(\tau^{\omega-1} + \frac{q}{2\tau})] \text{ flops.} \quad (16)$$

The constant in front of the leading contribution $t^{\omega+1}$ can be made much smaller than in **chebit** by choosing a large τ, and even the leading exponent $\omega+1$ can be lowered: take $\tau \simeq t^\alpha$, and choose α (in the range $0 \leq \alpha \leq 1$) so as to minimize the *maximum* exponent of time in Eq. (16). This gives $\alpha = 2\frac{1-\omega}{3-\omega}$, so that

$$\text{cost of } \mathbf{self\text{-}deco}(e_0, t, t^{1-\alpha}, \varepsilon) \simeq Q[(1 + D_\omega) t^{2\frac{1+\omega}{3-\omega}} + \text{lower-order terms}] \text{ flops.} \quad (17)$$

Then, when $\omega < 1$, this algorithm is characterized by a better computational complexity than the iterated Chebyshev algorithm.

5. Fast Self-Decomposition Generalized Chebyshev Series

The power of the self-decomposition is fully achieved when one notices that each of its steps provides a new set of generalized Bessel functions, which can be used for further integration: Letting $\tau = \frac{t}{2^k}$, with k a suitable integer, and evaluating preliminarily by **cheby** the generalized Bessel functions $\mathbf{J}_n(\mu;\tau)$, we can write

$$\mathbf{J}_k(\mu; 2\tau) = \sum_{n=0}^{\infty} \mathbf{J}_n(\mu;\tau)(e_k, p_n(\mu; J_\mu)\sum_{l=0}^{\infty} \mathbf{J}_l(\mu;\tau) e_l), \quad (18)$$

which provides the generalized Bessel functions at time 2τ. This result can now be used in the right-hand side to obtain $\mathbf{J}_k(\mu; 4\tau)$ in the left-hand side, and so on. Observe that

$$\mathbf{J}_k(\mu; 2\tau) = \sum_{n=0}^{\infty}\sum_{l=0}^{\infty} \mathbf{J}_n(\mu;\tau)\mathbf{J}_l(\mu;\tau) h_{k,n,l}, \quad (19)$$

where the matrix $h_{k,n,l}$ is defined by $h_{k,n,l} := (e_k, p_n(\mu; J_\mu) e_l)$ and can be computed as

$$h_{k,n,l} = \int d\mu(x) p_k(\mu, x) p_n(\mu; x) p_l(\mu; x), \quad (20)$$

so that the following properties can be verified:

Lemma 5.1. *There holds $h_{k,n,l} = 0$ if any two of the set of indices $\{k, n, l\}$ differ by an integer larger than the remaining index.*

Theorem 5.2. (Fast Generalized Chebyshev Series) *The doubly infinite sum in* Eq. (19) *converges absolutely for all values of k and τ.*

For the proofs of these results see [13]; they lead to the algorithm

fgenchs(e_0, t, k, ε):
$v_0 = e_0$, $\tau = t\, 2^{-k}$.

$v_1 = \mathbf{cheby}(v_0, \tau, \varepsilon)$, $\mathbf{J}_n(\mu; \tau) = (e_n, v_1)$, $\tau_1 = \tau$.
compute $N(\mu; \varepsilon, \tau_1)$
for $j := 1$ to $k - 1$ **do**
  $v_{j+1} = \sum_{n=0}^{N(\mu;\varepsilon,\tau_n)} \mathbf{J}_n(\mu; \tau) p_n(\mu; J_\mu) v_j$,
  $\mathbf{J}_l(\mu; 2\tau_j) = (e_l, v_j)$,
  compute $N(\mu; \varepsilon, \tau_j)$
  $\tau_{j+1} = 2\tau_j$.
end

The algorithmic cost, altogether, is

$$\text{cost of } \mathbf{fgenchs}(e_0, t, k, \varepsilon) \simeq Q[\tau^2 + E_\omega t^{2\omega} + qF_\omega t^\omega] \text{ flops}, \tag{21}$$

where $E_\omega := 3\frac{2^{2\omega}}{2^{2\omega}-1}$ and $F_\omega := \frac{2^\omega}{2^\omega-1}$. This equation permits us to rightly call this a *Fast Generalized Chebyshev Series*: comparison with the previous procedures shows that the computational complexity power-law growth of the number of flops is here the least. And indeed, being the square of that governing the growth of the effective basis, it is also the least theoretically achievable in a Chebyshev series approach.

6. Numerical Experiments

In this section we present a few results of our numerical experimentations of the techniques presented in the previous sections. We consider the orthogonal polynomials defined by the real, symmetric Jacobi matrix with zero diagonal elements, and off-diagonal elements taking on two possible values, W and S, arranged in the Fibonacci sequence $WSWWSWSW\ldots$. This can be generated automatically by the inflation rule $W \to WS$, $S \to W$ on the initial seed W. This Jacobi matrix is associated with a singular, multifractal measure. The effective dimension $N(\mu; \varepsilon, t)$ grows with exponent $\omega \simeq 0.843$. Even in such a simple almost-periodic system, an improvement of the Chebyshev series is therefore particularly needed. We have tested our algorithms and confirmed cost estimates and numerical stability. For instance, in Fig. 1 we compare the results obtained by **chebit** and **self-ref**, under identical conditions. The figure shows a linear error propagation.

Because of the zero diagonal elements, the spectral measure μ is symmetric with respect to the origin, and the even generalized Bessel functions are real, while the odd are purely imaginary. The graph of $\mathbf{J}_{120}(\mu; t)$ is displayed in Fig. 2, showing the *intermittent* character noticeable, for instance, in the irregular spacing of the zeros. Further experimentations are presented in [13].

7. Conclusions

We have presented a generalization of the standard Chebyshev series approach to the evaluation of the matrix exponential, which overcomes the inefficiency of the conventional algorithm in the case of singular continuous spectral measures.

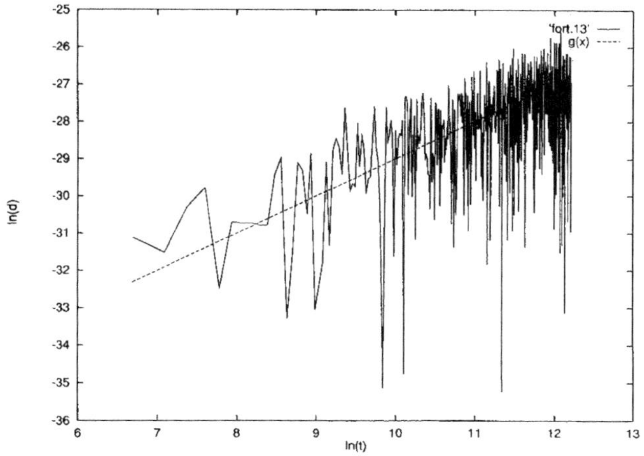

FIGURE 1. Logarithm of the absolute value of the difference between **chebit** and **self-ref** results, at fixed site e_{120}, versus the logarithm of time, for the Fibonacci Jacobi matrix, with $W = .15$, $S = .25$, for $\tau = 400$, $t = 2\times10^5$. The effective dimension $N(\mu;\varepsilon,t)$ is 30,000. The fitting line has unit exponent.

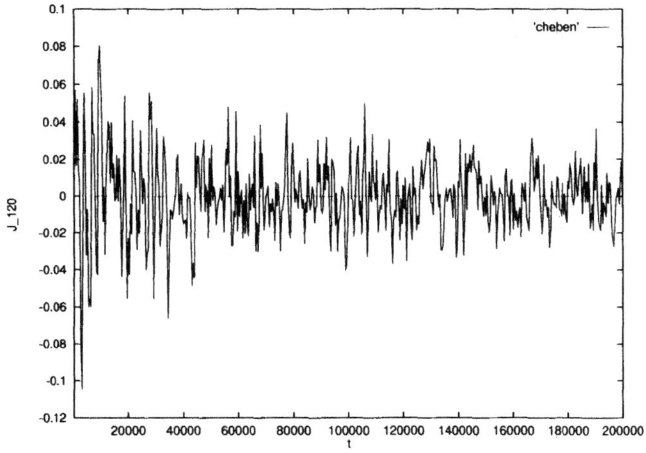

FIGURE 2. Generalized Bessel function $\mathbf{J}_{120}(\mu;t)$ versus time, for the Fibonacci Jacobi matrix, with $W = .15$, $S = .25$

This paper has been concerned with Jacobi matrices; an extension to more general systems is possible, via Krylov space techniques, as shown in [13].

We believe that the properties of the *generalized* Bessel functions defined herein are of utmost mathematical and physical significance. We hope that our techniques will provide reliable methods for their analysis.

References

[1] N.I. Akhiezer, *The classical moment problem and some related questions in analysis*, translated by N. Kemmer, Hafner, New York, 1965.

[2] M.F. Barnsley, J.S. Geronimo, and A.N. Harrington, *Infinite-dimensional Jacobi matrices associated with Julia sets*, Proc. Amer. Math. Soc., **88** (1983), 625–630.

[3] J. Béllissard, D. Bessis, and P. Moussa, *Chaotic states of almost periodic Schrödinger operators*, Phys. Rev. Lett., **49** (1982), 701–704.

[4] D. Bessis and G. Mantica, *Orthogonal polynomials associated to almost periodic Schrödinger operators. A trend towards random orthogonal polynomials*, J. Comput. Appl. Math., **48** (1993), 17–32.

[5] T. Geisel, R. Ketzmerick, and G. Petschel, *New class of level statistics in quantum systems with unbounded diffusion*, Phys. Rev. Lett., **66** (1991), 1651–1654; *Metamorphosis of a Cantor spectrum due to classical chaos*, ibid., **67** (1991), 3635–3638.

[6] G.H Golub and C.F. Van Loan, *Matrix computations*, 3d ed., John Hopkins Univ. Press, Baltimore, MD, 1996.

[7] I. Guarneri and G. Mantica, *Multifractal energy spectra and their dynamical implications*, Phys. Rev. Lett., **73** (1994), 3379–3382.

[8] K.A. Makarov, *Asymptotic expansions for Fourier transform of singular self-affine measures*, J. Math. Anal. Appl., **187** (1994), 259–286.

[9] G. Mantica, *A stable Stieltjes technique for computing orthogonal polynomials and Jacobi matrices associated with a class of singular measures*, Constr. Approx., **12** (1996), 509–530.

[10] G. Mantica, *Quantum intermittency in almost-periodic lattice systems derived from their spectral properties*, Physica D, **103** (1997), 576–589.

[11] G. Mantica, *Wave propagation in almost-periodic structures*, Physica D, **109** (1997), 113–127.

[12] G. Mantica, *On computing Jacobi matrices associated with recurrent and Möbius iterated function systems*, Dynamical Systems Preprint, Como, 1998.

[13] G. Mantica, *Algorithms for the quantum evolution generated by infinite matrices with singular continuous spectral measures*, Dynamical Systems Preprint, Como, 1998.

[14] C.B. Moler and C.F. Van Loan, *Nineteen dubious ways to compute the exponential of a matrix*, SIAM Rev., **20** (1978), 801–836.

[15] J.A. Shohat and J.D. Tamarkin, *The problem of moments*, American Mathematical Society Mathematical Surveys, vol. II, Amer. Math. Soc., New York, 1943.

[16] H. Tal-Ezer and R. Kosloff, *An accurate and efficient scheme for propagating the time dependent Schrödinger equation*, J. Chem. Phys., **81** (1984), 3967–3971.

Giorgio Mantica
International Center for the Study of Dynamical Systems
University of Milano at Como
via Lucini 3
I-22100 Como, Italy
INFM-INFN sez. di Milano
E-mail address: giorgio@fis.unico.it

International Series of Numerical Mathematics
Vol. 131,

On a Sequence of Fast Decreasing Polynomial Operators

Hrushikesh N. Mhaskar[1] and Jürgen Prestin

Abstract. Let f be a piecewise analytic function on the unit interval (respectively, the unit circle of the complex plane). Starting from the Chebyshev (respectively, Fourier) coefficients of f, we construct a sequence of fast decreasing polynomials (respectively, trigonometric polynomials) which "detect" the points where f fails to be analytic, provided f is not infinitely differentiable at these points.

1. Introduction

A fast decreasing sequence of polynomials is a sequence of polynomials which are "large" near some points, and decrease exponentially fast at other points. Such polynomials have been constructed by Gaier, Ivanov, Saff, Totik, and others ([1], [6] and references therein) in connection with the approximation of piecewise analytic functions on a real interval. The error in approximating such functions decays exponentially at the points of analyticity, and (typically) polynomially at the singularities; i.e., at the points where the function fails to be analytic.

In [2], we used segment approximation to **detect** the singularities of a piecewise analytic function on a compact, real interval, provided the number of such points is known in advance. A central observation in our approach is the fact that a piecewise analytic function on an interval (in the sense of [2]) necessarily has jump discontinuities in some derivative at the singularities. Motivated by the work [1] of Gaier, we define in this paper a sequence of fast decreasing polynomials which enables us to detect the location of those singularities of a function where it is not infinitely many times differentiable. Our approach also does not require the prior knowledge of the number of such singularities.

Our construction uses the Chebyshev coefficients of the function. We would like to observe here that the singularity or analyticity of a function is a local property, whereas each of the Chebyshev coefficients represents global information about the function. In [5], we have constructed polynomial operators, based on the Jacobi coefficients, that detect the points where the derivatives of a function

[1] This research was supported, in part, by the Alexander von Humboldt foundation and the U.S. Air Force Office of Scientific Research, Grant number F49620-97-1-0211.

have jump discontinuities. However, these operators do not define fast decreasing polynomials. In the case of Chebyshev polynomials, we are able to use function-theoretic methods to construct operators which yield fast decreasing polynomials.

Our techniques enable us to detect the singularities of a piecewise analytic function defined on the unit circle as well. In fact, the starting point of our research is to construct fast decreasing trigonometric polynomials for this purpose.

We discuss the construction of fast decreasing trigonometric polynomials in the next section. In §3, we point out the connection with Chebyshev polynomials. In §4, we present some numerical examples.

2. Functions on the Unit Circle

Let U denote the complex unit circle, $f : U \to \mathbb{C}$ be integrable with respect to the arclength of U. The Fourier coefficients of f are then defined by

$$\hat{f}_k := \frac{1}{2\pi i}\int_U \frac{f(\omega)}{\omega^{k+1}}d\omega, \qquad k = 0, \pm 1, \pm 2, \dots . \tag{1}$$

Suppose that both the power series $\sum_{k=0}^{\infty} \hat{f}_k z^k$ and $\sum_{k=1}^{\infty} \hat{f}_{-k} z^k$ have radius of convergence equal to 1. Then they define analytic functions on the open unit disc, and have singularities on U in the sense defined in Hille's book [3, §10.2, p.4 and §10.4, p.15]. We say that f is piecewise analytic if the total number of singularities on U of both of these power series is finite. The singularities of f are then defined to be the singularities of these power series. If $r \geq 0$ is an integer, we will say that f has a singularity of order r at a point w_0 if f is $r-1$ times continuously differentiable at w_0 with respect to the arclength, and the rth angular derivative (i.e., derivative with respect to the arclength) has a jump discontinuity at w_0.

Next, we define certain operators related to those defined by Gaier in [1]. For $m = 0, 1, \dots$, and $z \in \mathbb{C}$, we write

$$s_m(f, z) := \sum_{k=-m}^{m} \hat{f}_k z^k.$$

Our operators are defined by applying a summability method with the s_m's. Towards this end, we write for integers $\ell \geq 1$, $m \geq 0$,

$$h_{\ell,m} := \begin{cases} 0, & \text{if } 0 \leq m \leq \ell - 1, \\ -\dfrac{1}{\ell}, & \text{if } \ell \leq m \leq 2\ell - 1, \\ \dfrac{1}{2\ell}, & \text{if } 2\ell \leq m \leq 4\ell - 1, \\ 0, & \text{if } m \geq 4\ell, \end{cases}$$

and $h_{0,m} := 0$. For integer $n \geq 0$, the summability factors are defined by

$$g_{n,m} := 2^{-n} \sum_{\ell=0}^{n} \binom{n}{\ell} \{h_{\ell,4m} + h_{\ell,4m+1} + h_{\ell,4m+2} + h_{\ell,4m+3}\}$$
$$= 2^{-n} \sum_{\ell=m+1}^{n} \binom{n}{\ell} \{h_{\ell,4m} + h_{\ell,4m+1} + h_{\ell,4m+2} + h_{\ell,4m+3}\}. \quad (2)$$

Our operators are now defined for integer $n \geq 1$ by

$$\tau_n(f) := \sum_{m=0}^{n-1} g_{n,m} s_m(f) =: \sum_{k=-n+1}^{n-1} \tilde{g}_{n,|k|} \hat{f}_k z^k.$$

We observe that for each integer $n \geq 1$, τ_n is a linear operator defined on the class of all functions on U integrable with respect to the arclength, and for each such function f, $\tau_n(f)$ is a Laurent polynomial of degree at most $n-1$ (cf. Fig. 2).

For $f : U \to \mathbb{C}$, and $x \geq 0$, $\|f\|$ denotes $\sup_{z \in U} |f(z)|$, and we write

$$E_x(f) := \inf_{T \in \mathbb{H}_x} \|f - T\|,$$

where $\mathbb{H}_x$ denotes the class of all functions of the form $\sum_{|k| \leq x} a_k (\cdot)^k$. Restricted to U, elements of $\mathbb{H}_x$ are just trigonometric polynomials of order at most x in the argument variable.

For integer $r \geq 0$, and $z \in U$, we define

$$\Gamma_r(z) := \sum_{k=-\infty, k \neq 0}^{\infty} \frac{z^k}{(ik)^{r+1}}.$$

The function Γ_r is analytic on U, except at $z = 1$, where its rth angular derivative (i.e., derivative with respect to the arclength) has a jump discontinuity. Hence, if $f : U \to \mathbb{C}$ has a singularity of order r at $w_0 \in U$, then for a judiciously chosen complex constant $a(f)$, the rth angular derivative of $f - a(f)\Gamma_r(\cdot/w_0)$ is continuous at w_0.

In the sequel, we adopt the following notations and conventions. Letters $c, c_1, c_2, \ldots$ will denote positive constants depending on explicitly indicated quantities only. Their values may be different at different occurrences, even within the same formula. The main theorem of this section is the following.

Theorem 2.1. *Let $f : U \to \mathbb{C}$ be continuous. For each integer $n \geq 1$, there holds*

$$\|\tau_n(f)\| \leq c_1 e^{-cn} \|f\| + 6E_{n/12}(f). \quad (3)$$

If $w_0 \in U$, and f is analytic in the disc $\{z : |z - w_0| < d\}$ for some $0 < d := d(f, w_0) < 1$, then

$$|\tau_n(f, w_0)| \leq \frac{c(f)}{d^4} \left(1 - \frac{d^2}{900}\right)^{n/2}. \quad (4)$$

Suppose that f has only finitely many singularities, and that the only discontinuities in all angular derivatives of f at these singularities are jump discontinuities. If $r \ge 0$ is an integer, there is an arc $J := J_r(f)$ of the circle, such that if $w_0 \in U$ is a singularity of f of order r, then

$$n^r |\tau_n(f,w)| \ge c(f,r), \qquad (ww_0^{-1})^n \in J. \tag{5}$$

In order to prove Theorem 2.1, we need some preparation. First, we define the de la Vallée Poussin operators by $v_0(f,z) := \hat{f}_0$ and

$$v_\ell(f,z) := \frac{1}{\ell} \sum_{m=\ell}^{2\ell-1} s_m(f,z), \qquad \ell = 1, 2, \ldots .$$

We also need the Euler means of the difference of these operators. Thus, we define for $n = 0, 1, \ldots$

$$\tau_n^*(f,z) := 2^{-n} \sum_{\ell=0}^{n} \binom{n}{\ell} \Big(v_{2\ell}(f,z) - v_\ell(f,z) \Big).$$

Next, let

$$F(z) := f(z^4).$$

For $z = \rho e^{i\theta}$, we will write $z^{1/4} := \rho^{1/4} \exp(i\theta/4)$, $\theta \in (-\pi, \pi]$. If $z_0 \in U$, $|z - z_0^{1/4}| < \delta < 1$, then

$$|z^4 - z_0| < |z - z_0^{1/4}|(1 + |z|)(1 + |z|^2) < 15\delta.$$

Hence, if f is analytic in the disc of radius $d < 1$, centered at some $z_0 \in U$, then F is analytic in the disc of radius $d/15$ around $z_0^{1/4}$ (and hence, also in the discs of radius $d/15$ around $-z_0^{1/4}$ and $\pm i z_0^{1/4}$). Similarly, f has a singularity of order r at $z_0 \in U$ if and only if F has singularities of order r at $\pm z_0^{1/4}$ and $\pm i z_0^{1/4}$.

Further, we observe that for integer $m \ge 0$,

$$s_{4m}(F,z) = s_{4m+1}(F,z) = s_{4m+2}(F,z) = s_{4m+3}(F,z) = s_m(f,z^4).$$

Hence, for integer $\ell \ge 1$,

$$\begin{aligned} v_{2\ell}(F,z) - v_\ell(F,z) &= \sum_{m=0}^{4\ell-1} h_{\ell,m} s_m(F,z) \\ &= \sum_{m=0}^{\ell-1} \{h_{\ell,4m} + h_{\ell,4m+1} + h_{\ell,4m+2} + h_{\ell,4m+3}\} s_m(f,z^4) \\ &=: \sum_{k=-\ell+1}^{\ell-1} \tilde{h}_{\ell,|k|} \hat{f}_k z^{4k}. \end{aligned}$$

Consequently, we obtain that

$$\begin{aligned}
&\tau_n^*(F,z)\\
&= 2^{-n}\sum_{\ell=0}^{n}\binom{n}{\ell}\Big(v_{2\ell}(F,z)-v_\ell(F,z)\Big)\\
&= 2^{-n}\sum_{m=0}^{n-1}\Big\{\sum_{\ell=m+1}^{n}\binom{n}{\ell}\Big[h_{\ell,4m}+h_{\ell,4m+1}+h_{\ell,4m+2}+h_{\ell,4m+3}\Big]\Big\}s_m(f,z^4)\\
&= \tau_n(f,z^4).
\end{aligned}$$

Thus, Theorem 2.1 will be proved if we prove the following.

Theorem 2.2. *Let $F : U \to \mathbb{C}$ be continuous, and*

$$F(z) = F(-z) = F(iz) = F(-iz), \qquad z \in \mathbb{C}.$$

For each integer $n \geq 0$, there holds

$$\|\tau_n^*(F)\| \leq 0.15e^{-n/18}\|F\| + 6E_{n/3}(F). \tag{6}$$

If $w_0 \in U$, and F is analytic in the disc $\{z : |z - w_0| < d\}$ for some $0 < d := d(F, w_0) < 1$, then

$$|\tau_n^*(F,w_0)| \leq \frac{c(F)}{d^4}\left(1-\frac{d^2}{4}\right)^{n/2}. \tag{7}$$

Suppose that F has only finitely many singularities, and that the only discontinuities in all angular derivatives of F at these singularities are jump discontinuities. If $r \geq 0$ is an integer, there is an arc $J := J_r(F)$ of the circle, such that if $w_0 \in U$ is a singularity of F of order r, then

$$n^r|\tau_n^*(F,w)| \geq c(F,r), \qquad (ww_0^{-1})^n \in J. \tag{8}$$

Proof. It is well known that

$$\|v_\ell(F)\| \leq 3\|F\|, \quad \|v_\ell(F) - v_{2\ell}(F)\| \leq 6E_\ell(F), \quad \ell = 0, 1, \ldots.$$

Hence, (6) follows from an estimate for binomial sums (cf. [4]) and Stirling's inequality

$$2^{-n}\sum_{\ell=0}^{[n/3]-1}\binom{n}{\ell} \leq 2^{-n}\binom{n}{[n/3]} \leq \left(\frac{3}{2^{5/3}}\right)^n \frac{3e^{1/(12n)}}{2\sqrt{\pi n}} \leq 0.15e^{(\ln 3 - 5\ln 2)n},$$

where the last inequality is true for $n \geq 32$. For $n < 32$ the first sum can be estimated directly by computation.

To prove the next assertion of the theorem, we assume without loss of generality that $w_0 = 1$, and that F is analytic in the disc $\{z : |z-1| < d\}$ for some d, $0 < d < 0.7$. Following Gaier [1], we consider the arc γ of the circle $|z+1| = \sqrt{4-d^2}$ intercepted by U and inside U. Let w and W be the end points of γ in the upper and lower half plane respectively (cf. Fig. 1). Both w and W are on the arc from $e^{-i\pi/4}$ to $e^{i\pi/4}$. Let C_0 be the Jordan arc from $e^{-i\pi/4}$ to $e^{i\pi/4}$, following γ from W to w, and U otherwise. Let

$$C_1 := C_0 \cup (iC_0) \cup (-C_0) \cup (-iC_0)$$

and C_2 be the reflection of C_1 in the unit circle. We observe that the lens-shaped region between γ and its reflection in U lies in $\{z \,:\, |z-1| < d\}$, and hence, F is analytic in this region.

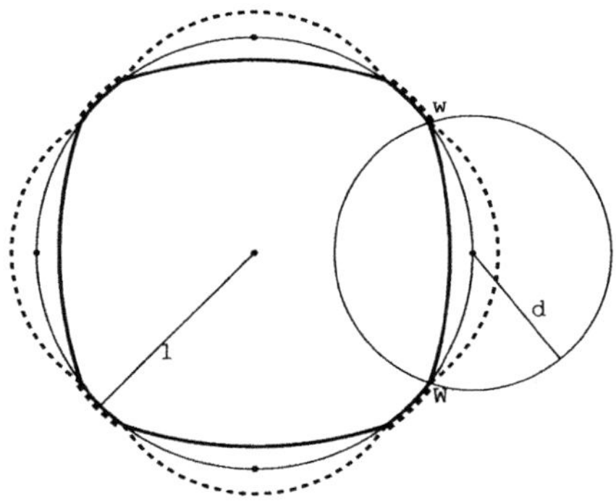

FIGURE 1. The unit circle with curves C_1 (thick line) and C_2 (dotted line)

Following Gaier [1], we obtain from (1) and Cauchy's formula that for integer $m \geq 0$,

$$s_m(F,1) = F(1) + \frac{1}{2\pi i}\oint_{C_1} \frac{F(\zeta)}{(\zeta-1)}\zeta^m d\zeta - \frac{1}{2\pi i}\oint_{C_2} \frac{F(\zeta)}{(\zeta-1)}\zeta^{-m-1} d\zeta,$$

and hence, after a little computation, that for integer $\ell \geq 1$,

$$\begin{aligned} v_{2\ell}(F,1) - v_\ell(F,1) = {} & \frac{1}{4\pi i}\oint_{C_1} \frac{F(\zeta)}{(\zeta-1)^2}\frac{\zeta^{4\ell} - 3\zeta^{2\ell} + 2\zeta^\ell}{\ell} d\zeta \\ & - \frac{1}{4\pi i}\oint_{C_2} \frac{F(\zeta)}{(\zeta-1)^2}\frac{\zeta^{-4\ell} - 3\zeta^{-2\ell} + 2\zeta^{-\ell}}{\ell} d\zeta, \end{aligned}$$

and for integer $n \geq 1$,

$$\tau_n^*(F,1) = \frac{1}{4\pi i}\oint_{C_1} \frac{F(\zeta)}{(\zeta-1)^2} H_n(\zeta) d\zeta - \frac{1}{4\pi i}\oint_{C_2} \frac{F(\zeta)}{(\zeta-1)^2} H_n(\zeta^{-1}) d\zeta, \tag{9}$$

where

$$H_n(z) := 2^{-n}\sum_{\ell=1}^{n} \binom{n}{\ell} \frac{z^{4\ell} - 3z^{2\ell} + 2z^\ell}{\ell}.$$

Writing $h_n(t) = t^{-1}((1+t)^n - 1)$, we observe that

$$H_n(z) = 2^{-n}\left\{\int_{z^2}^{z^4} h_n(t)dt - 2\int_z^{z^2} h_n(t)dt\right\},$$

where the integrations are over any paths joining the indicated end points. For $\zeta \in C_1$, we may choose these paths to lie entirely inside C_1 and outside the disc $\{z : |z| \le (\sqrt{4-d^2}-1)^{1/4}\}$. With this choice,

$$|h_n(t)| \le \frac{(4-d^2)^{n/2}+1}{(\sqrt{3}-1)^{1/4}}$$

for all t on all these paths. Therefore, for $\zeta \in C_1$,

$$|H_n(\zeta)| \le \frac{2\pi}{(\sqrt{3}-1)^{1/4}}\left\{\left(1-\frac{d^2}{4}\right)^{n/2}+2^{-n}\right\} \le 10.72\left(1-\frac{d^2}{4}\right)^{n/2}.$$

Along with (9), this implies

$$|\tau_n^*(F,1)| \le \frac{10.72}{(2-\sqrt{4-d^2})^2}\left(1-\frac{d^2}{4}\right)^{n/2} c(F),$$

which leads to (7).

Next, let w_0 be a singularity of F. Without loss of generality, we may again assume that $w_0 = 1$. Let R be the maximum of the orders of all the singularities of F. Our assumption on F implies that

$$F(z) = \omega \sum_{\nu=0}^{3} \Gamma_r(zi^\nu) + \Phi_1(z) + \Phi_2(z), \qquad z \in U,$$

where $\omega \in \mathbb{C}$, Φ_1 is a (finite) linear combination of terms of the form $\Gamma_k(zw_j^{-1})$, $w_j \ne 1$, $0 \le k \le R$, and Φ_2 has R continuous angular derivatives on U. In particular, Φ_1 is analytic at 1, and both Φ_1 and Φ_2 satisfy

$$\Phi_j(z) = \Phi_j(iz) = \Phi_j(-z) = \Phi_j(-iz), \quad j = 1, 2.$$

From (6) and Favard's Theorem, $\tau_n^*(\Phi_2, w) = o(n^{-R})$ for $w \in U$. From (7), $\tau_n^*(\Phi_1, w) = o(e^{-c(F)n})$ for w on some arc J_1 around 1. We observe that

$$\tau_n^*(\Gamma_r(i^\nu \cdot), w) = \tau_n^*(\Gamma_r, i^\nu w)$$

for $\nu = 0, 1, 2, 3$ and $w \in U$. Hence, the estimate (8) will be proved if we prove that

$$n^r|\tau_n^*(\Gamma_r, i^\nu w)| = \mathcal{O}(n^{-1}), \qquad \nu = 1, 2, 3, \tag{10}$$

and

$$n^r|\tau_n^*(\Gamma_r, w)| \ge c_r \tag{11}$$

for sufficiently large integer values of n, and w^n on some arc of U which intersects J_1.

We write $w =: e^{i\theta}$, $\tau_{n,r} := \tau_n^*(\Gamma_r)$, $v_{\ell,r} := v_\ell(\Gamma_r)$, and

$$h(t) := \frac{1}{\pi}\begin{cases} 0, & \text{if } t \in [0, \pi/2), \\ t^{-r-1}(2t-\pi), & \text{if } t \in [\pi/2, \pi), \\ t^{-r-1}(2\pi - t), & \text{if } t \in [\pi, 2\pi). \end{cases}$$

A simple summation by parts argument leads to the formula (for integer $\ell \ge 1$)

$$v_{2\ell,r}(w) - v_{\ell,r}(w) = 4\pi\left(\frac{2\pi}{4\ell}\right)^r \frac{1}{4\ell}\sum_{k=0}^{4\ell-1} h\left(\frac{2\pi k}{4\ell}\right)\cos(k\theta - (r+1)\pi/2).$$

Next, we recall that for any integer $p \ge 0$ and $\ell \ge 1$,

$$\ell^{-p} = \frac{\ell!}{(\ell+p)!}\left(1 + \mathcal{O}(\ell^{-1})\right) \le c(p)\frac{\ell!}{(\ell+p)!}\,. \tag{12}$$

Hence, an application of Lemma 3.2 of [5] gives

$$\begin{aligned} v_{2\ell,r}(w) - v_{\ell,r}(w) &= 2\left(\frac{2\pi}{4\ell}\right)^r\left\{\int_0^{2\pi} h(t)\cos(\ell\alpha t - \beta)dt + \mathcal{O}(\ell^{-1})\right\} \\ &= 2(\frac{\pi}{2})^r\frac{\ell!}{(\ell+r)!}\left\{\int_0^{2\pi} h(t)\cos(\ell\alpha t - \beta)dt + \mathcal{O}(\ell^{-1})\right\}, \end{aligned} \tag{13}$$

where $\alpha := 2\theta/\pi$, $\beta := (r+1)\pi/2$.

Let $\nu \in \{1,2,3\}$. Using this formula with $i^\nu w$ in place of w, we get

$$v_{2\ell,r}(i^\nu w) - v_{\ell,r}(i^\nu w) = 2(\frac{\pi}{2})^r\frac{\ell!}{(\ell+r)!}\left\{\int_0^{2\pi} h(t)\cos(\ell\alpha_\nu t - \beta)dt + \mathcal{O}(\ell^{-1})\right\},$$

where $\alpha_\nu = \alpha + \nu$. For $|\theta| < \pi/8$, we have $|\alpha_\nu| \ge c$. Hence, using integration by parts, we see that

$$\int_0^{2\pi} h(t)\cos(\ell\alpha_\nu t - \beta)dt = \mathcal{O}(\ell^{-1}). \tag{14}$$

Now, using (12), we obtain

$$\sum_{\ell=1}^{n}\binom{n}{\ell}\ell^{-p} \le c(p)\frac{n!}{(n+p)!}\sum_{\ell=1}^{n}\binom{n+p}{\ell+p} \le c_1(p)n^{-p}2^n. \tag{15}$$

Together with (13) and (14), this estimate (with $p = 1$) leads to (10).

Next, we deduce from (13) and (15) that

$$\begin{aligned} &\tau_{n,r}(w) \\ &= 2(\frac{\pi}{2})^r\frac{n!}{(n+r)!}\mathrm{Re}\left\{e^{-i\beta}\int_0^{2\pi} h(t)e^{-ir\alpha t}2^{-n}\sum_{l=r+1}^{n+r}\binom{n+r}{\ell}e^{i\alpha t\ell}dt \right.\\ &\qquad\qquad\left. + \mathcal{O}(n^{-1})\right\} \\ &= 2\pi^r\frac{n!}{(n+r)!}\mathrm{Re}\left\{e^{-i\beta}\int_0^{2\pi} h(t)e^{-ir\alpha t}((1+e^{i\alpha t})/2)^{n+r}dt \right.\\ &\qquad\qquad\left. + \mathcal{O}(n^{-1})\right\}. \end{aligned}$$

For any $A > 0$ and $|\alpha| \le A/n$, this yields

$$\begin{aligned}\tau_{n,r}(w) &= 2\pi^r \frac{n!}{(n+r)!}\mathrm{Re}\left\{e^{-i\beta}\int_0^{2\pi} h(t)e^{in\alpha t}dt + \mathcal{O}(n^{-1})\right\}\\ &= 2\pi^r \frac{n!}{(n+r)!}\left\{\int_0^{2\pi} h(t)\cos(n\alpha t-\beta)dt + \mathcal{O}(n^{-1})\right\},\end{aligned} \qquad (16)$$

where the unspecified constants may depend also on A. Now, the entire function

$$z \mapsto \int_0^{2\pi} h(t)\cos(zt-\beta)dt$$

is not identically equal to zero, and hence, every neighborhood of 0 contains a real interval I such that

$$\left|\int_0^{2\pi} h(t)\cos(zt-\beta)dt\right| \ge c_r$$

for all $z \in I$. Along with (16), this leads to (11). □

3. Functions on $[-1,1]$

Let $f : [-1,1] \to \mathbb{R}$. We will say that f is piecewise analytic if there exist points $-1 =: y_0 < y_1 < \cdots < y_s := 1$ such that f is analytic in each of the open intervals (y_{j-1}, y_j) for $j = 1, \ldots, s$. The points y_j will be called the singularities of f. This notion is more general than that used in [2], where the restriction of f to each of the *closed* intervals $[y_{j-1}, y_j]$ was assumed to be analytic. For the analysis in this section, we do have to assume the existence of all the one-sided derivatives of f at the points y_j, including at ± 1. Our objective is to detect the location of the points y_j that are in the open interval $(-1,1)$, given the sequence of Chebyshev coefficients of f (see (17) below for the definition). Unlike [2], we do not require that the number of these "interior singularities" be known in advance. We say that a point $y \in (-1,1)$ is a singularity of f of order r if f has r continuous derivatives in a deleted neighborhood of y, and the rth derivative of f has a jump discontinuity at y.

We recall that the Chebyshev polynomials are defined by the formula

$$T_n(\cos\theta) := \cos(n\theta), \qquad n = 0,1,2,\ldots, \ \theta \in [0,\pi].$$

They satisfy the following orthogonality relations:

$$\int_{-1}^{1} T_n(t)T_m(t)\frac{dt}{\sqrt{1-t^2}} = \begin{cases} \pi, & \text{if } n = m = 0,\\ \pi/2, & \text{if } n = m \ne 0,\\ 0, & \text{if } n \ne m.\end{cases}$$

If f is a continuous function on $[-1,1]$, we may define the Chebyshev coefficients for integer $k \ge 0$ by

$$f_k^T := \frac{1}{\sqrt{\pi}} \begin{cases} \displaystyle\int_{-1}^{1} f(t)(1-t^2)^{-1/2}dt, & \text{if } k=0, \\ \sqrt{2}\displaystyle\int_{-1}^{1} f(t)T_k(t)(1-t^2)^{-1/2}dt, & \text{if } k \neq 0. \end{cases} \tag{17}$$

The partial sums of the Chebyshev expansion of f are then defined for integer $m \ge 0$ and $x \in [-1,1]$ by

$$S_m(f,x) := \frac{1}{\sqrt{\pi}} f_0^T + \sqrt{\frac{2}{\pi}} \sum_{k=1}^{m} f_k^T T_k(x).$$

With the summability factors $g_{n,m}$ as in (2), we may now define the operators $\mathcal{T}_n$ by

$$\mathcal{T}_n(f,x) := \sum_{m=0}^{n-1} g_{n,m} S_m(f,x), \qquad n = 1,2,\dots, \ x \in [-1,1].$$

Finally, we write

$$\|f\|_I := \max_{x \in [-1,1]} |f(x)|,$$

and for $y \ge 0$,

$$\epsilon_y(f) := \min_{P \in \Pi_y} \|f - P\|_I,$$

where Π_y denotes the class of all algebraic polynomials of degree at most y.

Theorem 2.1 now leads to the following analogous result.

Theorem 3.1. *Let $f : [-1,1] \to \mathbb{R}$ be continuous. For each integer $n \ge 1$, there holds*

$$\|\mathcal{T}_n(f)\|_I \le c_1 e^{-cn} \|f\|_I + 6\epsilon_{n/12}(f). \tag{18}$$

If $x_0 \in [-1,1]$, and f is analytic in the disc $\{z \in \mathbb{C} : |z - x_0| < d\}$ for some $0 < d := d(f,x_0) < 1$, then

$$|\mathcal{T}_n(f,w_0)| \le \frac{c(f)}{d^4}(1 - c_1 d^2)^{n/2}. \tag{19}$$

Suppose that f fails to be analytic only at finitely many points, and that the only discontinuities in all derivatives of f at these points are jump discontinuities. We also assume that all the one-sided derivatives of f at ± 1 are defined. If $r \ge 0$ is an integer, there is a subinterval $J \subseteq [0,\pi]$ such that if $y =: \cos\phi \in (-1,1)$ is a singularity of f of order r, then

$$n^r |\mathcal{T}_n(f,\cos\theta)| \ge c(f,r), \qquad n|\theta - \phi| \in J. \tag{20}$$

Proof. We define $f^\circ : U \to \mathbb{R}$ by

$$f^\circ(e^{i\theta}) := f(\cos\theta), \qquad \theta \in [-\pi,\pi],$$

and consider the Joukowski transformation

$$z = \frac{1}{2}\left(w + \frac{1}{w}\right), \qquad |w| > 1,$$

that maps the exterior of $[-1, 1]$ conformally onto the exterior of the unit disc in the w-plane. It is easy to verify from the definitions that

$$S_m(f, \cos\theta) = s_m(f^\circ, e^{i\theta}), \qquad \theta \in [-\pi, \pi],$$

and hence, that

$$\mathcal{T}_n(f, \cos\theta) = \tau_n(f^\circ, e^{i\theta}), \qquad \theta \in [-\pi, \pi].$$

Together with (3), this implies (18).

We observe that

$$z - \cos\theta = \frac{1}{2w}(w - e^{i\theta})(w - e^{-i\theta}),$$

and that the right-hand side is a rational function of w with pole at the origin. Therefore, if f is analytic in the disc $\{z \in \mathbb{C} \ : \ |z - \cos\theta| < d\}$ for some $\theta \in (0, \pi)$ and $0 < d < 1$, then f° is analytic in discs of radius $d/2$ around $e^{i\theta}$ and $e^{-i\theta}$. Thus, (4) implies (19).

Our assumption about the one-sided derivatives of f implies that the only singularities of f° are at those points $e^{\pm i\theta}$ where f has a singularity at $\cos\theta$, $\theta \in (0, \pi)$. Moreover, all the angular derivatives of f° at these points will have at most jump discontinuities of the same order as the corresponding singularity of f. We have also assumed the existence of all the one-sided derivatives of f at ± 1. The function f° is then infinitely often differentiable at the points ± 1 on U. The estimate (5) then leads to (20). □

4. Numerical Results

We consider the function

$$f(z) := \frac{1}{2}\left\{\Gamma_1(ze^{0.8i}) + \Gamma_3(ze^{-i})\right\},$$

which has a singularity of order 1 at $z = z_1 = e^{-0.8i}$ and of order 3 at $z = z_3 = e^{i}$. In the following figures, the quantity $\arg z$ is on the x-axis, and the y-axis denotes the values of the operators in question. The results of our work [5] indicate that the operator

$$V_{1024}(f, z) := v_{512}(f, z) - v_{256}(f, z)$$

is also able to detect the singularity at z_1 and perhaps also at z_3. We observe that both $\tau_{1024}(f)$ and $V_{1024}(f)$ are in $\mathbb{H}_{1023}$, and the operator V_{1024} is obviously much easier to compute. On the other hand, Fig. 2 shows the summability factors $\tilde{g}_{1024,k}$ (respectively $\tilde{h}_{256,k}$) for the operators τ_{1024} (respectively V_{1024}). It is clear that while both of them are formally of the same order, the real contribution to τ_{1024} actually comes from a polynomial of order less than 700.

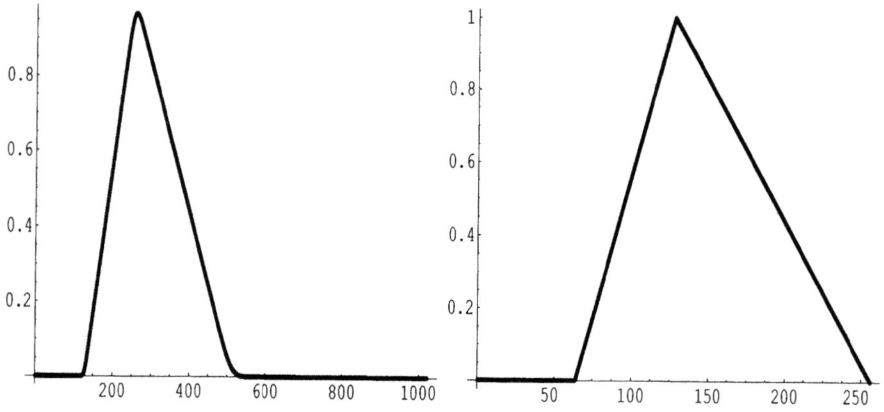

FIGURE 2. Left: $\tilde{g}_{1024,k}$, Right: $\tilde{h}_{256,k}$

Fig. 3 shows that τ_{1024} and V_{1024} are both sufficiently well localized for the detection of the singularity at z_1.

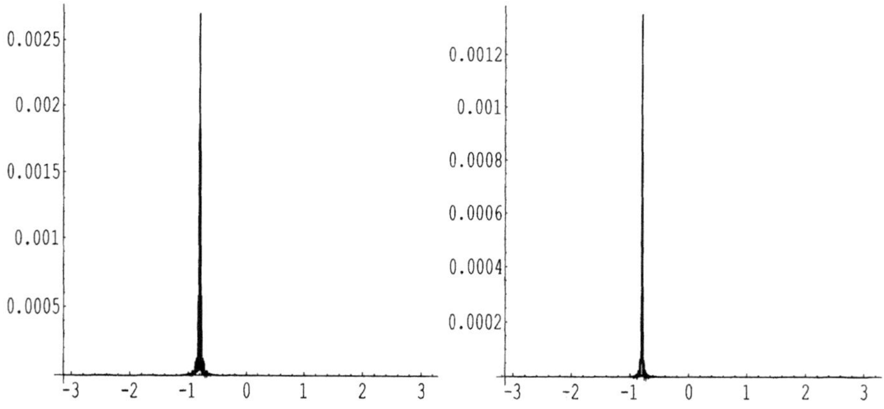

FIGURE 3. Left: $|\tau_{1024}(f)|$, Right: $|V_{1024}(f)|$ on $[-\pi, \pi]$

However, as Fig. 4 shows, in the presence of this singularity z_1, V_{1024} is not at all able to detect the higher-order singularity at z_3. The operator τ_{1024} is able to do so.

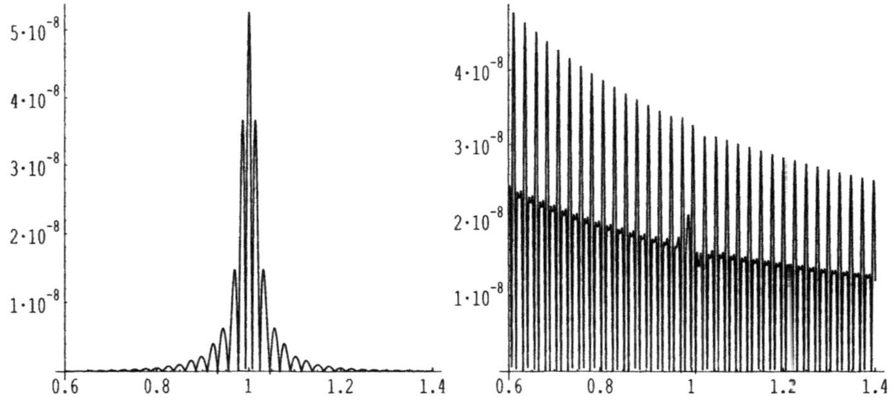

FIGURE 4. Left: $|\tau_{1024}(f)|$, Right: $|V_{1024}(f)|$ on $[0.6, 1.4]$

In Fig. 5, we examine the behavior of both near z_1. As we go away from this point, the operator τ_{1024} seems to decay slower in the beginning, but then decreases much more rapidly as the theory predicts.

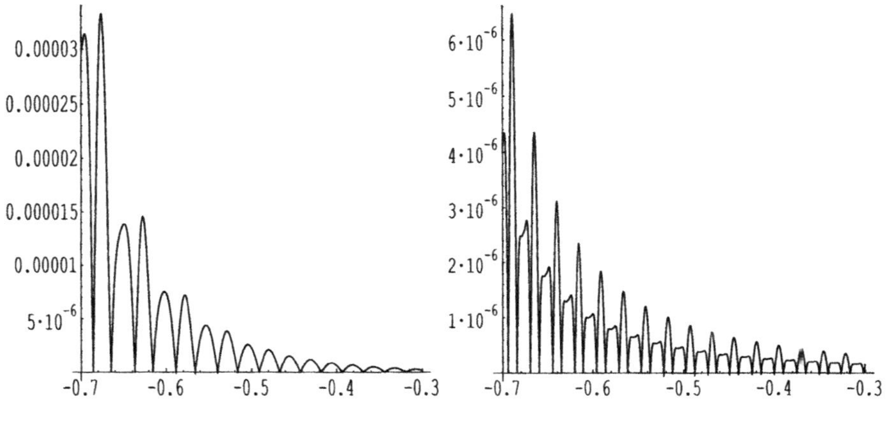

FIGURE 5. Left: $|\tau_{1024}(f)|$, Right: $|V_{1024}(f)|$ on $[-0.7, -0.3]$

In Fig. 6, it is clear that the oscillations in τ_{1024} have already "died" long before we come to z_3, while those of V_{1024} are still not sufficiently small to affect the behavior significantly at z_3.

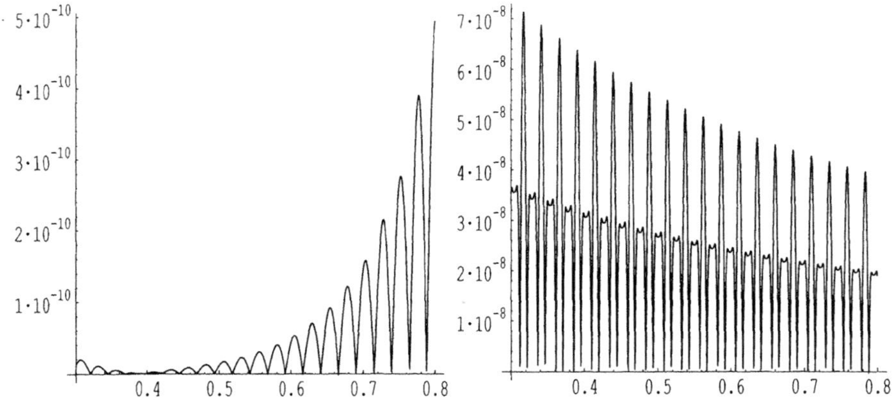

FIGURE 6. Left: $|\tau_{1024}(f)|$, Right: $|V_{1024}(f)|$ on $[0.3, 0.8]$

References

[1] D. Gaier, *Polynomial approximation of piecewise analytic functions,* J. Anal., **4** (1996), 67–79.

[2] R. Grothmann and H.N. Mhaskar, *Detection of singularities using segment approximation,* Math. Comp., **59** (1992), 533–540.

[3] E. Hille, *Analytic function theory, Vol. II,* Introductions to higher mathematics, Ginn and Co., Boston, 1962.

[4] R. Labahn, *On an interesting property for binomial coefficients,* Rostock. Math. Kolloq., **29** (1986), 21–24.

[5] H.N. Mhaskar and J. Prestin, *Polynomial frames for the detection of singularities,* Manuscript.

[6] E.B. Saff and V. Totik, *Logarithmic potentials with external fields,* Grundlehren der Mathematischen Wissenschaften **316**, Springer-Verlag, Berlin, 1997.

Hrushikesh N. Mhaskar
Department of Mathematics
California State University
Los Angeles, CA 90032, USA
E-mail address: hmhaska@calstatela.edu

Jürgen Prestin
GSF – National Research Institute for Environment and Health
Institute of Biomathematics and Biometrics
D-85764, Neuherberg, Germany
E-mail address: prestin@gsf.de

International Series of Numerical Mathematics
Vol. 131, © 1999 Birkhäuser Verlag Basel/Switzerland

Müntz Orthogonal Polynomials and Their Numerical Evaluation

Gradimir V. Milovanović[1]

Abstract. This paper is devoted to some classes of orthogonal Müntz polynomials on $(0,1)$, their connection with orthogonal rational functions, as well as their numerical computation. Also, we consider some important special cases of such polynomials. For evaluating Müntz polynomials we develop a numerical procedure based on numerical integration in the complex plane.

1. Introduction

Let $\Lambda = \{\lambda_0, \lambda_1, \lambda_2, \dots\}$ be a complex sequence. We adopt the following definition for x^λ:

$$x^\lambda = e^{\lambda \log x}, \qquad x \in (0,\infty),\ \lambda \in \mathbb{C},$$

and the value at $x=0$ to be the limit of x^λ as $x \to 0$ from $(0,\infty)$ whenever the limit exists, and consider Müntz polynomials as linear combinations of the Müntz system $\{x^{\lambda_0}, x^{\lambda_1}, \dots, x^{\lambda_n}\}$. By $M_n(\Lambda)$ we denote the set of all such polynomials, i.e.,

$$M_n(\Lambda) = \operatorname{span}\{x^{\lambda_0}, x^{\lambda_1}, \dots, x^{\lambda_n}\},$$

where the linear span is over the real (or complex) numbers. The union of all $M_n(\Lambda)$ is denoted by $M(\Lambda)$, i.e., $M(\Lambda) = \bigcup_{n=0}^{\infty} M_n(\Lambda)$.

Such generalized polynomials can be orthogonalized and applied to several approximation problems, including quadrature problems. The orthogonal Müntz systems were considered first by the Armenian mathematicians Badalyan [2] and Taslakyan [18]. Recently, they were investigated by McCarthy, Sayre and Shawyer [13] and more completely by Borwein, Erdélyi, and Zhang [4] (see also the recent book [3]).

In this paper we consider orthogonal Müntz polynomials and their numerical evaluation. Section 2 is devoted to some classes of orthogonal Müntz systems on $(0,1)$ and their connection with orthogonal rational functions. Numerical evaluation of Müntz polynomials is considered in §3.

[1]This work was partly supported by the Serbian Scientific Foundation under grant #04M03.

2. Orthogonal Müntz Polynomials

We investigate two classes of Müntz polynomials which are orthogonal with respect to some inner products. The first of them was introduced by Badalyan [2], and we refer to it as Müntz-Legendre polynomials. The second class was recently defined in [5] and [15].

2.1. Orthogonal Müntz-Legendre polynomials

Let the complex sequence $\Lambda = \{\lambda_0, \lambda_1, \lambda_2, \dots\}$ be such that $\operatorname{Re}(\lambda_k) > -1/2$ for every $k \in \mathbb{N}_0$ and let $\Lambda_n = \{\lambda_0, \lambda_1, \dots, \lambda_n\}$.

If Γ is a simple contour surrounding all the zeros of the denominator in the rational function

$$W_n(s) = \prod_{k=0}^{n-1} \frac{s + \bar{\lambda}_k + 1}{s - \lambda_k} \cdot \frac{1}{s - \lambda_n} \qquad (n \in \mathbb{N}_0), \tag{2.1}$$

then the Müntz-Legendre polynomials are defined by (see [2], [18], [13], [4], [3])

$$P_n(x) = P_n(x; \Lambda_n) = \frac{1}{2\pi i} \oint_{\Gamma} W_n(s) x^s \, ds. \tag{2.2}$$

In the case $n = 0$, an empty product in (2.1) should be taken to be equal to 1.

If $\lambda_k \neq \lambda_j$ $(k \neq j)$, by the Cauchy residue theorem, these generalized polynomials can be expressed in power form as

$$P_n(x) = \sum_{k=0}^{n} C_{nk} x^{\lambda_k}, \qquad C_{nk} = \frac{\prod\limits_{\nu=0}^{n-1} (\lambda_k + \bar{\lambda}_\nu + 1)}{\prod\limits_{\substack{\nu=0 \\ \nu \neq k}}^{n} (\lambda_k - \lambda_\nu)}. \tag{2.3}$$

For the Müntz-Legendre polynomials (2.2) the following orthogonality relation holds:

$$(P_n, P_m) = \int_0^1 P_n(x) \overline{P_m(x)} \, dx = \frac{\delta_{nm}}{\lambda_n + \bar{\lambda}_n + 1}.$$

Also, some recurrence relations exist, e.g.,

$$x P_n'(x) - x P_{n-1}'(x) = \lambda_n P_n(x) + (1 + \bar{\lambda}_{n-1}) P_{n-1}(x) \tag{2.4}$$

and

$$P_n(x) = P_{n-1}(x) - (\lambda_n + \bar{\lambda}_{n-1} + 1) x^{\lambda_n} \int_x^1 t^{-\lambda_n - 1} P_{n-1}(t) \, dt \qquad (x \in (0, 1]).$$

It is easy to prove that

$$P_n(1) = 1 \qquad \text{and} \qquad P_n'(1) = \lambda_n + \sum_{k=0}^{n-1} (\lambda_k + \bar{\lambda}_k + 1).$$

In the special case when $\lambda_0 = \lambda_1 = \cdots = \lambda$, (2.2) gives

$$P_n(x; \Lambda_n) = x^{\lambda} L_n(-(\lambda + \bar{\lambda} + 1) \log x),$$

where $L_n(x)$ is the Laguerre polynomial orthogonal with respect to e^{-x} on $[0, \infty)$ and such that $L_n(0) = 1$.

Taking $x = e^{-t}$, the Müntz-Legendre polynomials can be expressed in terms of a Laplace transform. Namely, we can prove:

Theorem 2.1. *If* $W_n(s)$ *is given by* (2.1) *and*

$$G_n(s) = -W_n(-s) = \prod_{k=0}^{n-1} \frac{s - (\bar{\lambda}_k + 1)}{s + \lambda_k} \cdot \frac{1}{s + \lambda_n},$$

then $P_n(e^{-t})$ *is the inverse Laplace transform of* $G_n(s)$, *i.e.*,

$$P_n(e^{-t}) = \mathcal{L}^{-1}[G_n(s)].$$

In the proof of this result we can take, for example, $\alpha > 1/2$, and then prove that

$$P_n(e^{-t}) = \frac{1}{2\pi i} \int_{\alpha - i\infty}^{\alpha + i\infty} G_n(s) e^{-st}\, ds.$$

An interesting question is connected with the zero distribution of the Müntz-Legendre polynomials for a real sequence Λ. A nice proof of the following result was given in [4].

Theorem 2.2. *For real numbers* $\lambda_\nu > -1/2$ $(\nu = 0, 1, \dots)$ *the Müntz-Legendre polynomial* $P_n(x; \Lambda_n)$ *has exactly* n *distinct zeros in* $(0, 1)$, *and it changes sign at each of these zeros. Furthermore, the zeros of the polynomials*

$$P_{n-1}(x; \Lambda_{n-1}) \qquad \text{and} \qquad P_n(x; \Lambda_n)$$

in $(0, 1)$ *strictly interlace.*

Now we consider the important special case where

$$\lambda_{2k} = \lambda_{2k+1} = k \qquad (k = 0, 1, \dots).$$

Namely, we take $\lambda_{2k} = k$ and $\lambda_{2k+1} = k + \varepsilon$ $(k = 0, 1, \dots)$, where ε decreases to zero. The corresponding limit process leads to orthogonal Müntz polynomials with logarithmic terms. Then, (2.1) becomes

$$W_n(s) = \begin{cases} \displaystyle\prod_{\nu=0}^{m-1} \left(\frac{s + \nu + 1}{s - \nu} \right)^2 \frac{1}{s - m}, & \text{when } n = 2m, \\ \displaystyle\prod_{\nu=0}^{m} \left(\frac{s + \nu + 1}{s - \nu} \right)^2 \frac{1}{s + m + 1}, & \text{when } n = 2m + 1. \end{cases}$$

Applying the Cauchy residue theorem to the integral in (2.2), with this rational function, we obtain the following representation for the corresponding Müntz polynomials:

$$P_n(x) = R_n(x) + S_n(x) \log x \qquad (n = 0, 1, \dots), \tag{2.5}$$

where $R_n(x)$ and $S_n(x)$ are algebraic polynomials of degree $[n/2]$ and $[(n-1)/2]$, respectively, i.e.,

$$R_n(x) = \sum_{\nu=0}^{[n/2]} a_\nu^{(n)} x^\nu, \qquad S_n(x) = \sum_{\nu=0}^{[(n-1)/2]} b_\nu^{(n)} x^\nu. \tag{2.6}$$

Notice that $P_n(1) = R_n(1) = 1$. The first few Müntz polynomials (2.5) are:

$$\begin{aligned}
P_0(x) &= \ \ 1,\\
P_1(x) &= \ \ 1 + \log x,\\
P_2(x) &= -3 + 4x - \log x,\\
P_3(x) &= \ \ 9 - 8x + 2(1+6x)\log x,\\
P_4(x) &= -11 - 24x + 36x^2 - 2(1+18x)\log x,\\
P_5(x) &= \ \ 19 + 276x - 294x^2 + 3(1 + 48x + 60x^2)\log x,\\
P_6(x) &= -21 - 768x + 390x^2 + 400x^3 - 3(1 + 96x + 300x^2)\log x.
\end{aligned}$$

The following theorem gives explicit expressions for the coefficients of the polynomials (2.6) for arbitrary n.

Theorem 2.3. *If n is an even number, $n = 2m$, we have*

$$a_\nu^{(2m)} = -\binom{m+\nu}{m}^2 \binom{m}{\nu}^2 \left[\frac{2m+1}{2\nu+1} + 2(m-\nu)\sum_{\substack{j=0\\ j\neq\nu}}^{m-1} \frac{2j+1}{(j-\nu)(j+\nu+1)}\right]$$

and

$$b_\nu^{(2m)} = -(m-\nu)\binom{m+\nu}{m}^2 \binom{m}{\nu}^2,$$

for each $0 \le \nu \le m-1$. *For* $\nu = m$ *we have*

$$a_m^{(2m)} = \binom{2m}{m}^2 \qquad \textit{and} \qquad b_m^{(2m)} = 0.$$

If n is an odd number, $n = 2m+1$, we have

$$a_\nu^{(2m+1)} = \binom{m+\nu}{m}^2 \binom{m}{\nu}^2 \left[\frac{2m+1}{2\nu+1} + 2(m+\nu+1)\sum_{\substack{j=0\\ j\neq\nu}}^{m} \frac{2j+1}{(j-\nu)(j+\nu+1)}\right]$$

and

$$b_\nu^{(2m+1)} = (m+\nu+1)\binom{m+\nu}{m}^2 \binom{m}{\nu}^2,$$

for each $0 \le \nu \le m$.

A simple proof of Theorem 2.3 can be obtained from the residue theorem. An explicit expression for $S_n(x)$, $n = 2m+1$, is given in [3, Theorem A.2.1].

These orthogonal Müntz polynomials can be used in the proof of the irrationality of $\zeta(3)$ and of other familiar numbers (see [3, pp. 372–381] and [19]).

Putting $\lambda_k + \beta/2$ instead of λ_k, $k = 0, 1, \ldots$, in the sequence Λ, we can define a kind of Müntz-Jacobi polynomials $P_n^{(\beta)}(x)$ by

$$P_n^{(\beta)}(x) = \frac{x^{-\beta/2}}{2\pi i} \oint_\Gamma W_n^{(\beta)}(s) x^s \, ds, \tag{2.7}$$

where

$$W_n^{(\beta)}(s) = \prod_{k=0}^{n-1} \frac{s + \bar{\lambda}_k + \beta/2 + 1}{s - \lambda_k - \beta/2} \cdot \frac{1}{s - \lambda_n - \beta/2}.$$

Then, the following result holds:

Theorem 2.4. *Let $\beta \in \mathbb{R}$ and $\operatorname{Re} \lambda_k > -(\beta + 1)/2$ for each $k \in \mathbb{N}_0$. Then*

$$(P_n^{(\beta)}, P_m^{(\beta)}) = \int_0^1 P_n^{(\beta)}(x) \overline{P_m^{(\beta)}(x)} \, x^\beta \, dx = \frac{\delta_{nm}}{\lambda_n + \bar{\lambda}_n + \beta + 1}.$$

The proof of this result and other properties of $P_n^{(\beta)}(x)$ will be given elsewhere.

In the special case $\lambda_k = k$ $(k = 0, 1, \ldots)$, the generalized polynomials (2.7) reduce to the classical Jacobi polynomials $\tilde{P}_n^{(0,\beta)}$ $(\beta > -1)$ shifted to $[0, 1]$. Then

$$\begin{aligned} P_n^{(\beta)}(x) = \tilde{P}_n^{(0,\beta)}(2x - 1) &= (-1)^n \binom{n+\beta}{n} {}_2F_1(-n, n+\beta+1; \beta+1; x) \\ &= (-1)^n \binom{n+\beta}{n} \sum_{k=0}^{n} \frac{(-n)_k (n+\beta+1)_k}{(\beta+1)_k} \cdot \frac{x^k}{k!} \\ &= \sum_{k=0}^{n} C_{nk}^{(\beta)} x^k, \end{aligned}$$

where

$$C_{nk}^{(\beta)} = \frac{(-1)^{n-k}}{k!(n-k)!} \prod_{\nu=0}^{n-1} (\beta + 1 + k + \nu).$$

Here, ${}_2F_1$ is the hypergeometric function, $(p)_k$ is defined by $(p)_k = \Gamma(p+k)/\Gamma(p)$, and Γ is the gamma function. For $\beta = 0$ these polynomials reduce to the Legendre polynomials shifted to $[0, 1]$.

Remark 2.5. *It would be interesting to construct the Müntz-Jacobi polynomials $P_n^{(\alpha,\beta)}(x)$ orthogonal with respect to the inner product*

$$(f, g) = \int_0^1 f(x) \overline{g(x)} \, (1 - x)^\alpha x^\beta \, dx.$$

2.2. Another class of orthogonal Müntz polynomials

Recently, we defined an external operation for the Müntz polynomials from $M(\Lambda)$ (see [5] and [15]). Namely, for $\alpha, \beta \in \mathbb{C}$ we define

$$x^\alpha \odot x^\beta = x^{\alpha\beta} \qquad (x \in (0,\infty)),$$

and then for polynomials $P \in M_n(\Lambda)$ and $Q \in M_m(\Lambda)$, i.e.,

$$P(x) = \sum_{i=0}^{n} p_i x^{\lambda_i} \qquad \text{and} \qquad Q(x) = \sum_{j=0}^{m} q_j x^{\lambda_j}, \tag{2.8}$$

we define

$$(P \odot Q)(x) = \sum_{i=0}^{n}\sum_{j=0}^{m} p_i q_j x^{\lambda_i \lambda_j}. \tag{2.9}$$

Under the restrictions that for each i and j we have

$$|\lambda_i| > 1, \qquad \operatorname{Re}(\lambda_i \bar{\lambda}_j - 1) > 0, \tag{2.10}$$

we can introduce a new inner product for Müntz polynomials (2.8) (see [15]),

$$[P,Q] = \int_0^1 (P \odot \overline{Q})(x)\,\frac{dx}{x^2}, \tag{2.11}$$

where $(P \odot Q)(x)$ is defined by (2.9).

Under the conditions (2.10), we defined (see [15]) the Müntz polynomials $Q_n(x) \equiv Q_n(x;\Lambda_n)$, $n = 0, 1, \ldots$, orthogonal with respect to the inner product (2.11). These polynomials are associated with the rational functions

$$W_n(s) = \frac{\prod_{\nu=0}^{n-1}(s - 1/\bar{\lambda}_\nu)}{\prod_{\nu=0}^{n}(s - \lambda_\nu)} \qquad (n = 0, 1, \ldots) \tag{2.12}$$

in the sense that

$$Q_n(x) = \frac{1}{2\pi i}\oint_\Gamma W_n(s) x^s\, ds, \tag{2.13}$$

where the simple contour Γ surrounds all the points λ_ν ($\nu = 0, 1, \ldots, n$). We note that the functions (2.12) form a system known as Malmquist system of rational functions (see Walsh [20, p. 305], Djrbashian [6]–[8]), which are orthogonal on the unit circle $|s| = 1$ with respect to the inner product

$$(u,v) = \frac{1}{2\pi i}\oint_{|s|=1} u(s)\overline{v(s)}\,\frac{ds}{s} = \frac{1}{2\pi}\int_{-\pi}^{\pi} u(e^{i\theta})\overline{v(e^{i\theta})}\, d\theta. \tag{2.14}$$

This generalizes the Szegő class of polynomials orthogonal on the unit circle (see Szegő [17, pp. 287–295]).

The following theorem gives the orthogonality relation for the polynomials $Q_n(x)$.

Theorem 2.6. *Under the conditions* (2.10) *on the sequence* Λ, *the Müntz polynomials* $Q_n(x)$, $n = 0, 1, \ldots$, *defined by* (2.13), *are orthogonal with respect to the inner product* (2.11), *i.e.,*

$$[Q_n, Q_m] = \frac{1}{(|\lambda_n|^2 - 1)|\lambda_0\lambda_1 \cdots \lambda_{n-1}|^2}\, \delta_{n,m}.$$

The proof of this theorem is based on the orthogonality of the Malmquist system of rational functions (2.12). Namely, we can prove that

$$[Q_n, Q_m] = (W_n, W_m),$$

where the inner products $[\cdot\,,\,\cdot]$ and $(\cdot\,,\,\cdot)$ are given by (2.11) and (2.14), respectively.

Assuming that $\lambda_i \neq \lambda_j$ $(i \neq j)$, we get a representation of (2.13) in the form

$$Q_n(x) = \sum_{k=0}^{n} A_{n,k} x^{\lambda_k}, \qquad A_{n,k} = \frac{\displaystyle\prod_{\nu=0}^{n-1} (\lambda_k - 1/\bar{\lambda}_\nu)}{\displaystyle\prod_{\substack{\nu=0\\ \nu\neq k}}^{n} (\lambda_k - \lambda_\nu)} \qquad (k = 0, 1, \ldots, n). \tag{2.15}$$

Now we mention some recurrence relations for the polynomials $Q_n(x)$.

Theorem 2.7. *Suppose that* Λ *is a complex sequence satisfying* (2.10). *Then the polynomials* $Q_n(x)$, *defined by* (2.13), *satisfy the following relations:*

$$xQ_n'(x) = xQ_{n-1}'(x) + \lambda_n Q_n(x) - (1/\bar{\lambda}_{n-1})Q_{n-1}(x),$$

$$xQ_n'(x) = \lambda_n Q_n(x) + \sum_{k=0}^{n-1} (\lambda_k - 1/\bar{\lambda}_k) Q_k(x),$$

$$xQ_n''(x) = (\lambda_n - 1)Q_n'(x) + \sum_{k=0}^{n-1} (\lambda_k - 1/\bar{\lambda}_k) Q_k'(x),$$

$$Q_n(1) = 1, \quad Q_n'(1) = \lambda_n + \sum_{k=0}^{n-1} (\lambda_k - 1/\bar{\lambda}_k),$$

$$Q_n(x) = Q_{n-1}(x) - (\lambda_n - 1/\bar{\lambda}_{n-1}) x^{\lambda_n} \int_x^1 t^{-\lambda_n - 1} Q_{n-1}(t)\, dt \quad (x \in (0, 1]).$$

One particular result for (2.13) when $\lambda_\nu \to \lambda$ for each ν, may be interesting:

Corollary 2.8. *Let* $Q_n(x)$ *be defined by* (2.13) *and let* $\lambda_0 = \lambda_1 = \cdots = \lambda_n = \lambda$. *Then*

$$Q_n(x) = x^\lambda L_n(-(\lambda - 1/\bar{\lambda}) \log x),$$

where $L_n(x)$ *is the Laguerre polynomial.*

Also, for a real sequence Λ such that

$$1 < \lambda_0 < \lambda_1 < \cdots \tag{2.16}$$

we have:

Theorem 2.9. *Let* Λ *be a real sequence satisfying* (2.16). *Then the polynomial* $Q_n(x)$, *defined by* (2.13), *has exactly* n *simple zeros in* $(0,1)$ *and no other zeros in* $[1,\infty)$.

3. Numerical Evaluation of Müntz Polynomials

A direct evaluation of Müntz polynomials $P_n(x)$ (or $Q_n(x)$) in the power form (2.3) (or (2.15)) can be problematic in finite arithmetic, especially when n is a large number and x is close to 1. The polynomial coefficients C_{nk} (or A_{nk}) become very large numbers when n increases, but their sums are always equal to 1. (Recall that $P_n(1) = 1$ and $Q_n(1) = 1$.) In order to illustrate this fact we consider a special class of Müntz polynomials determined by (2.5). Their coefficients are given in Theorem 2.3.

3.1. A special case of Müntz polynomials

Let $\mathbf{a}_n$ and $\mathbf{b}_n$ be the vectors of coefficients of the polynomials $R_n(x)$ and $S_n(x)$, defined by (2.6), i.e.,

$$\mathbf{a}_n = \begin{bmatrix} a_0^{(n)} & a_1^{(n)} & \cdots & a_{[n/2]}^{(n)} \end{bmatrix}^T, \qquad \mathbf{b}_n = \begin{bmatrix} b_0^{(n)} & b_1^{(n)} & \cdots & b_{[(n-1)/2]}^{(n)} \end{bmatrix}^T.$$

Using Theorem 2.3, we can calculate these vectors. For $n = 10$ and $n = 20$ we have

$$\mathbf{a}_{10} = \frac{1}{3}\begin{bmatrix} -134 & -52020 & -999810 & -1133440 & 1994895 & 190512 \end{bmatrix}^T,$$

$$\mathbf{b}_{10} = -5\begin{bmatrix} 1 & 720 & 26460 & 125440 & 79380 \end{bmatrix}^T$$

and

$$\mathbf{a}_{20} = \frac{1}{63}\begin{bmatrix} -7318 \\ -52049250 \\ -24527715300 \\ -2114001489600 \\ -48491344751850 \\ -337299299349012 \\ -625811341034880 \\ 163660745064960 \\ 674793629510715 \\ 173135700710830 \\ 2150491110768 \end{bmatrix}, \qquad \mathbf{b}_{20} = -10\begin{bmatrix} 1 \\ 10890 \\ 7056720 \\ 824503680 \\ 26512946460 \\ 286339821768 \\ 1131219048960 \\ 1633930721280 \\ 775478838420 \\ 85336948840 \end{bmatrix},$$

respectively. The absolute values of some coefficients increase very fast. For example, the vectors $\mathbf{a}_{30}$ and $\mathbf{b}_{30}$ are

$$\frac{1}{6006}\begin{bmatrix} -1189751 \\ -44524636800 \\ -114037231854000 \\ -57424250116396800 \\ -8712106595849556900 \\ -497343278165994868608 \\ -12111056103238468936560 \\ -134308661572934174177280 \\ -686397947699727236368395 \\ -1495971634625290991808640 \\ -818037563553451371290160 \\ 1213101132351099611692800 \\ 1461592959488813080647900 \\ 439304716456359806505600 \\ 33189655142727694141200 \\ 144513038735768102400 \end{bmatrix}, \quad -15\begin{bmatrix} 1 \\ 53760 \\ 176729280 \\ 110279070720 \\ 20527415148240 \\ 1445130026436096 \\ 44257107059605440 \\ 642345550526085120 \\ 4645714010933775780 \\ 17129875530027420160 \\ 32118516618801412800 \\ 29659336430994816000 \\ 12512532556825938000 \\ 2060815557834432000 \\ 90230418791064000 \end{bmatrix},$$

respectively.

Using Horner's scheme for evaluating the values of $P_n(x)$, written in the form (where $b_{n/2}^{(n)} = 0$ for n even)

$$P_n(x) = \sum_{\nu=0}^{[n/2]} c_\nu^{(n)} x^\nu, \qquad c_\nu^{(n)} = a_\nu^{(n)} + b_\nu^{(n)} \log x,$$

we obtain numerical results heavily affected by errors. Relative errors in the values of $P_n(x)$, for $n = 10(10)40$ and some selected values of x, obtained by using D-arithmetic (with machine precision $\approx 2.22 \times 10^{-16}$), are presented in Table 3.1. Numbers in parentheses indicate decimal exponents.

TABLE 3.1. Relative errors in the values $P_n(x)$ in D- (and Q-) arithmetic

x	$n = 10$	$n = 20$	$n = 30$	$n = 40$	Q-arith. ($n = 40$)
10^{-3}	3.08(−15)	9.92(−14)	2.31(−12)	1.03(−10)	
10^{-2}	2.84(−14)	2.44(−11)	5.72(−9)	1.12(−6)	
0.1	1.52(−12)	8.29(−7)	8.06(−3)	7.62(+1)	
0.2	8.38(−12)	3.21(−6)	4.10(−1)	2.43(+5)	9.05(−14)
0.5	3.42(−10)	3.83(−4)	1.93(+2)	5.49(+11)	5.08(−7)
0.9	5.34(−10)	1.11(−2)	7.28(+4)	9.24(+12)	4.95(−5)
1.0	2.13(−10)	5.13(−3)	4.89(+4)	4.81(+11)	8.41(−5)

As we can see, the values obtained for $n \geq 30$ are quite wrong, excluding cases when x is very close to zero. When $n = 10$ and $n = 20$, at $x = 1$ we lost approximatively 6 and 13 decimal digits, respectively. Also, when we used

Q-arithmetic (machine precision $\approx 1.93 \times 10^{-34}$ on the MICROVAX 3400), for $n = 40$ and $x = 1$ we lost about 30 digits (see the last column in Table 3.1). Notice that the shapes of the curves $y = R_n(x)$ and $y = S_n(x)\log x$ (Fig. 1) are very similar at first sight, but we know that the sum $R_n(x) + S_n(x)\log x$ represents the Müntz polynomial $P_n(x)$ which changes its sign n times on $[0, 1]$. Its zeros are more densely distributed around 0 than in other parts of the interval $[0, 1]$. In Figs. 2 and 3 we display $P_{20}(x)$ on the intervals $[0.05, 1]$ (14 zeros), $[10^{-3}, 0.05]$ (4 zeros), and $[0, 10^{-3}]$ (two zeros).

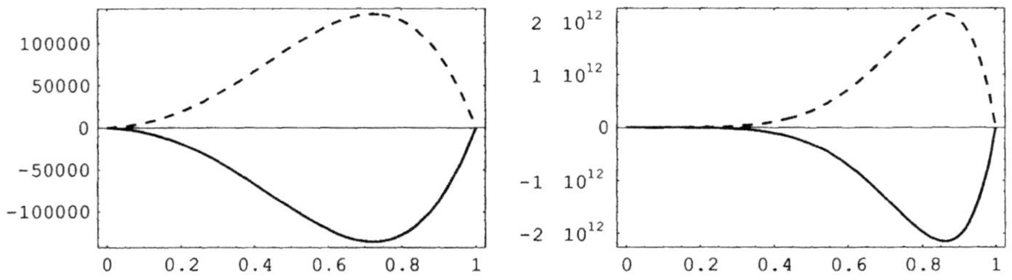

FIGURE 1. Graphics $x \mapsto R_n(x)$ (solid line) and $x \mapsto S_n(x)\log x$ (broken line) for $n = 10$ and $n = 20$

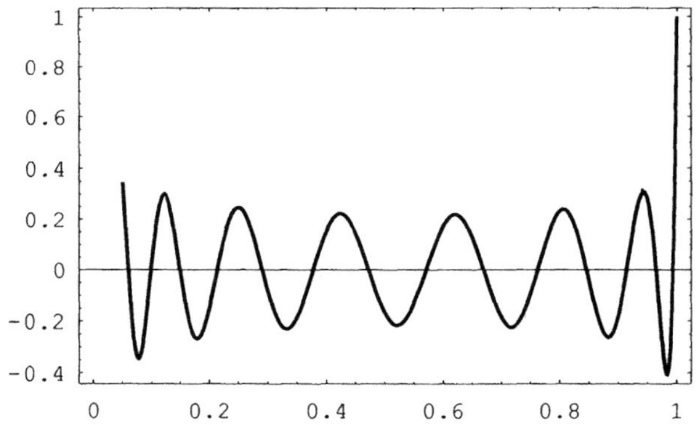

FIGURE 2. The Müntz polynomial $P_{20}(x) = R_{20}(x) + S_{20}(x)\log x$ on $[0.05, 1]$

Before concluding this subsection we mention that the Müntz polynomials (2.5) have a logarithmic behaviour in the neighbourhood of zero, i.e.,

$$P_{2m}(x) \sim -m\log x, \qquad P_{2m+1}(x) \sim (m+1)\log x \qquad (x \to 0+).$$

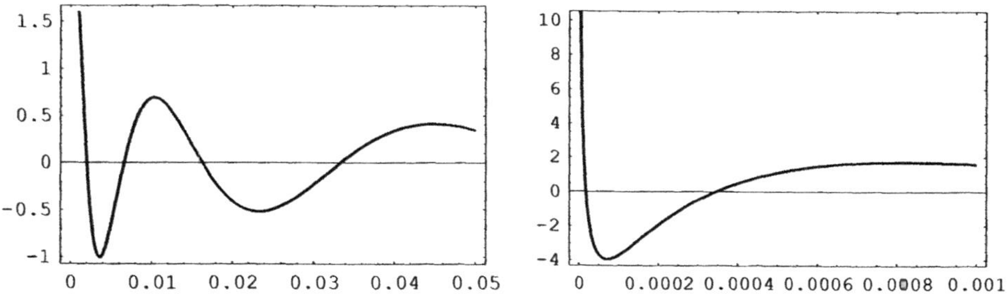

FIGURE 3. The Müntz polynomial $P_{20}(x)$ on $[0.001, 0.05]$ and $[0, 0.001]$

3.2. A numerical method for evaluating Müntz polynomials

In this subsection we give a stable numerical method for evaluating the values of the Müntz-Legendre polynomials defined by (2.1) and (2.2), i.e.,

$$P_n(x) = \frac{1}{2\pi i} \oint_\Gamma W_n(s) x^s \, ds, \qquad W_n(s) = \prod_{\nu=0}^{n-1} \frac{s + \bar{\lambda}_\nu + 1}{s - \lambda_\nu} \cdot \frac{1}{s - \lambda_n}. \tag{3.1}$$

For evaluating Müntz polynomials $Q_n(x)$, defined in §2.2, we can use the same procedure with the rational function (2.12).

Our method is based on a direct evaluation of the contour integral in (3.1). First we take the contour $\Gamma = \Gamma_R = C_R \cup L_R$ (see Fig. 4). Thus, C_R is a semicircle with center at $\sigma < -1/2$ and radius R such that all poles of $W_n(s)$ are inside the contour Γ_R, and L_R is the straight line $s = \sigma + it$, $-R \le t \le R$. (Notice that the function $W_n(s)$ has only real poles marked by crosses in Fig. 4 for $n = 5$.)

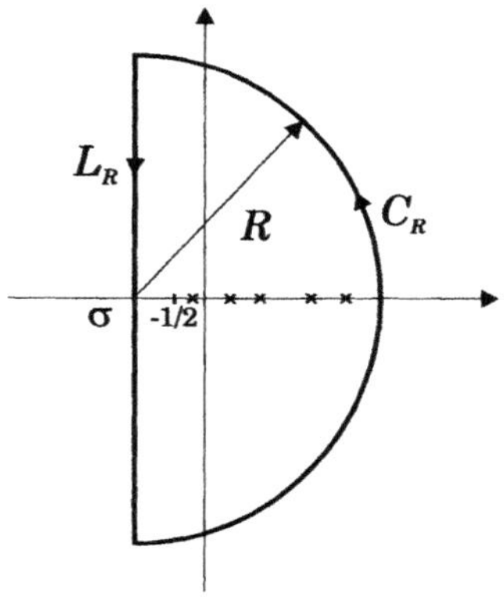

FIGURE 4. The contour of integration for the integral in (3.1)

Lemma 3.1. *We have* $\int_{C_R} W_n(s) x^s \, ds \to 0$ *when* $R \to \infty$.

Proof. Let $s \in C_R$, i.e., $s = \sigma + Re^{i\theta}$, $-\pi \le \theta \le \pi$. For a sufficiently large R, there exists a positive constant $M > 1$ such that $|W_n(s)| \le M/R$. Indeed, this follows from

$$|W_n(s)| = \left| \prod_{\nu=0}^{n-1} \frac{\sigma + \bar{\lambda}_\nu + Re^{i\theta} + 1}{\sigma - \lambda_\nu + Re^{i\theta}} \right| \cdot \left| \frac{1}{\sigma - \lambda_n + Re^{i\theta}} \right|$$

$$= \frac{1}{R} \cdot \frac{1}{\left| 1 + \dfrac{\sigma - \lambda_n}{R} e^{-i\theta} \right|} \prod_{\nu=0}^{n-1} \left| \frac{1 + \dfrac{\sigma + \bar{\lambda}_\nu + 1}{R} e^{-i\theta}}{1 + \dfrac{\sigma - \lambda_\nu}{R} e^{-i\theta}} \right|.$$

Now, we have

$$\left| \int_{C_R} W_n(s) x^s \, ds \right| \le \int_{-\pi/2}^{\pi/2} |W_n(s)| \cdot \left| e^{(\sigma + Re^{i\theta}) \log x} \right| R \, d\theta$$

$$\le M e^{\sigma \log x} \int_{-\pi/2}^{\pi/2} e^{-R \cos\theta \log(1/x)} \, d\theta.$$

Using the Jordan inequality $\cos\theta > 1 - 2\theta/\pi \quad (0 < \theta < \pi/2)$ and putting $\omega = \log(1/x) > 0$, we get

$$\left| \int_{C_R} W_n(s) x^s \, ds \right| \le 2M e^{-\sigma\omega} \int_0^{\pi/2} e^{-R\omega(1 - 2\theta/\pi)} \, d\theta = \frac{M\pi e^{-\sigma\omega}}{R\omega} \left(1 - e^{-R\omega}\right) \to 0$$

when $R \to \infty$. □

Thus, when $R \to \infty$, integration along the contour Γ_R reduces to integration over the line L_R, so that

$$P_n(x) = \frac{1}{2\pi i} \int_{\sigma + i\infty}^{\sigma - i\infty} W_n(s) x^s \, ds = -\frac{x^\sigma}{2\pi} \int_{-\infty}^{\infty} W_n(\sigma + it) e^{-i\omega t} \, dt,$$

i.e.,

$$P_n(x) = \frac{x^\sigma}{2\pi} \int_{-\infty}^{\infty} \left[-\frac{1}{\omega} W_n\left(\sigma - i\frac{t}{\omega}\right) \right] e^{it} \, dt.$$

Since

$$-\frac{1}{\omega} W_n\left(\sigma - i\frac{t}{\omega}\right) = -\frac{1}{\omega} \prod_{\nu=0}^{n-1} \frac{\sigma - it/\omega + \bar{\lambda}_\nu + 1}{\sigma - it/\omega - \lambda_\nu} \cdot \frac{1}{\sigma - it/\omega - \lambda_n} = \frac{1}{i} f_n(t; \omega),$$

where

$$f_n(t; \omega) = \prod_{\nu=0}^{n-1} \frac{t + i(\sigma + \bar{\lambda}_\nu + 1)\omega}{t + i(\sigma - \lambda_\nu)\omega} \cdot \frac{1}{t + i(\sigma - \lambda_n)\omega}, \tag{3.2}$$

we obtain

$$P_n(x) = \frac{x^\sigma}{2\pi i} \int_{-\infty}^{\infty} f_n(t; \omega) e^{it} \, dt.$$

This gives the following result:

Theorem 3.2. *Let $\sigma < -1/2$, $f_n(t;\omega)$ be defined by (3.2), and*

$$\varphi_n(t;\omega) = \frac{1}{2i}\Big(f_n(t;\omega)e^{it} + f_n(-t;\omega)e^{-it}\Big). \tag{3.3}$$

Then the Müntz-Legendre polynomials can be represented in the integral form

$$P_n(x) = \frac{x^\sigma}{\pi}\int_0^\infty \varphi_n(t;\omega)\,dt. \tag{3.4}$$

In the sequel we consider the case when the sequence Λ is real. An important corollary of Theorem 3.2 is the following result:

Theorem 3.3. *Let $\Lambda = \{\lambda_0, \lambda_1, \lambda_2, \dots\}$ be a real sequence such that $\lambda_k > -1/2$ for every $k \in \mathbb{N}_0$, $f_n(t;\omega)$ be defined by (3.2), and $\sigma < -1/2$. Then the Müntz-Legendre polynomials have the following integral representation:*

$$P_n(x) = \frac{x^\sigma}{\pi}\operatorname{Im}\left\{\int_0^\infty f_n(t;\omega)e^{it}\,dt\right\}. \tag{3.5}$$

In this case we have $f_n(-t;\omega) = \overline{f_n(t;\omega)}$, and then (3.3) becomes $\varphi_n(t;\omega) = \operatorname{Im}\{f_n(t;\omega)e^{it}\}$, and (3.4) gives (3.5). The poles of $f_n(t;\omega)$ are then purely imaginary, $i(\lambda_\nu - \sigma)\omega$, $\nu = 0, 1, \dots, n$, and located on the positive part of the imaginary axis, because of $\lambda_\nu > \sigma$ and $\omega = \log(1/x) > 0$.

In order to calculate the integral in (3.5) (or in (3.4)), we use the following idea on complex integration (see [14, Theorem 2.2]). We select a positive number $a = m\pi$ ($m \in \mathbb{N}$) and put

$$\begin{aligned}\int_0^\infty f_n(t;\omega)e^{it}\,dt &= \int_0^a f_n(t;\omega)e^{it}\,dt + \int_a^{+\infty} f_n(t;\omega)e^{it}\,dt \\ &= L_1(f_n(\,\cdot\,;\omega)) + L_2(f_n(\,\cdot\,;\omega)).\end{aligned}$$

Here,

$$L_1(f_n(\,\cdot\,;\omega)) = \int_0^{a=m\pi} f_n(t;\omega)e^{it}\,dt = \sum_{k=1}^m \int_{(k-1)\pi}^{k\pi} f_n(t;\omega)e^{it}\,dt,$$

i.e.,

$$L_1(f_n(\,\cdot\,;\omega)) = \pi\int_0^1\left[\sum_{k=1}^m f_n\big(\pi(\xi+k-1);\omega\big)e^{i\pi(\xi+k-1)}\right]d\xi, \tag{3.6}$$

and

$$L_2(f_n(\,\cdot\,;\omega)) = \int_a^{+\infty} f_n(t;\omega)e^{it}\,dt.$$

Since $f_n(z;\omega)$ is a holomorphic function in the region $D = \{z \in \mathbb{C} \mid \operatorname{Re} z \geq a > 0,\ \operatorname{Im} z \geq 0\}$ and $|f_n(z;\omega)| \leq A/|z|$ when $|z| \to +\infty$, for some positive constant A, we can prove

$$L_2(f_n(\,\cdot\,;\omega)) = ie^{ia}\int_0^{+\infty} f_n(a+iy;\omega)e^{-y}\,dy. \tag{3.7}$$

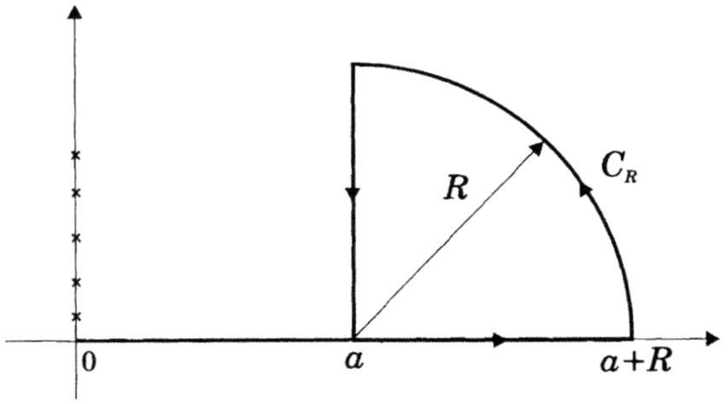

FIGURE 5. The contour of integration for the integral $L_2(f_n(\,\cdot\,;\omega))$

Indeed, if we take a closed contour of integration as in Fig. 5, consisting of the real segment $[a, a+R]$, the circular arc C_R, and the line segment joining the points $a+iR$ and a, we get, by Cauchy's residue theorem,

$$\int_a^{a+R} f_n(t;\omega)e^{it}\,dt + \int_0^{\pi/2} \left[f_n(z;\omega)e^{iz}\right]_{z=a+Re^{i\theta}} Rie^{i\theta}\,d\theta + i\int_R^0 f_n(a+iy;\omega)e^{i(a+iy)}\,dy = 0.$$

Using Jordan's lemma, we obtain the following estimate for the integral over the circular arc C_R,

$$\left|\int_0^{\pi/2} \left[f_n(z;\omega)e^{iz}\right]_{z=a+Re^{i\theta}} Rie^{i\theta}\,d\theta\right| \le \frac{\pi}{2}\cdot\frac{A}{\sqrt{a^2+R^2}}\left(1-e^{-R}\right) \to 0$$

when $R \to +\infty$. Thus we conclude that (3.7) holds.

Finally, for $a = m\pi$, (3.7) becomes

$$L_2(f_n(\,\cdot\,;\omega)) = (-1)^m \int_0^\infty \psi_n(y;\omega)e^{-y}\,dy, \tag{3.8}$$

where

$$\psi_n(y;\omega) = if_n(a+iy;\omega) = \prod_{\nu=0}^{n-1} \frac{y+(\sigma+\lambda_\nu+1)\omega - ia}{y+(\sigma-\lambda_\nu)\omega - ia}\cdot\frac{1}{y+(\sigma-\lambda_n)\omega-ia}.$$

Theorem 3.4. *Under the conditions of Theorem 3.3, the Müntz-Legendre polynomials have a computable representation*

$$P_n(x) = \frac{x^\sigma}{\pi}\mathrm{Im}\,\left\{L_1(f_n(\,\cdot\,;\omega)) + L_2(f_n(\,\cdot\,;\omega))\right\}, \tag{3.6}$$

where $L_1(f_n(\,\cdot\,;\omega))$ and $L_2(f_n(\,\cdot\,;\omega))$ are given by (3.6) and (3.8), respectively.

In the numerical implementation we use the Gauss-Legendre rule on $(0,1)$ and the Gauss-Laguerre rule for calculating $L_1(f_n(\cdot\,;\omega))$ and $L_2(f_n(\cdot\,;\omega))$, respectively. Numerical experiments show that a convenient choice for the parameter σ is $\lambda_{\min} - \pi/\omega$, where $\lambda_{\min} = \min\{\lambda_0, \lambda_1, \dots\}$.

In order to calculate the relative errors in Table 3.1, we used the previous numerical procedure for evaluating $P_n(x)$ with machine precision (in D-arithmetic).

Remark 3.5. *At the Oberwolfach Meeting on "Applications and Computation of Orthogonal Polynomials" (March, 1998), we also presented a stable numerical method for constructing the generalized Gaussian quadratures using orthogonal Müntz systems, but it is not included in this paper because of limited space. Some references in that direction are* [1], [9], [10], [11], [12], *and* [16].

Acknowledgments. The author is grateful to Professor Walter Gautschi for helpful comments.

References

[1] I.V. Andronov, *Application of orthogonal polynomials in the boundary-contact value problems,* VIII Simposium sobre polinomios ortogonales y sus aplicaciones (Sevilla, Sept. 22–26, 1997), Book of Abstracts, Facultad de Matemáticas, Universidad de Sevilla, 1997, p. 32.

[2] G.V. Badalyan, *Generalization of Legendre polynomials and some of their applications,* Akad. Nauk. Armyan. SSR Izv. Fiz.-Mat. Estest. Tekhn. Nauk, **8** (5) (1955), 1–28 and **9** (1) (1956), 3–22 (Russian, Armenian summary).

[3] P. Borwein and T. Erdélyi, *Polynomials and polynomial inequalities,* Graduate Texts in Mathematics **161**, Springer-Verlag, New York, 1995.

[4] P. Borwein, T. Erdélyi, and J. Zhang, *Müntz systems and orthogonal Müntz-Legendre polynomials,* Trans. Amer. Math. Soc., **342** (1994), 523–542.

[5] B. Danković, G.V. Milovanović, and S. Lj. Rančić, *Malmquist and Müntz orthogonal systems and applications,* in: Th.M. Rassias, Ed., *Inner product spaces and applications,* Pitman Res. Notes Math. Ser. **376**, Longman, Harlow, 1997, 22–41.

[6] M.M. Djrbashian, *Orthogonal systems of rational functions on the circle with given set of poles,* Dokl. Akad. Nauk SSSR, **147** (1962), 1278–1281.

[7] M.M. Djrbashian, *Orthogonal systems of rational functions on the circle,* Akad. Nauk. Armyan. SSR Izv. Mat., **1**(1) (1966), 3–24 and **1**(2) (1966), 106–125 (Russian).

[8] M.M. Djrbashian, *A survey on the theory of orthogonal systems and some open problems,* in: P. Nevai, Ed., *Orthogonal polynomials — theory and practice,* NATO Adv. Sci. Inst. Ser. C Math. Phys. Sci. **294**, Kluwer, Dordrecht, 1990, 135–146.

[9] W. Gautschi, *A survey of Gauss-Christoffel quadrature formulae,* in: P.L. Butzer and F. Fehér, Eds., *E.B. Christoffel,* Birkhäuser, Basel, 1981, 72–147.

[10] C.G. Harris and W.A.B. Evans, *Extension of numerical quadrature formulae to cater for end point singular behaviours over finite intervals,* Internat. J. Comput. Math., **6** (1977), 219–227.

[11] S. Karlin and W.J. Studden, *Tchebycheff systems with applications in analysis and statistics,* Pure and Applied Mathematics **15**, John Wiley Interscience, New York, 1966.

[12] J. Ma, V. Rokhlin, and S. Wandzura, *Generalized Gaussian quadrature rules for systems of arbitrary functions,* SIAM J. Numer. Anal., **33** (1996), 971–996.

[13] P.C. McCarthy, J.E. Sayre, and B.L.R. Shawyer, *Generalized Legendre polynomials,* J. Math. Anal. Appl., **177** (1993), 530–537.

[14] G.V. Milovanović, *Numerical calculation of integrals involving oscillatory and singular kernels and some applications of quadratures,* Comput. Math. Appl., **36** (1998), 19–39.

[15] G.V. Milovanović, B. Danković, and S. Lj. Rančić, *Some Müntz orthogonal systems,* J. Comput. Appl. Math., **99** (1998), 299–310.

[16] T.J. Stieltjes, *Sur une généralisation de la théorie des quadratures mécaniques,* C.R. Acad. Sci. Paris, **99** (1884), 850–851.

[17] G. Szegő, *Orthogonal polynomials,* Amer. Math. Soc. Colloq. Publ. **23**, 4th ed., Amer. Math. Soc., Providence, RI, 1975.

[18] A.K. Taslakyan, *Some properties of Legendre quasipolynomials with respect to a Müntz system,* Mathematics, Èrevan Univ., Èrevan, **2** (1984), 179–189 (Russian, Armenian summary).

[19] W. Van Assche, *Approximation theory and analytic number theory,* Special Functions and Differential Equations (Madras), 1997, to appear.

[20] J.L. Walsh, *Interpolation and approximation by rational functions in the complex plane,* 5th ed., Amer. Math. Soc. Colloq. Publ. **20**, Amer. Math. Soc., Providence, RI, 1969.

Gradimir V. Milovanović
Faculty of Electronic Engineering
Department of Mathematics
University of Niš
P.O. Box 73
18000 Niš
Serbia, Yugoslavia
E-mail address: grade@elfak.ni.ac.yu

International Series of Numerical Mathematics
Vol. 131,

Positivity of Gauss-Kronrod Formulae for a Certain Ultraspherical Weight Function

Knut Petras

Abstract. In this note, we show that for a sufficiently large number of nodes, the weights of the Gauss-Kronrod quadrature formulae with respect to the weight function $w(x) = (1-x^2)^{5/2}$ are positive. Since this weight function is a limit case in a certain sense, we are led to some conjectures concerning the positivity (and continuity) of Gauss-Kronrod quadrature for more general weight functions.

1. Introduction, Main Result, and Conjectures

The numerical calculation of integrals

$$I_w[f] = \int_{-1}^{1} f(x)w(x)\,dx, \qquad f \in C[-1,1],$$

where $w \in L_1[-1,1]$ is a nonnegative, not almost everywhere vanishing weight function, is usually done by using so-called quadrature formulae

$$Q_n[f] = \sum_{\nu=1}^{n} a_\nu f(x_\nu).$$

The corresponding error functional R_n is defined by

$$R_n[f] := I_w[f] - Q_n[f]$$

and the algebraic degree of exactness by

$$\deg(Q_n) = \sup\{d \mid R_n[\mathbb{P}_d] = 0\},$$

where $\mathbb{P}_k$ denotes the space of all polynomials of degree less than or equal to k.

A quadrature formula with very favourable properties (see, e.g., the books of Brass [1] and Davis and Rabinowitz [4] or the survey articles of Gautschi [8] and Brass, Fischer and Petras [3]) is the Gaussian quadrature formula

$$Q_n^G[f] = \sum_{\nu=1}^{n} a_\nu^G f(x_\nu^G),$$

which is defined uniquely by $\deg(R_n^G) = 2n-1$ (and there is no quadrature formula Q_n with higher algebraic degree of exactness). In order to estimate R_n^G, i.e., to have an indicator for the achieved accuracy, Kronrod [10] proposed the concept of an extension Q_m of the Gaussian formulae. This extension should fulfill certain requirements:

- It should use the already calculated function values $f(x_\nu^G)$.
- It should be a 'better' quadrature formula than Q_n^G. If one accepts the algebraic degree as a quality measure (and there are indeed results showing that this may be an indicator for the quality), then the extension should have a higher degree of exactness, so that $Q_m - Q_n^G$ might be a reasonable estimate for $R_n^G = I_w - Q_n^G$.
- It should use as few new function values as possible.

These requirements culminate in the following definition.

Definition 1.1. *The quadrature formula* Q_{2n+1}^{GK} *is called Gauss-Kronrod formula if it is of the form*

$$Q_{2n+1}^{GK}[f] = \sum_{\nu=1}^{n} A_\nu f(x_\nu^G) + \sum_{\mu=0}^{n} B_\mu f(y_\mu)$$

and satisfies $\deg(Q_{2n+1}^{GK}) \geq 3n+1$.

If a Gauss-Kronrod formula exists, then the additional nodes y_μ are the zeros of the uniquely defined so-called Stieltjes polynomial

$$E_{n+1}(x, w) = x^{n+1} + q(x), \qquad q \in \mathbb{P}_n,$$

which is orthogonal on $\mathbb{P}_n$ with respect to the (sign changing) weight function $p_n w$ (for a survey on Stieltjes polynomials and Gauss-Kronrod formulae, see Monegato [13]). Here and in the following,

$$p_n(x) = x^n + r(x), \qquad r \in \mathbb{P}_{n-1},$$

denotes the orthogonal polynomial with respect to the weight function w.

Definition 1.1 should yield a reasonable quadrature formula $Q_m = Q_{2n+1}^{GK}$:

1. Q_m should be a continuous linear functional on $C[-1,1]$. This property is equivalent to $x_\nu^G, y_\mu \in [-1,1]$, $\nu = 1, \dots, n$, $\mu = 0, \dots, n$. Since all nodes x_ν^G are automatically in $[-1,1]$, we have to require $y_\mu \in [-1,1]$, $\mu = 0, \dots, n$.
2. For several reasons, Q_m should be positive, i.e., all its weights A_ν, B_μ should be nonnegative. First of all, I_w is a positive functional and it is natural to approximate it by a positive functional. Secondly, positive quadrature formulae have minimal norm among all quadrature formulae of algebraic degree greater than or equal to 0 and the norm is a measure for the sensitivity to input errors. Finally, there are results which show that a high algebraic degree automatically implies favourable error estimates mainly if the quadrature formula is positive (see, e.g., Brass [2], Förster [6] or the author [19]).

Definition 1.2. *We say that a quadrature formula* Q_m *is continuous (positive) if it satisfies Property* 1 (*resp.* 2).

Since we have a sign changing weight function $p_n w$, the Gauss-Kronrod formulae are not necessarily continuous and positive. A necessary condition for positivity is the interlacing property

$$y_0 < x_1 < y_1 < x_2 < \cdots < x_n < y_n,$$

which is also sufficient for the positivity of the B_μ, $\mu = 0, \ldots, n$ (Monegato [11]).

For the standard weight function $w(x) \equiv 1$, Properties 1 and 2 have been proved by Szegö [20] and Monegato [12] respectively. The most frequently investigated generalization of the constant weight function is the ultraspherical,

$$w_\lambda(x) = \left(1 - x^2\right)^{\lambda - 1/2}, \qquad \lambda > -1/2.$$

Considering the whole class of these weight functions, we might get an idea for what kind of more general weight functions Gauss-Kronrod formulae are continuous and positive. The known results are:

Proposition 1.3.

a) *For $\lambda < 0$ and $n = 3$, there are nodes outside $[-1, 1]$* (Szegö [20]).

b) *For $\lambda \in [0, 1]$, we have continuous positive Gauss-Kronrod quadrature for all n* (Szegö [20] *and* Monegato [12]).

c) *For $\lambda \in (0, 2]$, we have continuous Gauss-Kronrod quadrature for all n. Furthermore, the interlacing property holds, and for each $\varepsilon > 0$ there is a number $n_0(\varepsilon)$ such that for $n > n_0(\varepsilon)$ all weights corresponding to nodes contained in $[\varepsilon - 1, 1 - \varepsilon]$ are positive* (Szegö [20] *and* Ehrich [5]).

d) *For $\lambda = 3$ and sufficiently large n, the Gauss-Kronrod quadrature formula is continuous and the nodes satisfy the interlacing property* (Peherstorfer and Petras [18]).

e) *For $\lambda > 3$ and sufficiently large n, the Gauss-Kronrod quadrature formulae are not continuous* (Peherstorfer and Petras [18]).

No concrete result for the case $\lambda \in (2, 3)$ is known, but Proposition 1.3 seems to indicate that we have continuity and interlacing also for these λ. If case d) could be characterized more completely, we could formulate a more complete conjecture about the behaviour for arbitrary λ. Our main result is the following.

Theorem. *For sufficiently large n, the Gauss-Kronrod formula Q_n^{GK} with respect to the weight function w_3 is continuous and positive.*

Numerical results (see Gautschi and Notaris [9]) indicate

Conjecture 1.4. *The statement of our theorem holds for all n.*

'Interpolating' between the cases $\lambda = 2$ and $\lambda = 3$, we get

Conjecture 1.5. *Ultraspherical Gauss-Kronrod quadrature is continuous and positive for all n if and only if $\lambda \in [0, 3]$.*

Gauss-Kronrod quadrature has also been considered for 'smooth' modifications of some special ultraspherical weight functions.

Proposition 1.6.

a) *For*

$$w^{(\pm)}(x) = \frac{\left(1-x^2\right)^{\pm 1/2}}{q(x)}, \qquad q \in \mathbb{P}_m, \quad q(x) > 0 \text{ for } x \in [-1,1],$$

Gauss-Kronrod quadrature is continuous and positive if $n \geq m$ *for* $w^{(+)}$ *and* $n \geq m+2$ *for* $w^{(-)}$ (Notaris [14] *and* Peherstorfer [15]).

b) *For*

$$w(x) = v(x)\sqrt{1-x^2}\,, \qquad v \in C^2[-1,1], \quad v(x) > 0 \text{ for } x \in [-1,1],$$

Gauss-Kronrod quadrature is continuous and positive for all sufficiently large n. *Similar results hold for weight functions of the form* $w(x) = v(x)/\sqrt{1-x^2}$ (*see* Peherstorfer [16] *and* [17]).

Of course, we may expect that 'smooth' modifications of the weight function do not disturb the properties of Gauss-Kronrod quadrature considerably.

Conjecture 1.7. *Proposition* 1.6b *also holds when* w_1 *or* w_0 *are replaced respectively by* w_λ, $\lambda \in [0,3]$. *Continuity of Gauss-Kronrod quadrature formulae does not hold for sufficiently large* n *and smoothly* (*in the sense of Proposition* 1.6b) *modified weight functions* w_λ, $\lambda \notin [0,3]$.

Conjecture 1.7 says that reasonable Gauss-Kronrod quadrature might be restricted to a certain boundary behaviour of the weight function.

The proof of the theorem is given in the next section. The first lemma below could be useful also for the more general case $w = w_\lambda$, $\lambda \in (2,3)$, while the second and third lemma might help to prove Conjecture 1.4.

2. Proof of the Main Result

We introduce the notation

$$E_{n+1}^{(\lambda)} := E_{n+1}(\,\cdot\,, w_\lambda) \qquad \text{and} \qquad \tilde{E}_{n+1}^{(\lambda)} := \frac{2^{n+1}\Gamma(n+\lambda+1)}{\sqrt{\pi}\,\Gamma(n+2\lambda)} E_{n+1}^{(\lambda)}$$

(the polynomials $\tilde{E}_{n+1}^{(\lambda)}$ are used by Ehrich [5] while the $E_{n+1}^{(\lambda)}$ are used in [18]). Furthermore, $P_n^{(\lambda)}$ denotes the nth ultraspherical polynomial (see Szegö [21, pp. 80 ff.]).

Lemma 2.1. *Denote by* $x_\nu = x_\nu^G$, $\nu = 1, \ldots, n$, *the nodes of the Gaussian formula with respect to the weight function* w_λ, $\lambda > 1$. *Let the zeros of* $\tilde{E}_{n+1}^{(\lambda)}$ *be in the interval* $[-1,1]$ *and interlace with the* x_ν. *If*

$$\left[\tilde{E}_{n+1}^{(\lambda)}(x_\nu)\right]^2 \geq \frac{2^{2-2\lambda}(n+\lambda)\Gamma(n+1)}{\pi\Gamma(n+2\lambda)}\left(1-x_\nu^2\right)^{1-\lambda}, \qquad \nu = 1, \ldots, n,$$

then the corresponding Gauss-Kronrod quadrature formula is positive.

Proof. We have

$$A_\nu = a_\nu^G + \frac{2^{2-2\lambda}\sqrt{\pi}}{\Gamma(\lambda) P_n^{(\lambda)'}(x_\nu)\tilde{E}_{n+1}^{(\lambda)}(x_\nu)}$$

(see, e.g., Ehrich [5]). The first summand on the right-hand side is positive while the second is negative by the interlacing property. Furthermore,

$$a_\nu^G = \frac{2^{2-2\lambda}\pi\Gamma(n+2\lambda)}{\Gamma(\lambda)^2\Gamma(n+1)(1-x_\nu^2)\left[P_n^{(\lambda)'}(x_\nu)\right]^2} > \frac{\pi}{n+\lambda}\left(1-x_\nu^2\right)^\lambda \tag{1}$$

(for the equation, see Szegö [21, Eq. (15.3.2)] and for the inequality, see Förster and Petras [7, Corollary 1]). The inequality $A_\nu > 0$ is thus equivalent to

$$\left|\frac{1}{\tilde{E}_{n+1}^{(\lambda)}(x_\nu)}\right| < \left|\frac{\sqrt{\pi}\,\Gamma(n+2\lambda)}{\Gamma(\lambda)\Gamma(n+1)(1-x_\nu^2)P_n^{(\lambda)'}(x_\nu)}\right|. \tag{2}$$

The inequality in (1) provides the bound

$$\left|\frac{1}{P_n^{(\lambda)'}(x_\nu)}\right| > \Gamma(\lambda)2^{\lambda-1}\sqrt{\frac{\Gamma(n+1)\left(1-x_\nu^2\right)^{\lambda+1}}{(n+\lambda)\Gamma(n+2\lambda)}}\ .$$

Inserting this on the right-hand side of (2), we obtain the lemma. □

Now we restrict consideration to the case $\lambda = 3$.

Lemma 2.2. *Let $x_\nu = -\cos\theta_\nu$, $\theta_\nu \in [0,\pi]$, and*

$$\delta_n(z) = z^4 - 2\frac{n+5}{n+2}z^2 + \frac{(n+5)(n+4)}{(n+2)(n+1)}\,.$$

Denote furthermore by y_k, $1 \le k \le 4$, the zeros of δ_n and define $g_n \in \mathbb{P}_3$ by the interpolatory conditions

$$2g_n\left(\frac{y_k^2+1}{2y_k}\right) = \delta_n(y_k^{-1})y_k^{-n-1}, \qquad 1 \le k \le 4.$$

Then we have for $\theta \in \{\theta_1, \dots, \theta_n\}$ that

$$\left[2^n\frac{(n+2)(n+1)}{(n+5)(n+4)}E_{n+1}^{(3)}(\cos\theta)\cdot|\delta_n(e^{i\theta})|^2 + g_n(\cos\theta)\right]^2$$
$$= \frac{16}{(n+2)^2(n+1)^2}\left\{\left[(n^2+6n+11)\sin^2\theta - 3\right]^2 + 9(n+3)^2\sin^2\theta\cos^2\theta\right\}.$$

Proof. We have

$$2^n \frac{(n+2)(n+1)}{(n+5)(n+4)} E_{n+1}^{(3)}(\cos\theta) \cdot |\delta_n(e^{i\theta})|^2 + g_n(\cos\theta)$$
$$= \left(-4\frac{n+4}{n+1}\sin^2\theta + \frac{12}{(n+2)(n+1)}\cos^2\theta\right)\cos(n+3)\theta$$
$$+6\frac{n+3}{(n+2)(n+1)}\sin 2\theta \sin(n+3)\theta \qquad (3)$$
$$= \frac{-4}{(n+2)(n+1)}\Big\{\left[(n^2+6n+11)\sin^2\theta - 3\right]\cos(n+3)\theta$$
$$-3(n+3)\sin\theta\cos\theta\sin(n+3)\theta\Big\}$$

(the first equation has been stated in the proof of Theorem 2 in [18], while the second follows by an elementary calculation). Furthermore,

$$\sin^5\theta \cdot P_n^{(3)}(\cos\theta)$$
$$= \frac{\text{const}}{(n+2)(n+1)}\Big\{\left[(n^2+6n+11)\sin^2\theta - 3\right]\sin(n+3)\theta \qquad (4)$$
$$+3(n+3)\sin\theta\cos\theta\cos(n+3)\theta\Big\}$$

(see also the proof of Theorem 2 in [18]). Adding the square of the right-hand side in (3) and of 4/const times the right-hand side in (4), we obtain the lemma, since $P_n^{(3)}(\cos\theta_\nu) = 0$. □

By elementary calculation, we obtain

$$|\delta_n(e^{i\theta})|^2 = 16\frac{(n+5)(n+4)}{(n+2)(n+1)} \cdot \left[\sin^4\theta + \frac{3\sin^2\theta}{(n+2)(n+4)} + \frac{9}{(n+1)(n+2)(n+4)(n+5)}\right].$$

The preceding lemmas yield, after some calculation, the following result.

Lemma 2.3. *Let the assumptions of Lemmas* 2.1 *and* 2.2 *be satisfied. Then the Gauss-Kronrod formula* Q_{2n+1}^{GK} *is positive if for all* $\theta \in \{\theta_1, \dots, \theta_n\}$

$$\frac{2}{(n+2)(n+1)}\sqrt{\left[(n^2+6n+11)\sin^2\theta - 3\right]^2 + 9(n+3)^2\sin^2\theta\cos^2\theta}$$
$$\geq \sqrt{\frac{(n+5)(n+4)}{(n+2)(n+1)}}\left[\sin^2\theta + \frac{3}{(n+2)(n+4)} + \frac{9\sin^{-2}\theta}{(n+1)(n+2)(n+4)(n+5)}\right] + \frac{|g_n(\cos\theta)|}{2}.$$

Proof of the theorem. We leave the path of explicit calculations and turn to the asymptotic behaviour. In the following, we always assume $\theta \in \{\theta_1, \dots, \theta_n\}$. Note that $\theta \geq \frac{11\pi}{6(n+3)}$ (see Peherstorfer and Petras [18, proof of Theorem 2]), which means

$$S(\theta) := n^2 \sin^2\theta \geq 33$$

for sufficiently large n. In [18, proof of Theorem 2] it has also been shown that

$$g_n(\cos\theta) = \frac{8}{e\sqrt{e}} \sin\left(\frac{\sqrt{3}}{2} - \frac{\pi}{6}\right) \sin^2\theta - \frac{6}{n^2} \sin\left(\frac{\sqrt{3}}{2} + \frac{\pi}{6}\right) \cdot \begin{Bmatrix} 1 & \text{if } n \text{ is odd} \\ \cos\theta & \text{if } n \text{ is even} \end{Bmatrix} + O\left(\frac{\sin^2\theta}{n} + \frac{1}{n^3}\right).$$

Since the two main terms on the right-hand side of this equation have opposite sign and the modulus of the first one exceeds asymptotically that of the second, we have for sufficiently large n that

$$|g_n(\cos\theta)| \leq \frac{8\sin\left(\frac{\sqrt{3}}{2} - \frac{\pi}{6}\right)}{e\sqrt{e}} \sin^2\theta + O\left(\frac{\sin^2\theta}{n} + \frac{1}{n^3}\right) \leq 0.6 \sin^2\theta.$$

Using this and squaring both sides of the inequality in Lemma 2.3, we see that we have to satisfy

$$4\sin^4\theta + \frac{12}{n^2}\sin^2\theta + \frac{36}{n^4} + O\left(\sum_{k=0}^{2} \frac{\sin^{4-2k}\theta}{n^{1+2k}}\right)$$

$$\geq 1.69 \sin^4\theta + \frac{7.8}{n^2}\sin^2\theta + \frac{32.4}{n^4} + \frac{54}{n^6 \sin^2\theta} + \frac{81}{n^8 \sin^4\theta} + O\left(\sum_{k=0}^{4} \frac{\sin^{4-2k}\theta}{n^{1+2k}}\right),$$

i.e.,

$$2.31 + \frac{4.2}{S(\theta)} + \frac{3.6}{S^2(\theta)} \geq \frac{54}{S^3(\theta)} + \frac{81}{S^4(\theta)} + \frac{1}{n} O\left(\sum_{k=0}^{4} S^{-k}(\theta)\right).$$

Since we may assume $S(\theta) \geq 33$ (see above), the theorem is proved. □

References

[1] H. Brass, *Quadraturverfahren,* Studia Mathematica, Skript 3, Vandenhoeck & Ruprecht, Göttingen, 1977.

[2] H. Brass, *Bounds for Peano kernels,* in: H. Brass and G. Hämmerlin, Eds., *Numerical integration IV*, Internat. Ser. Numer. Math. **112**, Birkhäuser, Basel, 1993, 39–55.

[3] H. Brass, J.-W. Fischer, and K. Petras, *The Gaussian quadrature method,* Abh. Braunschweig. Wiss. Ges., **47** (1996), 115–150.

[4] P.J. Davis and P. Rabinowitz, *Methods of numerical integration,* 2d ed., Academic Press, Orlando, 1984.

[5] S. Ehrich, *Asymptotic properties of Stieltjes polynomials and Gauss-Kronrod quadrature formulae,* J. Approx. Theory, **82** (1995), 287–303.

[6] K.-J. Förster, *On quadrature formulae near Gaussian quadrature,* in: T.O. Espelid and A. Genz, Eds., *Numerical integration*, NATO Adv. Sci. Inst. Ser. C Math. Phys. Sci. **357**, Kluwer, Dordrecht, 1992, 45–54.

[7] K.-J. Förster and K. Petras, *On estimates for the weights in Gaussian quadrature in the ultraspherical case,* Math. Comp., **55** (1990), 243–264.

[8] W. Gautschi, *A survey of Gauss-Christoffel quadrature formulae,* in: P.L. Butzer and F. Fehér, Eds., *E.B. Christoffel*, Birkhäuser, Basel, 1981, 72–147.

[9] W. Gautschi and S.E. Notaris, *An algebraic study of Gauss-Kronrod quadrature formulae for Jacobi weight functions,* Math. Comp., **51** (1988), 231–248.

[10] A.S. Kronrod, *Nodes and weights for quadrature formulae. Sixteen-place tables* (Russian), Nauka, Moscow, 1964; English transl., Consultants Bureau, New York, 1965.

[11] G. Monegato, *A note on extended Gaussian quadrature rules,* Math. Comp., **30** (1976), 812–817.

[12] G. Monegato, *Positivity of the weights of extended Gauss-Legendre quadrature rules,* Math. Comp., **32** (1978), 243–245.

[13] G. Monegato, *Stieltjes polynomials and related quadrature rules,* SIAM Rev., **24** (1982), 137–158.

[14] S.E. Notaris, *Gauss-Kronrod quadrature formulae for weight functions of Bernstein-Szegö type II,* J. Comput. Appl. Math., **29** (1990), 161–169.

[15] F. Peherstorfer, *Weight functions admitting repeated positive Kronrod quadrature,* BIT, **30** (1990), 145–151.

[16] F. Peherstorfer, *On the asymptotic behaviour of functions of the second kind and Stieltjes polynomials and on the Gauss-Kronrod quadrature formulas,* J. Approx. Theory, **70** (1992), 156–190.

[17] F. Peherstorfer, *Stieltjes polynomials and functions of the second kind,* J. Comput. Appl. Math., **65** (1995), 319–338.

[18] F. Peherstorfer and K. Petras, *Ultraspherical Gauss-Kronrod quadrature is not possible for $\lambda > 3$,* submitted for publication.

[19] K. Petras, *One sided L1-approximation and bounds for Peano kernels,* in: T.O. Espelid and A. Genz, Eds., *Numerical integration*, NATO Adv. Sci. Inst. Ser. C Math. Phys. Sci. **357**, Kluwer, Dordrecht, 1992, 165–174.

[20] G. Szegö, *Über gewisse Polynome, die zu einer oszillierenden Belegungsfunktion gehören,* Math. Ann., **110** (1935), 501–513.

[21] G. Szegö, *Orthogonal polynomials,* Amer. Math. Soc. Colloq. Publ. **23**, Amer. Math. Soc., Providence, RI, 1939.

Knut Petras
Mathematisches Institut
Ludwig-Maximilians-Universität München
Theresienstraße 39
D-80333 München, Germany
E-mail address: petras@rz.mathematik.uni-muenchen.de

International Series of Numerical Mathematics
Vol. 131, © 1999 Birkhäuser Verlag Basel/Switzerland

A Christoffel-Darboux-Type Formula for Szegő Polynomials and Polynomial Evaluation

Michael-Ralf Skrzipek

Abstract. Polynomials Φ_n, $n \geq 0$, orthogonal on the complex unit circle satisfy a recurrence relation. If we shift their recurrence coefficients, we obtain the associated polynomials, which can be modified by changing the initialization. We prove a dual recurrence relation and a mixed Christoffel-Darboux-type formula, which expresses the derivative of an orthogonal polynomial in terms of orthogonal polynomials and the modified associated polynomials.

We exhibit connections between polynomial evaluation and these modified associated polynomials by showing, for example, how a polynomial q_n expanded in terms of the Φ_ν, $\nu = 0, \dots, n$, and its derivatives can be evaluated.

1. Introduction

For a system of real monic orthogonal polynomials $\{P_n\}_{n \geq -1}$, $P_{-1} = 0$, and its system of associated polynomials $\{P_n^{(k)}\}_{n \geq -1}$, $k \geq 0$, the following formula [8] is very useful:

$$\begin{aligned}\sum_{j=0}^{n} P_{n-j}^{(j+1)}(x) P_j(y) &= \begin{cases} \dfrac{P_{n+1}(x) - P_{n+1}(y)}{x-y}, & x \neq y, \\ P'_{n+1}(x), & x = y, \end{cases} \\ &= \sum_{j=0}^{n} P_{n-j}^{(j+1)}(y) P_j(x). \end{aligned} \tag{1.1}$$

With this formula we have shown some connections between associated polynomials and polynomial evaluation. Furthermore, we have derived an algorithm for evaluating derivatives of linear combinations of real orthogonal polynomials [6].

The proof of (1.1) uses a dual recurrence relation for real orthogonal polynomials. For Szegő polynomials Φ_n, $n \geq 0$, a dual recurrence formula is valid, too [5]. But with this one, we cannot derive a simple formula such as (1.1). This has the consequence that our abovementioned results for the real case cannot be

---

1991 *Mathematics Subject Classification.* 42C15, 65B10.

Key words and phrases. Orthogonal polynomials, Szegő polynomials, associated polynomials, dual recurrence relation, Christoffel-Darboux formula, Clenshaw algorithm.

carried over directly to the complex case. Thus we make a compromise. We want to write the difference quotient resp. the derivative of Φ_n as a linear combination of products of polynomials which should be comparable to the left side of (1.1) in its simplicity. This can be done if we modify the associated Szegő polynomials.

As an application we derive in §§4 and 5 algorithms for evaluating polynomials expanded in terms of (Bernstein-) Szegő polynomials, and their derivatives. These results generalize the Horner scheme for $\Phi_\nu(z) = z^\nu$. Because we use the Geronimus recurrence relation, our results are closely related to the Clenshaw algorithm for the real case (see, e.g., [6], [9]). (Both algorithms use a three-term recurrence relation.)

We first need some results about associated Szegő polynomials and a modification thereof. After some preliminaries in §2 we derive in §3 a dual recursion and a mixed Christoffel-Darboux formula for these polynomials. In §4 we give connections between these formulas and the evaluation of polynomials and their derivatives. The final §5 comments on details of implementation.

2. Recurrence Relations for Szegő Polynomials

Let $\{\Phi_n\}_{n\in\mathbb{N}_0}$ be a sequence of polynomials orthogonal on the unit circle $\mathbb{T} := \{z \in \mathbb{C} \; : \; |z| = 1\}$ with respect to μ, i.e.,

$$\int_{\mathbb{T}} \Phi_n(z)\overline{\Phi_m(z)}\, d\mu(\theta) = g_m \delta_{m,n}\,, \quad z = \exp(i\theta), \quad g_m \neq 0, \quad m,n \in \mathbb{N}_0.$$

These polynomials are called Szegő polynomials. Here, μ denotes a positive measure on $[-\pi, \pi)$ with infinite support.

By $\Phi_n^\star(z) := z^n\overline{\Phi}_n(1/z)$ we denote the reciprocal polynomials. Then $\{(\Phi_n, \Phi_n^\star)^T\}_{n\geq 0}$ is a solution of the recurrence (cf. [2], [7])

$$\begin{pmatrix} u_{n+1} \\ v_{n+1} \end{pmatrix} = A_n(z) \begin{pmatrix} u_n \\ v_n \end{pmatrix}, \qquad A_n(z) = \begin{pmatrix} z & -a_n \\ -\overline{a}_n z & 1 \end{pmatrix} \tag{2.1}$$

for $n \geq 0$, with $(\Phi_0, \Phi_0^\star)^T = (1,1)^T$ and $|a_n| < 1$ for all $n \in \mathbb{N}_0$ (see, e.g., [2, p. 5 and the remark on p. 3]). Obviously, we have $a_n = -\Phi_{n+1}(0)$. If $a_\nu = 0$ for all $\nu > j$ for some $j \geq -1$, we call these polynomials Bernstein-Szegő polynomials.

The k-associated polynomials $\Phi_n^{(k)}$ are defined by a k-shift in the reflection coefficients a_n of (2.1), i.e., they satisfy

$$\begin{pmatrix} u_{n+1} \\ v_{n+1} \end{pmatrix} = A_{n+k}(z) \begin{pmatrix} u_n \\ v_n \end{pmatrix} \tag{2.2}$$

with initial value $(\Phi_0^{(k)}, \Phi_0^{(k)^\star})^T = (1,1)^T$. For $k = 0$ we omit the superscripts. The polynomials Φ_n and $\Phi_{n-k}^{(k)}$ for $n \geq k$ satisfy the same recurrence (2.1).

If $a_{n+k-1} = \cdots = a_{n+k-m} = 0 \neq a_{n+k-m-1}$, $n - m \geq 0$, we have from (2.2)

$$\begin{pmatrix} u_n \\ v_n \end{pmatrix} = \prod_{\nu=1}^{m} A_{n+k-\nu}(z) \begin{pmatrix} u_{n-m} \\ v_{n-m} \end{pmatrix} = \begin{pmatrix} z^m u_{n-m} \\ v_{n-m} \end{pmatrix}. \tag{2.3}$$

The system (2.1) of two recurrences for $(\Phi_n, \Phi_n^\star)^T$ can be reformulated as a three-term recursion for calculating Φ_{n+1} [2, p. 4],

$$a_{n-1}u_{n+1}(z) = (za_{n-1} + a_n)u_n(z) - a_n(1 - |a_{n-1}|^2)zu_{n-1}(z), \quad n = 1, 2, \ldots . \tag{2.4}$$

The elements of $\{\Phi_n\}_{n\geq 0}$ satisfy this so-called Geronimus recurrence (2.4) with initial values $\Phi_0(z) := 1$, $\Phi_1(z) = z - a_0$. The k-associated version is given by

$$a_{n+k-1}u_{n+1}(z) = (za_{n+k-1} + a_{n+k})u_n(z) - a_{n+k}(1 - |a_{n+k-1}|^2)zu_{n-1}(z), \tag{2.5}$$

for $n \geq 1$, which is satisfied by the elements of $\{\Phi_n^{(k)}\}_{n\geq 0}$ with initial values $\Phi_0^{(k)}(z) := 1$, $\Phi_1^{(k)}(z) = z - a_k$, $k \geq 0$. Obviously, (2.5) breaks down if $a_{n+k-1} = 0$. In this sense, (2.4) and (2.5) are "weaker" than (2.1) resp. (2.2). But by using (2.3) the proof of (2.4) in [2] can be modified to obtain a formula also in this case.

3. A Modification of Associated Polynomials

Inspection of the proof of (1.1) given in [8] suggests that an analogous formula for Szegő polynomials may be valid if the recurrence (2.1) resp. (2.4) could be started one step earlier. Of course, this cannot be done in (2.1), because reciprocal polynomials are not defined for polynomial degree -1. But we can take (2.4). Since $a_{-1} = -1$, (2.4) remains formally true for $n = 0$, although u_{-1} is not defined. We use this sometimes for the purpose of simplifying formulas.

If (2.5) is started at $n = 0$, with $\widehat{\Phi}_{-1}^{(k)}(z) = 0$, $\widehat{\Phi}_0^{(k)}(z) = 1$, then for $a_{k-1} \neq 0$ one gets a sequence $\{\widehat{\Phi}_n^{(k)}\}_{n\geq -1}$ with

$$a_{k-1}\widehat{\Phi}_{n+1}^{(k)}(0) = a_{n+k} \quad \text{or} \quad -\hat{a}_{n+k} := \widehat{\Phi}_{n+1}^{(k)}(0) = \frac{a_{n+k}}{a_{k-1}} .$$

Since $\Phi_{n+1}^{(k)}(0) = -a_{n+k}$, $|a_{k-1}| < 1$, $k > 0$, we have $\widehat{\Phi}_n^{(k)} \neq \Phi_n^{(k)}$ for $k > 0$, $n \in \mathbb{N}$. If $a_{k-1} = 0$, then for $n = 0$ we have $0 = a_{k-1}\widehat{\Phi}_1^{(k)}(0) = a_k$. Increasing k, we would have $a_\nu = 0$ for all $\nu \geq k-1$, which is only valid if $\Phi_{k+\nu} = z^{\nu+1}\Phi_{k-1}$, i.e., for Bernstein-Szegő polynomials. In this case $\widehat{\Phi}_n^{(k)}$ is not defined. Thus, in the following we use $\widehat{\Phi}_n^{(k)}$ only if $a_{k-1} \neq 0$. To derive the desired Christoffel-Darboux-type identity, we need the following lemma.

Lemma 3.1. [5] *If $a_0 \neq 0$ and $l \geq 1$, we have*

$$\widehat{\Phi}_{l-1}^{(1)}(z)\Phi_{l-1}(z) - \widehat{\Phi}_{l-2}^{(1)}(z)\Phi_l(z) = \frac{a_{l-1}}{a_0} z^{l-1} \prod_{\nu=0}^{l-2} \left(1 - |a_\nu|^2\right) . \tag{3.1}$$

Remark. *There holds $\prod_{\nu=0}^{l-2}(1 - |a_\nu|^2) = \kappa_0^2/\kappa_{l-1}^2$, where κ_m denotes the leading coefficient of the orthonormal polynomial $\kappa_m\Phi_m$ [3].*

If $a_0 \neq 0$, then by Lemma 3.1 the Wronski determinant of two solutions $u_n = \Phi_n$, $v_n = \widehat{\Phi}^{(1)}_{n-1}$ of (2.4),

$$\begin{vmatrix} u_n(z) & v_n(z) \\ u_{n+1}(z) & v_{n+1}(z) \end{vmatrix} = \Phi_n(z)\widehat{\Phi}^{(1)}_n(z) - \Phi_{n+1}(z)\widehat{\Phi}^{(1)}_{n-1}(z)$$
$$= \begin{cases} 1, & n = 0, \\ \frac{a_n}{a_0} z^n \prod_{\nu=0}^{n-1}(1 - |a_\nu|^2), & a_0 \neq 0,\ n \geq 1, \end{cases}$$

is nonzero for $z \neq 0 \neq a_n$. This means that u_n and v_n are linearly independent in this case, and every solution of the recurrence (2.4) can be written as a linear combination of u_n and v_n:

$$\widehat{\Phi}^{(l)}_{n-l} = A\Phi_n + B\widehat{\Phi}^{(1)}_{n-1}, \quad n \geq l-1,$$

where A and B can be determined by setting $n = l$ resp. $n = l - 1$.

We obtain

$$\widehat{\Phi}^{(l)}_{n-l} = \frac{\Phi_{l-1}\widehat{\Phi}^{(1)}_{n-1} - \widehat{\Phi}^{(1)}_{l-2}\Phi_n}{\Phi_{l-1}\widehat{\Phi}^{(1)}_{l-1} - \widehat{\Phi}^{(1)}_{l-2}\Phi_l}.$$

If we choose $l = 2$ and use Lemma 3.1, we get

$$\Phi_n(z) = \Phi_1(z)\widehat{\Phi}^{(1)}_{n-1}(z) - \frac{a_1}{a_0}(1 - |a_0|^2)z\widehat{\Phi}^{(2)}_{n-2}(z) \text{ for } n \geq 1,\ a_0 \neq 0.$$

For $a_k \neq 0$ we can formulate the k-associated version of this formula. We get

$$\Phi^{(k)}_n(z) = \Phi^{(k)}_1(z)\widehat{\Phi}^{(k+1)}_{n-1}(z) - \frac{a_{k+1}}{a_k}(1 - |a_k|^2)z\widehat{\Phi}^{(k+2)}_{n-2}(z), \quad n \geq 1. \tag{3.2}$$

If we use $\Phi_\nu = \widehat{\Phi}^{(0)}_\nu$, $\nu \in \mathbb{N}_0$, and remember that $\widehat{\Phi}^{(k)}_n$ can be used only for $a_{k-1} \neq 0$, we get in the same manner

$$\widehat{\Phi}^{(k)}_n(z) = \widehat{\Phi}^{(k)}_1(z)\widehat{\Phi}^{(k+1)}_{n-1}(z) - \frac{a_{k+1}}{a_k}(1 - |a_k|^2)z\widehat{\Phi}^{(k+2)}_{n-2}(z), \quad n \geq 1, \tag{3.3}$$

if $a_{k-1}a_k a_{k+1} \neq 0$. In this case we call (3.3) a dual recurrence to (2.5).

We now substitute (3.3) in (3.2). We obtain

$$\begin{aligned} \widehat{\Phi}^{(k)}_n &= \widehat{\Phi}^{(k)}_1\widehat{\Phi}^{(k+1)}_{n-1} + \Phi^{(k)}_n - \Phi^{(k)}_1\widehat{\Phi}^{(k+1)}_{n-1} \\ &= (\widehat{\Phi}^{(k)}_1 - \Phi^{(k)}_1)\widehat{\Phi}^{(k+1)}_{n-1} + \Phi^{(k)}_n. \end{aligned}$$

This kind of substitution can be repeated for $\widehat{\Phi}^{(k+1)}_{n-1}, \widehat{\Phi}^{(k+2)}_{n-2}, \ldots$. By induction we get

$$\widehat{\Phi}^{(k)}_n = \sum_{\nu=0}^{p-1}\prod_{\mu=0}^{\nu-1}(\widehat{\Phi}^{(\mu+k)}_1 - \Phi^{(\mu+k)}_1)\Phi^{(\nu+k)}_{n-\nu} + \prod_{\mu=0}^{p-1}(\widehat{\Phi}^{(\mu+k)}_1 - \Phi^{(\mu+k)}_1)\widehat{\Phi}^{(p+k)}_{n-p}$$

for $p = 0, \dots, n$. Since $\prod_{\mu=0}^{\nu-1}(\widehat{\Phi}_1^{(\mu+k)} - \Phi_1^{(\mu+k)}) = \frac{a_{\nu+k-1}}{a_{k-1}} \prod_{\mu=0}^{\nu-1}(1 + a_{\mu+k-1})$, we have for $p = n$

$$\widehat{\Phi}_n^{(k)} = \frac{1}{a_{k-1}} \sum_{\nu=0}^{n} a_{\nu+k-1} \prod_{\mu=0}^{\nu-1} (1 + a_{\mu+k-1}) \Phi_{n-\nu}^{(\nu+k)}.$$

This shows again that $\widehat{\Phi}_n^{(k)}$ is only defined if $a_{k-1} \neq 0$.

With (2.4) and (3.3) we can prove a Christoffel-Darboux-type formula:

Theorem 3.1. 1) *If $a_k \neq 0$ for $k = -1, \dots, n$, then*

$$\sum_{k=0}^{n} \left(\Phi_k(w) - \frac{a_k}{a_{k-1}} (1 - |a_{k-1}|^2) \Phi_{k-1}(w) \right) \widehat{\Phi}_{n-k}^{(k+1)}(z) = \begin{cases} \dfrac{\Phi_{n+1}(w) - \Phi_{n+1}(z)}{w - z} & \text{for } w \neq z, \\ \dfrac{d}{dw} \Phi_{n+1}(w) & \text{for } w = z. \end{cases} \tag{3.4}$$

2) *If $a_{-1}, \dots, a_j \neq 0$, $a_{j+1} = \cdots = a_n = 0$ for some j, $-1 \leq j \leq n$, then*

$$\sum_{k=0}^{j} \left(\Phi_k(w) - \frac{a_k}{a_{k-1}} (1 - |a_{k-1}|^2) \Phi_{k-1}(w) \right) \widehat{\Phi}_{n-k}^{(k+1)}(z) + \sum_{k=j+1}^{n} \Phi_k(w) \Phi_{n-k}^{(k+1)}(z) = \begin{cases} \dfrac{\Phi_{n+1}(w) - \Phi_{n+1}(z)}{w - z} & \text{for } w \neq z, \\ \dfrac{d}{dw} \Phi_{n+1}(w) & \text{for } w = z. \end{cases} \tag{3.5}$$

Proof. From (2.3) we have for $n = k \geq 1$

$$z a_{k-1} \Phi_k(z) = a_{k-1} \Phi_{k+1}(z) - a_k \Phi_k(z) + a_k (1 - |a_{k-1}|^2) z \Phi_{k-1}(z), \tag{3.6}$$

which also holds for $k = 0$ since $a_{-1} = -1$. From (3.3), if $a_k \neq 0$ and if n is replaced by $n - k + 1$, we get

$$\begin{aligned} w a_{k-1} \widehat{\Phi}_{n-k}^{(k+1)}(w) &= a_{k-1} \widehat{\Phi}_{n-k+1}^{(k)}(w) - a_k \widehat{\Phi}_{n-k}^{(k+1)}(w) \\ &+ a_{k-1} \frac{a_{k+1}}{a_k} (1 - |a_k|^2) w \widehat{\Phi}_{n-k-1}^{(k+2)}(w). \end{aligned}$$

From these formulas we obtain for $a_{k-1} a_k \neq 0$

$$\begin{aligned} (z - w) \Phi_k(z) \widehat{\Phi}_{n-k}^{(k+1)}(w) &= \Phi_{k+1}(z) \widehat{\Phi}_{n-k}^{(k+1)}(w) - \Phi_k(z) \widehat{\Phi}_{n-(k-1)}^{(k)}(w) \\ + \frac{a_k}{a_{k-1}} (1 - |a_{k-1}|^2) z \Phi_{k-1}(z) \widehat{\Phi}_{n-k}^{(k+1)}(w) &- \frac{a_{k+1}}{a_k} (1 - |a_k|^2) w \Phi_k(z) \widehat{\Phi}_{n-(k+1)}^{(k+2)}(w). \end{aligned}$$

Let $a_0, \dots, a_n \neq 0$. Since $-a_{-1} = \Phi_0(0) = 1$, $\widehat{\Phi}_{-1}^{(k+1)}(z) = 0$, $\Phi_\nu^{(0)} = \Phi_\nu$, $\nu \in \mathbb{N}_0$, we get

$$(z-w)\sum_{k=0}^{n} \Phi_k(z)\widehat{\Phi}_{n-k}^{(k+1)}(w) = \Phi_{n+1}(z) - \Phi_{n+1}(w)$$
$$+(z-w)\sum_{k=0}^{n} \frac{a_k}{a_{k-1}}(1-|a_{k-1}|^2)\Phi_{k-1}(z)\widehat{\Phi}_{n-k}^{(k+1)}(w),$$

from which (3.4) follows.

Now let $a_{-1}, \dots, a_j \neq 0$, $a_{j+1} = \dots = a_n = 0$, where $-1 \leq j \leq n$. With

$$\Phi_{j+1+l}(z) = z^l \Phi_{j+1}(z), \quad 0 \leq l \leq n-j-1,$$
$$\Phi_{n-k}^{(k+1)}(z) = z^{n-k} \quad \text{for} \quad k = j+1, \dots, n,$$

we obtain

$$\sum_{k=j+1}^{n} \Phi_k(z)\Phi_{n-k}^{(k+1)}(w) = \Phi_{j+1}(z) \sum_{k=j+1}^{n} z^{k-j-1}\Phi_{n-k}^{(k+1)}(w)$$
$$= \Phi_{j+1}(z) \sum_{k=j+1}^{n} z^{k-j-1} w^{n-k}.$$

With $\widehat{\Phi}_{n-k}^{(k+1)}(z) = z^{n-j}\widehat{\Phi}_{j-k}^{(k+1)}(z)$ for $n, j \geq k$, we have for $z \neq w$

$$\sum_{k=0}^{j} \left(\Phi_k(z) - \frac{a_k}{a_{k-1}}(1-|a_{k-1}|^2)\Phi_{k-1}(z)\right) \widehat{\Phi}_{n-k}^{(k+1)}(w) + \sum_{k=j+1}^{n} \Phi_k(z)\Phi_{n-k}^{(k+1)}(w)$$
$$= w^{n-j}\frac{\Phi_{j+1}(z) - \Phi_{j+1}(w)}{z-w} + \Phi_{j+1}(z) z^{-j-1} w^n \sum_{k=j+1}^{n} \left(\frac{z}{w}\right)^k$$
$$= \dots = \frac{z^{n-j}\Phi_{j+1}(z) - w^{n-j}\Phi_{j+1}(w)}{z-w}$$
$$= \frac{\Phi_{n+1}(z) - \Phi_{n+1}(w)}{z-w}.$$

For $z = w$ we get

$$\sum_{k=0}^{j} \left(\Phi_k(z) - \frac{a_k}{a_{k-1}}(1-|a_{k-1}|^2)\Phi_{k-1}(z)\right) \widehat{\Phi}_{n-k}^{(k+1)}(z) + \sum_{k=j+1}^{n} \Phi_k(z)\Phi_{n-k}^{(k+1)}(z)$$
$$= z^{n-j}\Phi'_{j+1}(z) + \Phi_{j+1}(z)(n-j)z^{n-j-1}$$
$$= \frac{d}{dz}\left(z^{n-j}\Phi_{j+1}(z)\right) = \Phi'_{n+1}(z),$$

which proves the theorem. □

4. Evaluation of Polynomials and Their Derivatives

Let

$$q_n = \sum_{\nu=0}^{n} \alpha_\nu \Phi_\nu \tag{4.1}$$

be a polynomial expanded in terms of Szegő polynomials Φ_ν. Besides the Horner scheme for $\Phi_\nu(z) = z^\nu$, q_n can be evaluated at $z \in \mathbb{C}$ by using (2.1) [1]. As an application of the (modified) associated polynomials $\widehat{\Phi}_n^{(k)}$, we show that they can be used to derive a polynomial evaluation based on the Geronimus recurrence. As we have seen in the previous section, the use of $\widehat{\Phi}_n^{(k)}$ is problematic if a_{k-1} has a small modulus. This exactly reflects the behaviour of the Geronimus recurrence (2.4). In this section we want to get more insight into this behaviour.

For simplicity, and to make the main ideas more transparent, we assume from now on that $a_{-1}, \dots, a_j \neq 0$, $a_{j+1} = \cdots = a_{n-1} = 0$, for some j, $-1 \le j \le n-1$. If all $a_\nu \neq 0$, we set $j := n$. The general case of arbitrary reflection coefficients a_ν, $|a_\nu| < 1$, can be handled in a similar, though more technical, manner by using the first component of (2.3) instead of (2.4).

We define the following quantities:

$$\begin{aligned}
a_{n+1}^{(1)} &:= 0,\\
a_{n-\nu}^{(1)} &:= a_{n-\nu+1}^{(1)} z + \alpha_{n-\nu}\,, \quad \nu = 0, \dots, n-j-1,\\
a_j^{(1)} &:= a_{j+1}^{(1)}\left(z + \frac{a_j}{a_{j-1}}\right) + \alpha_j,\\
a_{j-\nu}^{(1)} &:= a_{j-\nu+1}^{(1)}\left(z + \frac{a_{j-\nu}}{a_{j-\nu-1}}\right) - a_{j-\nu+2}^{(1)} \frac{a_{j-\nu+1}}{a_{j-\nu}}(1 - |a_{j-\nu}|^2) z + \alpha_{j-\nu},\\
&\quad \nu = 1, \dots, j.
\end{aligned} \tag{4.2}$$

Theorem 4.1. *Let $q_n := \sum_{\nu=0}^{n} \alpha_\nu \Phi_\nu$. Defining $a_\nu^{(1)}$, $\nu = n+1, \dots, 0$, as in (4.2), we have*

$$a_{n-\nu}^{(1)} = \begin{cases} \sum_{\mu=0}^{\nu} \alpha_{n-\nu+\mu} \widehat{\Phi}_\mu^{(n-\nu)}(z), & n-\nu \le j+1,\\ \sum_{\mu=0}^{\nu} \alpha_{n-\nu+\mu} \Phi_\mu^{(n-\nu)}(z), & \text{otherwise.} \end{cases} \tag{4.3}$$

In particular, for $\nu = n$ we have

$$a_0^{(1)} = q_n(z). \tag{4.4}$$

Proof. For $\nu = 0$ we see immediately that $a_n^{(1)} = \alpha_n = \alpha_n \Phi_0^{(n)}(z)$. For $\nu = 1$ we have

Case 1: $n-2 \ge j$, i.e., $a_{n-1} = 0$.

For $n-2 = j$ we have

$$\begin{aligned}
a_{n-1}^{(1)} &= a_n^{(1)} z + \alpha_{n-1} = \alpha_n z \widehat{\Phi}_0^{(n-1)}(z) + \alpha_{n-1} \widehat{\Phi}_0^{(n-2)}(z)\\
&= \alpha_n \widehat{\Phi}_1^{(n-1)}(z) + \alpha_{n-1} \widehat{\Phi}_0^{(n-2)}(z).
\end{aligned}$$

If $n-2>j$, we have

$$\begin{aligned} a_{n-1}^{(1)} &= a_n^{(1)} z + \alpha_{n-1} = \alpha_n z \Phi_0^{(n-1)}(z) + \alpha_{n-1}\Phi_0^{(n-2)}(z) \\ &= \alpha_n \Phi_1^{(n-1)}(z) + \alpha_{n-1}\Phi_0^{(n-2)}(z). \end{aligned}$$

Case 2: $n-1=j$.
Using the dual recurrence formula (3.3), we get

$$\begin{aligned} a_{n-1}^{(1)} = a_n^{(1)} \left(z + \frac{a_{n-1}}{a_{n-2}}\right) + \alpha_{n-1} &= \alpha_n \widehat{\Phi}_0^{(n)}(z) \left(z + \frac{a_{n-1}}{a_{n-2}}\right) + \alpha_{n-1} \\ &= \alpha_n \widehat{\Phi}_1^{(n-1)}(z) + \alpha_{n-1}\widehat{\Phi}_0^{(n)}(z). \end{aligned}$$

Case 3: $n-1<j$.
This case can be reduced to Case 2 since $a_{n+1}^{(1)}=0$:

$$a_{n-1}^{(1)} = a_n^{(1)} \left(z + \frac{a_{n-1}}{a_{n-2}}\right) - a_{n+1}^{(1)} \frac{a_n}{a_{n-1}}(1-|a_{n-1}|^2)z + \alpha_{n-1}.$$

Now suppose the theorem is true for $a_{n-\eta}^{(1)}$, $\eta=0,\dots,\nu$. We proceed by induction from ν to $\nu+1<n$.

Case 1: For $n-\nu-1\geq j+1$ we have $\Phi_\mu^{(n-\nu)}(z)=z^\mu$, $\mu\geq 0$, and thus

$$\begin{aligned} a_{n-\nu-1}^{(1)} &= a_{n-\nu}^{(1)} z + \alpha_{n-\nu-1} = \sum_{\mu=0}^{\nu} \alpha_{n-\nu+\mu} z \Phi_\mu^{(n-\nu)}(z) + \alpha_{n-\nu-1} \\ &= \sum_{\mu=0}^{\nu} \alpha_{n-\nu+\mu} z^{\mu+1} + \alpha_{n-\nu-1} = \sum_{\mu=0}^{\nu} \alpha_{n-\nu+\mu} \Phi_{\mu+1}^{(n-\nu-1)}(z) + \alpha_{n-\nu-1} \\ &= \sum_{\mu=0}^{\nu+1} \alpha_{n+\mu-(\nu+1)} \Phi_\mu^{(n-(\nu+1))}(z). \end{aligned}$$

Case 2: $n-\nu=j+1$. From (3.3) we have

$$\begin{aligned} a_j^{(1)} &= a_{n-\nu-1}^{(1)} = a_{j+1}^{(1)} \left(z + \tfrac{a_j}{a_{j-1}}\right) + \alpha_j \\ &= \sum_{\mu=0}^{\nu} \alpha_{n-\nu+\mu} \widehat{\Phi}_\mu^{(n-\nu)}(z) \left(z + \frac{a_j}{a_{j-1}}\right) + \alpha_j = \sum_{\mu=0}^{\nu} \alpha_{n-\nu+\mu} \widehat{\Phi}_{\mu+1}^{(n-\nu-1)}(z) + \alpha_j \\ &= \sum_{\mu=0}^{\nu+1} \alpha_{n-\mu-(\nu+1)} \widehat{\Phi}_\mu^{(n-(\nu+1))}(z). \end{aligned}$$

Case 3: $n - \nu < j + 1$. Here we have

$$
\begin{aligned}
a^{(1)}_{j-(\nu+1)} &= a^{(1)}_{j-\nu}\left(z + \frac{a_{j-\nu-1}}{a_{j-\nu-2}}\right) - a^{(1)}_{j-\nu+1}\frac{a_{j-\nu}}{a_{j-\nu-1}}(1 - |a_{j-\nu-1}|^2)z + \alpha_{j-\nu-1} \\
&= \sum_{\mu=0}^{\nu} \alpha_{j-\nu+\mu}\widehat{\Phi}^{(j-\nu)}_{\mu}(z)\left(z + \frac{a_{j-\nu-1}}{a_{j-\nu-2}}\right) \\
&\quad - \sum_{\mu=0}^{\nu-1} \alpha_{j-\nu+1-\mu}\widehat{\Phi}^{(j-\nu+1)}_{\mu}(z)\left(\frac{a_{j-\nu}}{a_{j-\nu-1}}\left(1 - |a_{j-\nu-1}|^2\right)z\right) + \alpha_{j-\nu-1} \\
&= \alpha_{j-\nu-1} + \sum_{\mu=1}^{\nu} \alpha_{j-\nu+\mu}\left[\widehat{\Phi}^{(j-\nu)}_{\mu}(z)\left(z + \frac{a_{j-\nu-1}}{a_{j-\nu-2}}\right)\right. \\
&\quad \left. - \widehat{\Phi}^{(j-\nu+1)}_{\mu-1}(z)\left(\frac{a_{j-\nu}}{a_{j-\nu-1}}(1 - |a_{j-\nu-1}|^2)z\right)\right] + \alpha_{j-\nu}\left(z + \frac{a_{j-\nu-1}}{a_{j-\nu-2}}\right) \\
&= \sum_{\mu=1}^{\nu} \alpha_{j-\nu+\mu}\widehat{\Phi}^{(j-\nu-1)}_{\mu+1}(z) + \alpha_{j-\nu}\widehat{\Phi}^{(j-\nu-1)}_{1}(z) + \alpha_{j-\nu-1} \\
&\qquad \text{(by the dual recurrence (3.3) for } \widehat{\Phi}^{(j-\nu-1)}_{\mu+1}\text{)} \\
&= \sum_{\mu=2}^{\nu+1} \alpha_{j-\nu-1+\mu}\widehat{\Phi}^{(j-\nu-1)}_{\mu}(z) + \alpha_{j-\nu}\widehat{\Phi}^{(j-\nu-1)}_{1}(z) + \alpha_{j-\nu-1} \\
&= \sum_{\mu=0}^{\nu+1} \alpha_{j+\mu-(\nu+1)}\widehat{\Phi}^{(j-(\nu+1))}_{\mu}(z).
\end{aligned}
$$

□

The following theorem shows that we obtain values of derivatives of q_n if we use the dual recurrence relation (3.3) and the Christoffel-Darboux identities of Theorem 3.1 based on the Geronimus recurrence (2.4).

Theorem 4.2. *Let* $q_{n-k,k} := \sum_{\nu=0}^{n-k} \alpha^{(k)}_{\nu}\Phi_{\nu}$ *be given for* $k = 0$ *with* $\alpha^{(0)}_{\nu} := \alpha_{\nu}$. *If we apply* (4.2) *at* $z = w$ *for* $k = 0, \dots, n$, *and start with* $q_{n,0}$, *we get sequences of numbers* $a^{(k+1)}_{n-k}, \dots, a^{(k+1)}_{0}$ *with*

$$
a^{(k+1)}_{0} = \frac{1}{k!}\frac{d^k}{dz^k}q_n(z)\Big|_{z=w}. \tag{4.5}
$$

Here, if $k \geq 1$, *the polynomials* $q_{n-k,k}$ *to which* (4.2) *is applied repeatedly are defined by*

$$
\alpha^{(k)}_{\nu} := \begin{cases} a^{(k)}_{\nu+1} - a^{(k)}_{\nu+2}\frac{a_{\nu+1}}{a_{\nu}}(1 - |a_{\nu}|^2) & \text{for } 0 \leq \nu \leq j-1, \\ a^{(k)}_{\nu+1} & \text{for } j \leq \nu \leq n-k. \end{cases}
$$

Proof. We first assume $a_0, \dots, a_{n-1} \neq 0$. If $z \neq w$ then we have by Theorem 3.1,

$$
\begin{aligned}
&q_n(z) - q_n(w) \\
&= (z-w)\sum_{\nu=0}^{n} \alpha_\nu \sum_{k=0}^{\nu-1}\left(\Phi_k(z) - \frac{a_k}{a_{k-1}}(1-|a_{k-1}|^2)\Phi_{k-1}(z)\right)\widehat{\Phi}^{(k+1)}_{\nu-1-k}(w) \\
&= (z-w)\sum_{k=0}^{n-1}\left(\Phi_k(z) - \frac{a_k}{a_{k-1}}(1-|a_{k-1}|^2)\Phi_{k-1}(z)\right)\sum_{\nu=k+1}^{n}\alpha_\nu \widehat{\Phi}^{(k+1)}_{\nu-(k+1)}(w) \\
&= (z-w)\sum_{k=0}^{n-1} a^{(1)}_{k+1}\left(\Phi_k(z) - \frac{a_k}{a_{k-1}}(1-|a_{k-1}|^2)\Phi_{k-1}(z)\right).
\end{aligned}
$$

Here, we have used the fact that the sum in (3.4) is symmetric with respect to z and w, and we have abbreviated $\sum_{\nu=0}^{n-k-1}\alpha_{\nu+k+1}\widehat{\Phi}^{(k+1)}_\nu(w) =: a^{(1)}_{k+1,n}(w) =: a^{(1)}_{k+1}$ as in (4.3), where z is replaced by w.

With $a_{-1} = -1$ we define $\alpha^{(1)}_k := a^{(1)}_{k+1} - a^{(1)}_{k+2}\frac{a_{k+1}}{a_k}(1-|a_k|^2)$. Then we have

$$
q_n(z) - q_n(w) = (z-w)\sum_{k=0}^{n-1}\alpha^{(1)}_k \Phi_k(z) =: (z-w)q_{n-1,1}(z).
$$

If $a_0, \dots, a_j \neq 0 = a_{j+1} = \dots = a_{n-1}$, we can use the result just proved up to polynomial degree $j+1$ and obtain

$$
\begin{aligned}
&q_n(z) - q_n(w) = q_{j+1}(z) - q_{j+1}(w) + \sum_{\nu=j+2}^{n}\alpha_\nu(\Phi_\nu(z) - \Phi_\nu(w)) \\
&= q_{j+1}(z) - q_{j+1}(w) \\
&+(z-w)\sum_{\nu=j+2}^{n}\alpha_\nu\left\{\left[\sum_{k=0}^{j}\left(\Phi_k(z) - \frac{a_k}{a_{k-1}}(1-|a_{k-1}|^2)\Phi_{k-1}(z)\right)\widehat{\Phi}^{(k+1)}_{\nu-k-1}(w)\right]\right. \\
&\left.+\sum_{k=j+1}^{\nu-1}\Phi_k(z)\Phi^{(k+1)}_{\nu-k-1}(w)\right\} \\
&= q_{j+1}(z) - q_{j+1}(w) \\
&+(z-w)\left[\sum_{k=0}^{j}\left(\Phi_k(z) - \frac{a_k}{a_{k-1}}(1-|a_{k-1}|^2)\Phi_k(z)\right)\sum_{\nu=j+2}^{n}\alpha_\nu\widehat{\Phi}^{(k+1)}_{\nu-k-1}(w)\right. \\
&\left.+\sum_{k=j+1}^{n-1}\Phi_k(z)\sum_{\nu=k+1}^{n}\alpha_\nu\Phi^{(k+1)}_{\nu-k-1}(w)\right].
\end{aligned}
$$

With $a_{k+1,j}^{(1)} + \sum_{\nu=j+2}^{n} \alpha_\nu \widehat{\Phi}_{\nu-k-1}^{(k+1)}(w) = \sum_{\nu=0}^{n-k-1} \alpha_{\nu+k+1} \widehat{\Phi}_\nu^{(k+1)}(w) = a_{k+1,n}^{(1)}$ and $a_{n,n-1}^{(1)} = 0$ we have

$$\begin{aligned} q_n(z) - q_n(w) &= (z-w)\left\{ \sum_{k=0}^{j} a_{k+1,n}^{(1)} \left(\Phi_k(z) - \frac{a_k}{a_{k-1}}(1-|a_{k-1}|^2)\Phi_{k-1}(z) \right) \right. \\ &\qquad \left. + \sum_{k=j+1}^{n-1} a_{k+1,n}^{(1)} \Phi_k(z) \right\} \\ &= (z-w)\left\{ \sum_{k=0}^{n-1} a_{k+1,n}^{(1)} \Phi_k(z) - \sum_{k=0}^{j-1} a_{k+2,n}^{(1)} \frac{a_{k+1}}{a_k}(1-|a_k|^2)\Phi_k(z) \right\} \\ &=: (z-w) q_{n-1,1}(z), \end{aligned}$$

where

$$q_{n-1,1}(z) := \sum_{k=0}^{n-1} \alpha_k^{(1)} \Phi_k(z)$$

and

$$\alpha_k^{(1)} := \begin{cases} a_{k+1}^{(1)} - a_{k+2}^{(1)} \frac{a_{k+1}}{a_k}(1-|a_k|^2) & \text{for } 0 \le k \le j-1, \\ a_{k+1}^{(1)} & \text{for } j \le k \le n-1. \end{cases}$$

For a fixed chosen w and $z \to w$ we get $q_n'(w) = \sum_{k=0}^{n-1} \alpha_k^{(1)} \Phi_k(w) = q_{n-1,1}(w)$. Thus we have $q_n(z) = (z-w)q_{n-1,1}(z) + a_0^{(1)}$, where $a_0^{(1)} := q_n(w)$. By induction, using the Euclidean algorithm, we obtain

$$q_{n-k,k}(z) = (z-w) q_{n-k-1,k+1}(z) + a_0^{(k+1)}, \quad k = 0, \dots, n,$$

with

$$q_{n-k-1,k+1}(z) := \sum_{\nu=0}^{n-(k+1)} \alpha_\nu^{(k+1)} \Phi_\nu(z),$$

$$\alpha_\nu^{(k+1)} := \begin{cases} a_{\nu+1}^{(k+1)} - a_{\nu+2}^{(k+1)} \frac{a_{\nu+1}}{a_\nu}(1-|a_\nu|^2) & \text{for } 0 \le \nu \le j-1, \\ a_{\nu+1}^{(k+1)} & \text{for } j \le \nu \le n-(k+1), \end{cases} \tag{4.6}$$

where $a_\nu^{(k+1)}$, $\nu = n-k-1, \dots, 0$, are determined by (4.2) applied to $q_{n-k,k}$ at $z = w$, i.e.,

$$\begin{aligned} a_{n-k+1}^{(k+1)} &:= 0, \\ a_{n-k-\nu}^{(k+1)} &:= a_{n-k-\nu+1}^{(k+1)} w + \alpha_{n-k-\nu}^{(k)}, \quad \nu = 0, \dots, n-k-j-1, \\ a_j^{(k+1)} &:= a_{j+1}^{(k+1)} \left(w + \frac{a_j}{a_{j-1}} \right) + \alpha_j^{(k)}, \\ a_{j-\nu}^{(k+1)} &:= a_{j-\nu+1}^{(k+1)} \left(w + \frac{a_{j-\nu}}{a_{j-\nu-1}} \right) - a_{j-\nu+2}^{(k+1)} \frac{a_{j-\nu+1}}{a_{j-\nu}}(1-|a_{j-\nu}|^2) w + \alpha_{j-\nu}^{(k)}, \\ &\qquad \nu = 1, \dots, j, \end{aligned}$$

for $k = 0, \dots, n$. In particular, we have $a_0^{(k+1)} = q_{n-k,k}(w)$ and $q_{0,n} = \alpha_0^{(n)}\Phi_0 \equiv a_0^{(n+1)}$ by Theorem 4.1. Thus we get

$$q_n(z) = (z-w)q_{n-1,1}(z) + a_0^{(1)} = (z-w)^2 q_{n-2,2}(z) + (z-w)a_0^{(2)} + a_0^{(1)} = \cdots$$
$$= \sum_{k=0}^{n} a_0^{(k+1)}(z-w)^k.$$

By the identity theorem for power series we get the result of Theorem 4.2. □

The reader should carefully note the difference between the quantities a_ν, $\alpha_\nu =: \alpha_\nu^{(0)}$, $a_\nu^{(k+1)}$, and $\alpha_\nu^{(k+1)}$: the a_ν are the reflection coefficients of the orthogonal polynomials given in (2.1); the α_ν are the Fourier coefficients of q_n from (4.1). For a fixed k, by the Clenshaw-type algorithm (4.2), we calculate $a_\nu^{(k+1)}$, $\nu = n-k-1, \dots, 0$. If we increase k we need the Fourier coefficients $\alpha_\nu^{(k+1)}$ of $q_{n-k-1,k+1}$ given in (4.6).

5. Some Remarks on Implementation

First we formulate an algorithm to calculate values of a polynomial and its derivatives, which follows directly from the work in §4:

Algorithm 1.

Given: *A polynomial* $q_{n,0} := \sum_{\nu=0}^{n} \alpha_\nu^{(0)}\Phi_\nu$, *where the* Φ_ν *satisfy the recurrence* (2.1).

Objective: *Calculate* $\left.\frac{d^k}{dz^k} q_{n,0}(z)\right|_{z=w}$, $k = 0, \dots, n$.

Calculation: *(Iteration)* **For** $k := 0$ **to** n **do**

1) *Evaluate* $q_{n-k,k}(w) := \sum_{\nu=0}^{n-k} \alpha_\nu^{(k)}\Phi_\nu(w)$ *by using* (4.2). *We get a sequence* $a_{n-k-1}^{(k+1)}, \dots, a_0^{(k+1)}$.

2) *Store* $k!a_0^{(k+1)} = \frac{d^k}{dz^k} q_n(z)|_{z=w}$ *in a suitable manner.*

3) If $k < n$ **then** *calculate* $\alpha_\nu^{(k+1)}$, $\nu = 0, \dots, n-(k+1)$, *as defined in* (4.6).

By using the Euclidean algorithm we have shown that values of derivatives of $q_{n,0}$ can be calculated by a Clenshaw-like algorithm. Corresponding results based on the Szegő recurrence (2.1) are, however, too complicated for practical use [5].

As a consequence of using the Geronimus recurrence (2.4) there appear instabilities in (4.2) if some reflection coefficients have small moduli. They can be reduced by the following considerations. Again, we want to evaluate q_n given in (4.1) at $z \in \mathbb{C}$. Define $\tilde{\alpha}_\nu$ by

$$\alpha_\nu = \tilde{\alpha}_\nu \prod_{k=-1}^{\nu-2} a_k, \quad \nu = 0, \dots, j+1, \tag{5.1}$$

and

$$
\begin{aligned}
\tilde{a}_{j+1}^{(1)} &:= \tilde{\alpha}_{j+1}, \\
\tilde{a}_{j}^{(1)} &:= \tilde{a}_{j+1}^{(1)}(za_{j-1} + a_j) + \tilde{\alpha}_j, \\
\tilde{a}_{j-\nu+1}^{(1)} &:= \tilde{a}_{j-\nu+2}^{(1)}(a_{j-\nu}z + a_{j-\nu+1}) - \tilde{a}_{j-\nu+3}^{(1)}a_{j-\nu+2}a_{j-\nu}(1-|a_{j-\nu+1}|^2)z \\
&\quad + \tilde{\alpha}_{j-\nu+1}, \quad \nu = 2, \dots, j+1.
\end{aligned}
\tag{5.2}
$$

By induction we can prove

Theorem 5.1. *With $\tilde{\alpha}_\nu$ from (5.1) and $\tilde{a}_{j-\nu}^{(1)}$ from (5.2) we have*

$$
a_{j-\nu+1}^{(1)} = \tilde{a}_{j-\nu+1}^{(1)} \prod_{k=-1}^{j-\nu-1} a_k, \quad \nu = 0, \dots, j+1, \tag{5.3}
$$

where a_k denote the reflection coefficients from (2.1) resp. (2.4).

With $\beta_{j+1} := \beta_{j+1}(z) := \sum_{\nu=0}^{n-j-1} \alpha_{\nu+j+1} z^\nu$ we have

$$
\begin{aligned}
\sum_{\nu=0}^{n} \alpha_\nu \Phi_\nu(z) &= \sum_{\nu=0}^{j} \alpha_\nu \Phi_\nu(z) + \sum_{\nu=j+1}^{n} \alpha_\nu \Phi_\nu(z) \\
&= \sum_{\nu=0}^{j} \alpha_\nu \Phi_\nu(z) + \Phi_{j+1}(z) \sum_{\nu=j+1}^{n} \alpha_\nu z^{\nu-j-1} \\
&= \sum_{\nu=0}^{j} \alpha_\nu \Phi_\nu(z) + \beta_{j+1}\Phi_{j+1}(z).
\end{aligned}
$$

This yields an algorithm for evaluating q_n in which the instabilities of (4.2) appear only in the initialization (calculation of $z_\nu := 1/\prod_{k=-1}^{\nu} a_k$, $a_k \neq 0$). This can be handled numerically by using the fact that real numbers are approximated by rational numbers on a computer. We omit details here.

Algorithm 2.
Calculate β_{j+1} with the Horner scheme.

$\tilde{a}_0 := \alpha_0;$

$z_{-1} := -1;$

For $\nu := 1$ **to** j **do**
Begin

$\tilde{\alpha}_\nu := \alpha_\nu z_{\nu-2};$

$z_{\nu-1} := z_{\nu-2}/a_{\nu-1};$

End;

$\tilde{\alpha}_{j+1} := \beta_{j+1} z_{j-1};$

$\tilde{a}^{(1)}_{j+2} := \tilde{a}^{(1)}_{j+3} := 0;$

For $\nu := 0$ **to** $j+1$ **do**

$$\tilde{a}^{(1)}_{j-\nu+1} := \tilde{a}^{(1)}_{j-\nu+2}(a_{j-\nu}z + a_{j-\nu+1}) - \tilde{a}^{(1)}_{j-\nu+3}a_{j-\nu+2}a_{j-\nu}(1-|a_{j-\nu+1}|^2)z + \tilde{\alpha}_{j-\nu+1}.$$

From (4.4) and (5.3) we conclude $\tilde{a}^{(1)}_0 = q_n(z)$.

References

[1] G.S. Ammar, W.B. Gragg, and L. Reichel, *An analogue for Szegő polynomials of the Clenshaw algorithm*, J. Comput. Appl. Math., **46** (1993), 211–216.

[2] Ya.L. Geronimus, *Polynomials orthogonal on a circle and their applications*, Amer. Math. Soc. Translation, 1954.

[3] F. Peherstorfer, *On the asymptotic behaviour of functions of the second kind and Stieltjes polynomials and on Gauss-Kronrod quadrature formulas*, J. Approx. Theory, **70** (1992), 156–190.

[4] F. Peherstorfer, *A special class of polynomials orthogonal on the unit circle including the associated polynomials*, Constr. Approx., **12** (1996), 161–185.

[5] M.-R. Skrzipek, *Dual recurrence and Christoffel-Darboux-type formulas for orthogonal polynomials*, in: M.D. Buhmann, M.W. Müller, M. Felten, and D.H. Mache, Eds., *Proceedings of the IDoMAT 98 conference, Witten, Germany, Feb. 23 – 27, 1998*, Birkhäuser, Basel, to appear.

[6] M.-R. Skrzipek, *Polynomial evaluation and associated polynomials*, Numer. Math., **79** (1998), 601–613.

[7] G. Szegő, *Orthogonal polynomials*, Amer. Math. Colloq. Publ. **23**, 4th ed., Amer. Math. Soc., Providence, RI, 1975.

[8] W. Van Assche, *Orthogonal polynomials, associated polynomials and functions of the second kind*, J. Comput. Appl. Math., **37** (1991) 237–249.

[9] J. Wimp, *Computation with recurrence relations*, Applicable Mathematics Series, Pitman, Boston, 1984.

Michael-Ralf Skrzipek
Fernuniversität Gesamthochschule in Hagen
FB Mathematik
Postfach 940
D-58084 Hagen, Germany
E-mail address: michael.skrzipek@fernuni-hagen.de

International Series of Numerical Mathematics
Vol. 131, © 1999 Birkhäuser Verlag Basel/Switzerland

Applications of Tensor-Valued Tri-Variate Hermite Polynomials and Spherical Harmonics in the Kinetic Theory of Gases

Kurt Suchy

Abstract. An infinite (coupled) system of integro-differential balance equations for tensor-valued functions is very important in the kinetic theory of gases. For the complicated calculation of the integral part, the introduction of tensor-valued Hermite polynomials in velocity space is most useful. Various relations for these polynomials, including their connection with tensor-valued spherical harmonics, are presented, which are necessary to evaluate the infinite system for the velocity averages of the polynomials.

1. Tensor-Valued Functions as Physical Quantities

In the kinetic theory of gases and liquids, equations are derived connecting vector- and tensor-valued quantities. An important vector-valued quantity is the average velocity

$$\bar{\boldsymbol{c}}_j := \frac{m_j}{\rho_j} \int d^3c_j f_j(\boldsymbol{r}, t, \boldsymbol{c}_j)\boldsymbol{c}_j \tag{1}$$

of particles of species j with mass m_j, mass density ρ_j, and velocity distribution function $f_j(\boldsymbol{r}, t, \boldsymbol{c}_j)$, which depends on the vector-valued position $\boldsymbol{r}$ and time t.

An important (second-order) tensor-valued quantity is the dyadic partial pressure

$$\mathsf{p}_j := \rho_j \int d^3c_j f_j(\boldsymbol{r}, t, \boldsymbol{c}_j)\boldsymbol{C}_j\boldsymbol{C}_j \tag{2}$$

with the peculiar velocity

$$\boldsymbol{C}_j := \boldsymbol{c}_j - \boldsymbol{v}, \tag{3}$$

the mean-mass velocity

$$\boldsymbol{v} := \frac{1}{\rho}\sum_j \rho_j \bar{\boldsymbol{c}}_j, \tag{4}$$

and the total mass density

$$\rho := \sum_j \rho_j. \tag{5}$$

We have written the dyadic product $\boldsymbol{C}_j\boldsymbol{C}_j =: \boldsymbol{C}_j^2$ in (2) according to Gibbs ([7, art. 99]) without any multiplication sign, which is the standard notation in physics.

Another (third-order) tensor-valued quantity is the symmetric triadic

$$q_j^{(3)} := \rho_j \int d^3c_j f_j(\boldsymbol{r}, t, \boldsymbol{c}_j)\boldsymbol{C}_j\boldsymbol{C}_j\boldsymbol{C}_j =: \rho_j \int d^3c_j f_j(\boldsymbol{r}, t, \boldsymbol{c}_j)\boldsymbol{C}_j^3. \tag{6}$$

Half of its trace,

$$\boldsymbol{q}_j := \frac{1}{2}\rho_j \int d^3c_j f_j(\boldsymbol{r}, t, \boldsymbol{c}_j) C_j^2 \boldsymbol{C}_j, \tag{7}$$

is the energy flux of the particles of species j.

The vector-valued diffusion flux

$$\rho_j \overline{\boldsymbol{C}_j} := m_j \int d^3c_j f_j(\boldsymbol{r}, t, \boldsymbol{c}_j)\boldsymbol{C}_j, \tag{8}$$

the partial pressure p_j (2), and the energy flux triadic $q_j^{(3)}$ (6) are special cases of the tensor-valued velocity moments

$$\overline{\boldsymbol{C}_j^n} := \frac{m_j}{\rho_j} \int d^3c_j f_j(\boldsymbol{r}, t, \boldsymbol{c}_j)\boldsymbol{C}_j^n \tag{9}$$

with $n = 1, 2, 3$. Therefore, equations connecting these moments are desirable.

2. Balance Equations for Velocity Averaged Properties

To establish these equations, Enskog's ([3, Eqs. (8), (10)]) balance equation

$$\frac{\partial}{\partial t}\rho_j\overline{\phi}_j + \boldsymbol{\nabla}\cdot\left(\boldsymbol{v}\rho_j\overline{\phi}_j + \rho_j\overline{\boldsymbol{C}_j\phi_j}\right) - \rho_j \overline{\left(\frac{D}{Dt} + \boldsymbol{C}_j\cdot\boldsymbol{\nabla}\right)\phi_j\Big|_{\boldsymbol{C}_j}}$$

$$+\rho_j \overline{\left(\frac{D\boldsymbol{v}}{Dt} + \boldsymbol{C}_j\cdot\boldsymbol{\nabla}\boldsymbol{v} - \boldsymbol{a} + \boldsymbol{b}\times\boldsymbol{C}_j\right)\cdot\frac{\partial}{\partial\boldsymbol{C}_j}\phi_j} = \rho_j\overline{\frac{\delta\phi_j}{\delta t}} \tag{10}$$

for the velocity average

$$\overline{\phi}_j(\boldsymbol{r}, t) := \frac{m_j}{\rho_j}\int d^3c_j f_j(\boldsymbol{r}, t, \boldsymbol{c}_j)\phi_j(\boldsymbol{r}, t, \boldsymbol{C}_j) \tag{11}$$

of a property $\phi_j(\boldsymbol{r}, r, \boldsymbol{c}_j)$ was specialized by Johnston ([11, Eq. (6.1)]) for $\phi_j(\boldsymbol{r}, t, \boldsymbol{C}_j) = \boldsymbol{C}_j^n$ to

$$\frac{\partial}{\partial t}\rho_j\overline{\boldsymbol{C}_j^n} + \boldsymbol{\nabla}\cdot\left(\boldsymbol{v}\rho_j\overline{\boldsymbol{C}_j^n} + \rho_j\overline{\boldsymbol{C}_j^{n+1}}\right)$$

$$+n\rho_j \left(\widehat{\frac{D\boldsymbol{v}}{Dt}\overline{\boldsymbol{C}_j^{n-1}}} + \widehat{\overline{\boldsymbol{C}_j^n} \cdot \boldsymbol{\nabla v}} - \widehat{\boldsymbol{a}\overline{\boldsymbol{C}_j^{n-1}}} + \widehat{\boldsymbol{b} \times \overline{\boldsymbol{C}_j^n}} \right) = \rho_j \frac{\overline{\delta \boldsymbol{C}_j^n}}{\delta t}. \tag{12}$$

Here, $\boldsymbol{\nabla} := \partial/\partial \boldsymbol{r}$ is the vector-valued differential operator in position space, $D/Dt := \partial/\partial t + \boldsymbol{v} \cdot \boldsymbol{\nabla}$ is the convective derivative, $\boldsymbol{a}$ is the acceleration due to an external force independent of $\boldsymbol{C}_j$ (e.g. gravitational or electrical field), and $\boldsymbol{C}_j \times \boldsymbol{b}$ is due to a Coriolis force or a magnetic field. The symmetry symbol means

$$\widehat{\boldsymbol{xy}} := \frac{1}{2}(\boldsymbol{xy} + \boldsymbol{yx}), \qquad \widehat{\boldsymbol{xyy}} := \frac{1}{3}(\boldsymbol{xyy} + \boldsymbol{yxy} + \boldsymbol{yyx}), \quad \text{etc.}, \tag{13}$$

and the collision integral $\overline{\delta\phi_j/\delta t}$ for elastic collisions between monatomic particles of species j and k reads

$$\frac{\overline{\delta \phi_j}}{\delta t} = \sum_k \overline{\left(\frac{\delta \phi_j}{\delta t}\right)_k} \tag{14}$$

with

$$\overline{\left(\frac{\delta \phi_j}{\delta t}\right)_k} = \frac{m_j}{\rho_j} \int d^3 c_j \int d^3 c_k f_j f_k \left|\boldsymbol{C}_j - \boldsymbol{C}_k\right| \oint d^2 \Omega'_{jk} \sigma_{jk} (\phi'_j - \phi_j). \tag{15}$$

Here, σ_{jk} is the differential cross section for collisions between particles of species j and k, the primed quantities refer to values after a collision, and $d^2\Omega'_{jk}$ is the solid angle element for the relative velocity $\boldsymbol{c}_{jk} := \boldsymbol{c}_j - \boldsymbol{c}_k = \boldsymbol{C}_j - \boldsymbol{C}_k$ after the collision.

3. Balance Equations for Hermitian Moments

To calculate the collision integrals (15), assumptions must be made for the velocity distribution functions f_j and f_k. A most effective kind of ansatz is an expansion in tensor-valued tri-variate Hermite polynomials (Grad [9, Eqs. (4.4) to (4.8)]; Suchy [15, Eqs. (20), (22)])

$$f_j = \frac{\rho_j}{m_j} w_j \sum_{p=0}^{\infty} \frac{1}{p!} \left(\frac{m_j}{2k_B T}\right)^p \overline{\boldsymbol{He}_j^{(p)}} \underset{p}{\cdot} \boldsymbol{He}_j^{(p)} \tag{16}$$

with the local Maxwell distribution

$$w_j := \left(\frac{m_j}{2\pi k_B T}\right)^{3/2} \exp\left(-\frac{m_j C_j^2}{2k_B T}\right) \tag{17}$$

as weight function. Here, $T(\boldsymbol{r}, t)$ is the local temperature, k_B is Boltzmann's constant, and $\underset{p}{\cdot}$ means the p-fold contraction (scalar product) with the innermost indices contracted first (McCourt et al. [13, §20.4.1]).

Because of the orthogonality relation (Grad [8, Eq. (19)]; Erdélyi et al. [4, Eq. 12.9.1])

$$\int d^3C_j w_j He_j^{(p)} He_j^{(q)} \underset{q}{\cdot} T^{(q)} = \delta_{pq} \left(\frac{k_BT}{m_j}\right)^p p! \widehat{T^{(q)}} \tag{18}$$

(with $T^{(q)}$ an arbitrary tensor), the expansion coefficients $\overline{He_j^{(p)}}$ in (16) are just the velocity averages

$$\overline{He_j^{(p)}} = \frac{m_j}{\rho_j} \int d^3C_j f_j(\boldsymbol{r}, t, \boldsymbol{C}_j) He_j^{(p)} \left(\frac{k_BT}{m_j}; \boldsymbol{C}_j\right) \tag{19}$$

of the generalized Hermite polynomials (Suchy [14, Eq. (1.7)])

$$He_j^{(n)} \left(\frac{k_BT}{m_j}; \boldsymbol{C}_j\right) := \exp\left(\frac{m_j C_j^2}{2k_BT}\right) \left(\frac{-k_BT}{m_j} \frac{\partial}{\partial \boldsymbol{C}_j}\right) \exp\left(-\frac{m_j C_j^2}{2k_BT}\right). \tag{20}$$

The first polynomials are

$$He_j^{(0)} = 1, \qquad He_j^{(1)} = \boldsymbol{C}_j, \qquad He_j^{(2)} = \boldsymbol{C_j}^2 - \frac{k_BT}{m_j}\mathsf{I}, \tag{21}$$

$$He_j^{(3)} = \boldsymbol{C}_j^3 - 3\frac{k_BT}{m_j} \widehat{\boldsymbol{C}_j \mathsf{I}}$$

with the second-order unit tensor $\mathsf{I} = I^{(2)}$. Their velocity averages

$$\rho_j \overline{He_j^{(1)}} = \rho_j \overline{\boldsymbol{C}_j}, \qquad \rho_j \overline{He_j^{(2)}} = \mathsf{p}_j - p_j \mathsf{I},$$

$$\rho_j \overline{He_j^{(3)}} = q_j^{(3)} - 3\frac{k_BT}{m_j} \rho_j \widehat{\overline{\boldsymbol{C}} \mathsf{I}} =: Q_j^{(3)} \tag{22}$$

are directly related to the diffusion flux (8), the partial pressure p_j(2), the hydrostatic pressure $p_j := \rho_j k_BT/m_j$, and the heat flux, cf. (7),

$$\boldsymbol{Q}_j := \frac{1}{2}\mathsf{I} : Q_j^{(3)} = \boldsymbol{q}_j - \frac{5}{2} p_j \overline{\boldsymbol{C}}_j. \tag{23}$$

Therefore, it suggests itself to specialize the property $\phi_j(\boldsymbol{r}, t, \boldsymbol{c}_j)$ in Enskog's balance equation (10) not to the dyadic powers $\boldsymbol{C}_j^n$ as in (12), but to the generalized Hermite polynomials (20). Instead of seven terms on the left-hand side of (12) for $\overline{\boldsymbol{C}_j^n}$ there will be twelve terms for $\overline{He_j^{(n)}}$ (Suchy [14, Eq. (4.2)]; [16, Eq. (4.6)]), but the collision integrals on the right-hand side can now be evaluated (Suchy [15, Eqs. (24), (106)]). They are bilinear in the velocity averages $\overline{He_j^{(p)}}$ and $\overline{He_k^{(q)}}$.

4. Relations Between Hermite Polynomials

For the evaluation of the left-hand side of the balance equation (10) for $\overline{He_j^{(n)}}$ several relations are employed, which are specializations and generalizations of known results to the generalized Hermite polynomials (20).

Differentiation of the definition (20) of $He_j^{(n)}$ yields the ascending operation (Grad [8, Eq. (15)])

$$\left(\boldsymbol{C}_j - \frac{k_B T}{m_j}\frac{\partial}{\partial \boldsymbol{C}_j}\right) He_j^{(n)} = He_j^{(n+1)}. \tag{24}$$

Operating with $\boldsymbol{a} \cdot \partial/\partial \boldsymbol{C}_j$ on the definition (20), where $\boldsymbol{a}$ is an arbitrary constant vector, yields, after some intermediate steps, the descending operation (Bell [1, Eq. (1.9)])

$$\boldsymbol{a} \cdot \frac{\partial}{\partial \boldsymbol{C}_j} He_j^{(n)} = n\,\widehat{\boldsymbol{a} He_j^{(n-1)}}. \tag{25}$$

Scalar multiplication of (24) with $\boldsymbol{a}$ and use of (25) leads to the recurrence relation (Bell [1, Eq. (1.8)])

$$\boldsymbol{a} \cdot He_j^{(n+1)} = \boldsymbol{a} \cdot \boldsymbol{C}_j He_j^{(n)} - n\frac{k_B T}{m_j}\widehat{\boldsymbol{a} He_j^{(n-1)}}. \tag{26}$$

Combination of the descending operation with the recurrence relation (26) for $n-1$ instead of n leads, with $\boldsymbol{a}$ and $\boldsymbol{b}$ arbitrary constant vectors, to

$$\boldsymbol{ab} : \boldsymbol{C}_j \frac{\partial}{\partial \boldsymbol{C}_j} He_j^{(n)} = n\,\widehat{\boldsymbol{ab} \cdot He_j^{(n)}} + n(n-1)\frac{k_B T}{m_j}\widehat{\boldsymbol{ab} He_j^{(n-2)}}. \tag{27}$$

Replacing the dyadic $\boldsymbol{ab}$ by the unit dyadic I, we obtain

$$\boldsymbol{C}_j \cdot \frac{\partial}{\partial \boldsymbol{C}_j} He_j^{(n)} = n He_j^{(n)} + n(n-1)\frac{k_B T}{m_j}\widehat{\mathsf{I} He_j^{(n-2)}}. \tag{28}$$

Replacing $\boldsymbol{ab}$ in (27) by the antisymmetric dyadic $\boldsymbol{a} \times \mathsf{I}$ leads, after some intermediate steps, to

$$\boldsymbol{a} \times \boldsymbol{C}_j \cdot \frac{\partial}{\partial \boldsymbol{C}_j} He_j^{(n)} = n\,\widehat{\boldsymbol{a} \times He_j^{(n)}}. \tag{29}$$

With the ascending operation (24) Grad ([8, Eqs. (16), (24)]) proved by induction the explicit expression

$$He_j^{(n)}\left(\frac{k_B T}{m_j}; \boldsymbol{C}_j\right) = \sum_{m=0}^{[n/2]} \frac{(-k_B T/m_j)^m n!}{(n-2m)!(2m)!!}\widehat{\mathsf{I}^m \boldsymbol{C}_j^{n-2m}}. \tag{30}$$

Differentiating this with respect to k_BT/m_j, replacing m by $m+1$, using $(2m)!! = 2^m m!$ and $n!/(n-2)! = n(n-1)$, we obtain (Dattoli et al. [2, Eq. (10)])

$$\frac{m_j}{k_B}\frac{\partial}{\partial T}He_j^{(n)} = -\frac{n(n-1)}{2}\widehat{\mathbf{I}He_j^{(n-2)}}. \tag{31}$$

Multiplying this with $\boldsymbol{a}\cdot\boldsymbol{C}_j.$ and using the recurrence relation (26) with $n-2$ instead of n, we obtain

$$\boldsymbol{a}\cdot\boldsymbol{C}_j\frac{m_j}{k_B}\frac{\partial}{\partial T}He_j^{(n)} = -\frac{n(n-1)}{2}\left[\widehat{\mathbf{I}\boldsymbol{a}\cdot He_j^{(n-1)}}+\frac{k_BT}{m_j}(n-2)\,\widehat{\mathbf{I}\boldsymbol{a}He_j^{(n-3)}}\right]. \tag{32}$$

5. Important Auxiliary Formulas for the Collision Integrals

The evaluation of the collision integrals (15) for $\phi_j = He_j^{(n)}$ is much more involved than that of the left-hand side of the balance equations (10). The reason for this is mainly the appearance of the relative velocity $\boldsymbol{c}_{jk} := \boldsymbol{c}_j - \boldsymbol{c}_k = \boldsymbol{C}_j - \boldsymbol{C}_k$, not only explicitly with its modulus $|\boldsymbol{c}_{jk}|$, but also via the differential cross section σ_{jk}, which depends on $|\boldsymbol{c}_{jk}|$ and on the deflection angle $\chi :=$ arccos $(\hat{\boldsymbol{c}}_{jk}\cdot\hat{\boldsymbol{c}}_{jk})$ of the binary collision. (Here, $\hat{\boldsymbol{c}} := \boldsymbol{c}/|\boldsymbol{c}| = \boldsymbol{c}/c$ is the unit vector in the direction of the vector $\boldsymbol{c}$.) It is therefore necessary to express the peculiar velocities $\boldsymbol{C}_j$ and $\boldsymbol{C}_k$ as linear combinations of the relative velocity $\boldsymbol{c}_{jk}$ and the center-of-mass velocity $\boldsymbol{C}_{jk}$ relative to the mean-mass velocity $\boldsymbol{v}$ (4). Because of the factors m_j/k_BT of C_j^2 and m_k/k_BT of C_k^2 in the arguments of the weight functions w_j and w_k (17), it is useful to include these factors in the expressions for the linear combinations of the arguments of $He_j^{(p)}$ and $He_k^{(q)}$ (Kumar [12, Eqs. (83), (84), (87)]).

For the rather involved calculations, the following mathematical relations are needed.

The addition theorem is (Erdélyi et al. [4, Eq. 10.13.40])

$$He^{(n)}(1;\alpha\boldsymbol{x}+\beta\boldsymbol{y}) = \sum_{m=0}^{n}\binom{n}{m}\alpha^m\beta^{n-m}\,\widehat{He^{(m)}(1;\boldsymbol{x})He^{(n-m)}(1;\boldsymbol{y})} \tag{33}$$

for $\alpha^2+\beta^2=1$.

The generalized orthonormality relation reads (Erdélyi et al. [5, Eq. 16.5.15])

$$T^{(l)}\,{}_i\int d^3y\frac{\exp(-y^2/2\alpha)}{(2\pi\alpha)^{3/2}}\,\frac{He^{(l)}(\alpha;\boldsymbol{y})}{\alpha^{l/2}}\,\frac{He^{(m)}(\alpha;\boldsymbol{y})}{\alpha^{m/2}}\,\frac{He^{(n)}(\alpha;\boldsymbol{y})}{\alpha^{n/2}}\,{}_{\dot n}T^{(n)}$$

$$= \begin{cases} \widehat{T^{(l)}}\,{}_{\frac{l-\dot m+n}{2}}\,\widehat{T^{(n)}}\quad \dfrac{l!m!n!}{\frac{l-m+n}{2}!\,\frac{m-n+l}{2}!\,\frac{n-l+m}{2}!} & \text{for } l-m+n \text{ even,}\\ 0^{(m)} & \text{for } l-m+n \text{ odd.}\end{cases} \tag{34}$$

Symmetric and tracefree tensor-valued spherical harmonics are defined as (Wallace [17, p. 104])

$$\frac{x^{2l-1}}{(2l-1)!!}\left(\frac{-\partial}{\partial \boldsymbol{x}}\right)^{l}\frac{1}{x} =: Y^{(l)}(\boldsymbol{x}) = \boldsymbol{x}^{l}Y^{(l)}(\hat{\boldsymbol{x}}) \tag{35}$$

with $(2l-1)!! := (2l-1)(2l-3)\cdots 1$. Examples are

$$Y^{(0)} = 1, \qquad Y^{(1)}(\boldsymbol{x}) = \boldsymbol{x}, \qquad Y^{(2)}(\boldsymbol{x}) = \boldsymbol{x}^2 - \frac{x^2}{3}\mathbf{I},$$

$$Y^{(3)}(\boldsymbol{x}) = \boldsymbol{x}^3 - \frac{3x^2}{5}\widehat{\boldsymbol{x}\mathbf{I}}. \tag{36}$$

The orthogonality relation for spherical harmonics, with $T^{(m)}$ an arbitrary tensor of order m and $\overset{\circ}{\widehat{T^{(m)}}}$ its completely symmetric and completely tracefree part, reads (Wilhelm and Winkler [19, Eq. (51)])

$$\oint d^2\Omega Y^{(l)}(\hat{\boldsymbol{x}})Y^{(m)}(\hat{\boldsymbol{x}})_{\dot{m}}T^{(m)} = 4\pi\,\delta_{lm}\frac{l!}{(2l+1)!!}\overset{\circ}{\widehat{T^{(m)}}}. \tag{37}$$

The connection between the Hermite polynomials $He^{(m)}(\alpha;\boldsymbol{x})$ (20) and the spherical harmonics $Y^{(l)}(\boldsymbol{x})$ involves generalized Laguerre polynomials

$$L_n^{\alpha}(x) = \frac{e^x}{n!x^{\alpha}}\left(\frac{d}{dx}\right)^{n}(e^{-x}x^{n+\alpha}) \tag{38}$$

and reads (Ikenberry [10, Eqs. (2), (5), (11), (31)])

$$He^{(n)}(\alpha;\boldsymbol{x}) = n!\sum_{m}^{n}(-\alpha)^{\frac{n-m}{2}}\frac{(2m+1)!!}{m!(n+m+1)!!}L_{\frac{n-m}{2}}^{m+\frac{1}{2}}\left(\frac{x^2}{2\alpha}\right)\widehat{\mathbf{I}^{\frac{n-m}{2}}Y^{(m)}(\boldsymbol{x})}, \tag{39}$$

where the sum over m means $m = 0, 2, \ldots, n$ for n even, and $m = 1, 3, \ldots, n$ for n odd. (Hence $n - m$ is always even.)

A multiplication theorem expressing the product of two generalized Laguerre polynomials as a linear combination of such polynomials is (Feldheim [6, Eq. (3.4)])

$$L_m^{\alpha}(x)L_n^{\alpha}(x) = \sum_{s=|m-n|}^{m+n} a(m,n;\alpha,s)L_s^{\alpha}(x), \tag{40}$$

with the coefficients (Weinert [18, Eqs. (A3.3), (A3.15), (A3.17)])

$$a(m,n;\alpha,s)$$

$$= (-1)^{m+n+s}\frac{\binom{n+\alpha}{n}}{\binom{s+\alpha}{s}}\sum_{k=|m-n|}^{m+n}\binom{m+\alpha}{\frac{m-n+k}{2}}\binom{n}{\frac{m+n-k}{2}}\binom{m+n-k}{s-k}$$

$$= a(n,m;\alpha,s). \tag{41}$$

The summation index k runs through all even/odd numbers between $|m-n|$ and $m+n$ if $m+n$ is even/odd.

References

[1] E.T. Bell, *Exponential polynomials*, Ann. of Math., **35** (1934), 258–277.

[2] G. Dattoli, C. Chiccoli, S. Lorenzutta, G. Maino, and A. Torre, *Generalized Bessel functions and generalized Hermite polynomials*, J. Math. Anal. Appl., **178** (1993), 509–516.

[3] D. Enskog, *Bemerkungen zu einer Fundamentalgleichung in der kinetischen Gastheorie*, Phys. Z., **12** (1911) 533–539.

[4] A. Erdélyi, W. Magnus, F. Oberhettinger, and F.G. Tricomi, *Higher transcendental functions*, Vol. 2, McGraw-Hill, New York, 1953.

[5] A. Erdélyi, W. Magnus, F. Oberhettinger, and F.G. Tricomi, *Tables of integral transforms*, Vol. 2, McGraw-Hill, New York, 1954.

[6] E. Feldheim, *Quelques nouvelles relations pour les polynomes d'Hermite*, J. London Math. Soc., **13** (1938), 22–29.

[7] J.W. Gibbs, *Vector analysis*, Dover, New York, 1909.

[8] H. Grad, *Note on N-dimensional Hermite polynomials*, Comm. Pure Appl. Math., **2** (1949), 325–330.

[9] H. Grad, *On the kinetic theory of rarefied gases*, Comm. Pure Appl. Math., **2** (1949), 331–407.

[10] E. Ikenberry, *Representation of Grad's Hermite polynomials as sums of products of Sonine polynomials and solid spherical harmonics,* Arch. Rational Mech. Anal., **9** (1962), 255–259.

[11] T.W. Johnston, *Waves in warm quiescent plasmas*, Canad. J. Phys., **40** (1962), 1208–1244.

[12] K. Kumar, *Polynomial expansions in kinetic theory of gases*, Ann. Physics, **37** (1966), 113–141.

[13] F.R.W. McCourt, J.J.M. Beenakker, W.E. Köhler, and I. Kuščer, *Nonequilibrium phenomena in polyatomic gases*, Vol.2, Clarendon, Oxford, 1991.

[14] K. Suchy, *Balance equations for Hermitian moments in mixtures of rarefied gases*, in: J. Harvey and G. Lord, Eds., *Proc. 19th Internat. Sympos. Rarefied Gas Dynamics*, Oxford University Press, Oxford, 1995, 230–236.

[15] K. Suchy, *Collision integrals for tensorial Hermite polynomials in mixtures of rarefied gases and plasmas*, Austral. J. Phys., **49** (1996), 1135–1157.

[16] K. Suchy, *A short history of balance equations for rarefied monatomic gases and plasmas*, Contin. Mech. Thermodyn., **10** (1998), 55–60.

[17] P.R. Wallace, *Angular distribution of neutrons inside a scattering and absorbing medium*, Canad. J. Research Sect. A, **26** (1948), 99–114.

[18] U. Weinert, *Darstellung des Stoßoperators der kinetischen Gleichung für Orthogonalentwicklungen der Geschwindigkeitsverteilungsfunktionen*, Thesis, University of Düsseldorf, 1975,

[19] J. Wilhelm and R. Winkler, *Tensorielle Kugelflächenfunktionsentwicklung in der Boltzmann-Gleichung für das Lorentz-Plasma bzw. -Gas*, Beitr. Plasmaphysik, **8** (1968), 167–188.

Kurt Suchy
Institut für Theoretische Physik
Universität Düsseldorf
Universitätsstr. 1
D-40225 Düsseldorf, Germany
fax: +49 211 311 3117

International Series of Numerical Mathematics
Vol. 131, © 1999 Birkhäuser Verlag Basel/Switzerland

Indeterminate Moment Problems and a Conjecture on the Growth of the Entire Functions in the Nevanlinna Parametrization

Galliano Valent

Abstract. Combining previous results on a quartic indeterminate Stieltjes moment problem and new results for a cubic one, we present a conjecture on the order, type, and Phragmen-Lindelöf indicator for a large class of indeterminate Stieltjes moment problems.

1. Introduction and Background Material

The study of indeterminate moment problems using Nevanlinna parametrization has been intensively developed in recent years, either for q-polynomials or for polynomials related to birth and death processes [6], [5], [8].

To set the stage, define polynomials $P_n(x)$ and $Q_n(x)$ as solutions of the three-term recurrence relation

$$xP_n(x) = b_n P_{n+1}(x) + a_n P_n(x) + b_{n-1} P_{n-1}(x), \qquad n \geq 1, \tag{1}$$

with the initial conditions

$$P_0(x) = 1, \quad P_1(x) = \frac{x - a_0}{b_0}; \qquad Q_0(x) = 1, \quad Q_1(x) = \frac{1}{b_0}, \tag{2}$$

and with the restrictions $a_n \in \mathbb{R}$ and $b_n > 0$ for $n \geq 0$.

The vectors $P = (P_0, P_1, \ldots, P_n, \ldots)$ and $Q = (Q_0, Q_1, \ldots, Q_n, \ldots)$ are linearly independent eigenvectors of the Jacobi matrix

$$J = \begin{pmatrix} a_0 & b_0 & 0 & 0 & \cdots \\ b_0 & a_1 & b_1 & 0 & \cdots \\ 0 & b_1 & a_2 & b_2 & \cdots \\ \cdots & \cdots & \cdots & \cdots & \cdots \\ \cdots & \cdots & \cdots & \cdots & \cdots \end{pmatrix}, \qquad J\,P(x) = x\,P(x). \tag{3}$$

If the corresponding Hamburger (resp. Stieltjes) problem is indeterminate, we will denote it by indet H (resp. indet S).

For an indet H problem, the object of greatest interest is the Nevanlinna matrix of entire functions [1]

$$\mathcal{N}(z)=\begin{pmatrix} A(z) & C(z) \\ B(z) & D(z) \end{pmatrix}, \qquad \det \mathcal{N}(z)=1, \tag{4}$$

given by

$$\begin{aligned} A(z) &= z\sum_{n\geq 0} Q_n(0)Q_n(z), & C(z) &= 1+z\sum_{n\geq 0} P_n(0)Q_n(z), \\ B(z) &= -1+z\sum_{n\geq 0} Q_n(0)P_n(z), & D(z) &= z\sum_{n\geq 0} P_n(0)P_n(z). \end{aligned} \tag{5}$$

All orthogonality measures ν_ϕ for which

$$\int P_m(x)\,P_n(x)\,d\nu_\phi(x)=\delta_{mn} \tag{6}$$

are parametrized according to

$$\int \frac{d\nu_\phi(u)}{z-u}=\frac{A(z)\phi(z)-C(z)}{B(z)\phi(z)-D(z)}, \qquad z\in\mathbb{C}\backslash\mathbb{R}, \tag{7}$$

where $\phi(z)\in\mathcal{P}\cup\{\infty\}$ and $\mathcal{P}$ is the space of Pick functions, i.e., holomorphic functions from $\mathbb{C}\backslash\mathbb{R}\to\mathbb{C}$ such that

$$\phi(\overline{z})=\overline{\phi}(z) \quad \text{and} \quad \operatorname{Im}\phi(z)\geq 0 \quad \text{for} \quad \operatorname{Im} z>0.$$

The Nevanlinna extremal measures correspond to the choice

$$\phi=t, \qquad t\in\mathbb{R}\backslash\{\infty\}.$$

The corresponding measures ν_t are purely discrete and are precisely the ones for which the polynomials $P_n(x)$ are dense in $L^2(\nu)$ by a theorem of M. Riesz [13]. Indeed, the ν_t are the spectral measures of the self-adjoint extensions of the Jacobi matrix.

If the moment problem is also indet S, then the quantity

$$\alpha=\lim_{n\to\infty}\frac{P_n(0)}{Q_n(0)}<0, \tag{8}$$

which is of prime importance, is finite. Indeed, one has [6], [5]

$$\text{supp }(\nu_t)\subseteq[0,+\infty) \quad \text{iff} \quad t\in[\alpha,0].$$

It was proved in [12] that $t=\alpha$ corresponds to the Friedrichs extension of the Jacobi matrix (3).

A natural question about the entire functions appearing in the Nevanlinna matrix is their growth properties at infinity (see [11] for basic information). These are best described in terms of their order ρ,

$$\rho(f)=\limsup_{r\to\infty}\frac{\ln\ln M(r)}{\ln r}, \qquad M(r)=\max_{|z|=r}|f(z)|, \tag{9}$$

and the Phragmen-Lindelöf indicator

$$h_f(\theta) = \limsup_{r\to\infty} \frac{\ln|f(r\,e^{i\theta})|}{r^{\rho(f)}}, \qquad \theta \in [0, 2\pi], \tag{10}$$

from which one deduces the type by

$$\sigma(f) \equiv \limsup_{r\to\infty} \frac{\ln M(r)}{r^{\rho(f)}} = \sup_{\theta\in[0,2\pi]} |h_f(\theta)|. \tag{11}$$

Roughly speaking, for large r the behaviour of $\ln M(r)$ is given by $\sigma\, r^{\rho}$.

Berg and Pedersen have proved in [4] that all elements of the Nevanlinna matrix have the same order, Phragmen-Lindelöf indicator, and type. One can therefore speak of the order, Phragmen-Lindelöf indicator, and type of the Nevanlinna matrix itself.

2. Growth Properties of the Nevanlinna Matrix for a Quartic Case

If the coefficients a_n and b_n can be written as

$$a_n = \lambda_n + \mu_n, \qquad b_n = \sqrt{\lambda_n \mu_{n+1}}, \quad n \geq 0,$$

with the positivity constraints

$$\lambda_n > 0, \quad n \geq 0; \qquad \text{and} \qquad \mu_0 \geq 0, \quad \mu_n > 0, \quad n \geq 1, \tag{12}$$

one can switch from the polynomials $P_n(x)$ defined by (1) to the polynomials $F_n(x)$ according to

$$P_n(x) = \frac{(-1)^n}{\sqrt{\pi_n}} F_n(x), \qquad \pi_0 = 1, \quad \pi_n = \frac{\lambda_0 \cdots \lambda_{n-1}}{\mu_1 \cdots \mu_n}, \quad n \geq 1. \tag{13}$$

The resulting recurrence relation then is

$$\begin{gathered}(\lambda_n + \mu_n - x)F_n(x) = \mu_{n+1}F_{n+1}(x) + \lambda_{n-1}F_{n-1}(x), \qquad n \geq 1, \\ F_0(x) = 1, \qquad F_1(x) = \frac{\lambda_0 + \mu_0 - x}{\mu_1},\end{gathered} \tag{14}$$

and the orthogonality relation

$$\int F_m(x)\, F_n(x)\, d\nu_\phi(x) = \frac{1}{\pi_n}\, \delta_{mn}.$$

According to [9], [7], the polynomials $F_n(x)$ describe a birth and death process with birth (resp. death) transition rates λ_n (resp. μ_n).

For $\mu_0 = 0$ the moment problem is indet S [5] if and only if

$$\sum_{n\geq 1} \left(\pi_n + \frac{1}{\mu_n\, \pi_n}\right) < \infty. \tag{15}$$

The quantity α can then be determined [6], [5] by

$$-\frac{1}{\alpha}=\sum_{n\geq 1}\frac{1}{\mu_n\,\pi_n}. \tag{16}$$

In this framework, by means of the generating functions obtained in [14], the full Nevanlinna matrix for the case

$$\lambda_n=(4n+1)(4n+2)^2(4n+3), \qquad \mu_n=(4n-1)(4n)^2(4n+1),$$

which is indet S (therefore indet H), was worked out. The simplest elements are the functions B and D, which were found to be

$$D(z)=\frac{4}{B(1/2,1/2)}z^{1/2}\delta_2(\theta_4(1)z^{1/4}), \quad B(z)-\frac{1}{\alpha}D(z)=-\delta_0(\theta_4(1)z^{1/4}), \tag{17}$$

with the notations

$$B(1/2,1/2)=\pi,\ \theta_4(1)=\int_0^1\frac{du}{\sqrt{1-u^4}},\ \delta_l(u)=\sum_{n\geq 0}(-1)^n\frac{u^{4n+l}}{(4n+l)!},\ l=0,2. \tag{18}$$

The order, the Phragmen-Lindelöf indicator, and the type were found to be

$$\rho=\frac{1}{4},\quad h(\theta)=\frac{\sigma}{\sqrt{2}}(\cos\theta/4+\sin\theta/4),\quad \sigma=\int_0^1\frac{du}{\sqrt{1-u^4}}. \tag{19}$$

As a last piece of information, the moments were shown to behave for large n as

$$s_n\equiv\int x^n\,d\nu\sim\frac{4(4n+1)!}{[\sqrt{2}\theta_4(1)]^{4n+2}}. \tag{20}$$

These results, except for the large n behaviour of the moments, were shown to hold true [15] also for the associated polynomials with

$$\lambda_n=(4n+4c+1)(4n+4c+2)^2(4n+4c+3), \quad \mu_n=(4n+4c-1)(4n+4c)^2(4n+4c+1)$$

for $c\geq 0$. This was an indication that the growth properties of the Nevanlinna matrix could be expected to be independent of the parameters appearing in λ_n and μ_n as long as the problem is indeterminate and $\lim_{n\to\infty}\frac{\lambda_n}{\mu_n}=1$, and they still hold for $c>\frac{1}{4}$, which corresponds to $\mu_0>0$.

All these particular examples, as well as potential physical applications, indicate that the whole family of transition rates

$$\lambda_n=(n+b_1)\cdots(n+b_p), \qquad \mu_n=(n+d_1)\cdots(n+d_p), \quad p=3,4,\ldots$$

is of great interest. In order to have a better insight, we will compute the Nevanlinna matrix for a new cubic example, after the following remarks:

1. It is difficult to make a good guess of the general structure of the Phragmen-Lindelöf indicator on the basis of this single result. We will see that the new cubic example is of some help.

2. We concentrate on the case where λ_n and μ_n are polynomials, largely because of applications. Observe that $p = 1,\ 2$ are excluded because they lead only to det S moment problems.
3. We stick to the situation where
$$\lim_{n\to\infty} \frac{\mu_n}{\lambda_n} = k^2$$
is equal to 1. For $k^2 \neq 1$ the moment problem is again det S.
4. One should be careful about the type, which does depend on the overall normalization of the coefficients λ_n and μ_n.

3. Growth Properties of the Nevanlinna Matrix for a Cubic Case

Consider the following rates
$$\lambda_n = (3n+1)(3n+2)^2, \qquad \mu_n = (3n)^2(3n+1). \tag{21}$$
The corresponding moment problem is indet S, and we would like to present the computations of the functions $D(z)$ and $B(z)$. This can be achieved by using the relations proved in [15], which we state as
$$D(z) = z\sum_{n\geq 0} F_n(z), \qquad B(z) - \frac{1}{\alpha}D(z) = -1 + z\sum_{n\geq 0}\frac{\tilde{F}_n(z)}{\tilde{\mu}_0}. \tag{22}$$
The polynomials F_n satisfy the recurrence relation (14) with coefficients $\lambda_n,\ \mu_n$ while the $\tilde{F}_n$ satisfy the same recurrence relation with $\tilde{\lambda}_n = \mu_{n+1},\ \tilde{\mu}_n = \lambda_n$, the so-called dual process in the terminology of Karlin and McGregor [9]. It follows that knowledge of a generating function for these polynomials is in principle sufficient to determine $D(z)$ and $B(z)$.

The technique to be used is a variant of the cubic decomposition [3]. We first define nearly monic polynomials $G_n(x)$ by
$$G_0(x) = 1, \qquad G_n(x) = \mu_1\cdots\mu_n\, F_n(x), \quad n\geq 1. \tag{23}$$
They satisfy the recurrence relation
$$\begin{gathered}(\lambda_n + \mu_n - x)G_n(x) = G_{n+1}(x) + \lambda_{n-1}\mu_n F_{n-1}(x), \qquad n\geq 1,\\ G_0(x) = 1, \qquad G_1(x) = \lambda_0 + \mu_0 - x.\end{gathered} \tag{24}$$
In order to solve it, we use the following device; consider the system
$$\begin{aligned}&d_{3n+1}(jx^{1/3}) = jx^{1/3}d_{3n}(jx^{1/3}), \quad n\geq 0,\\ &d_{3n+2}(jx^{1/3}) = jx^{1/3}d_{3n+1}(jx^{1/3}) + \mu_n d_{3n-1}(jx^{1/3}), \quad n\geq 0, \qquad j = e^{i\pi/3},\\ &d_{3n+3}(jx^{1/3}) = jx^{1/3}d_{3n+2}(j\rho) + \lambda_n d_{3n}(jx^{1/3}), \quad n\geq 0\end{aligned} \tag{25}$$
with the initial conditions
$$d_0(jx^{1/3}) = 1, \qquad d_{-1}(jx^{1/3}) = 0.$$

As will be apparent later on, the precise choice of the cubic root of x is immaterial. Then one can check that

$$G_n(x) = d_{3n}(jx^{1/3}) \equiv \frac{d_{3n+1}(jx^{1/3})}{jx^{1/3}}, \qquad n \geq 0, \tag{26}$$

is a solution of the recurrence (24) with the rates λ_n, μ_n given in (21), while

$$\tilde{G}_n(x) = \frac{d_{3n+2}(jx^{1/3})}{(jx^{1/3})^2}, \qquad n \geq 0, \tag{27}$$

is a solution of its dual process

$$\tilde{\lambda}_n = \mu_{n+1} = (3n+3)^2(3n+4), \qquad \tilde{\mu}_n = \lambda_n = (3n+1)(3n+2)^2.$$

To obtain a solution of (25), we define the functions

$$\frac{\sigma_l(u)}{j^l} = \sum_{n\geq 0}(-1)^l \frac{u^{3n+l}}{(3n+l)!}, \qquad l = 0, 1, 2, \tag{28}$$

which are solutions of the differential system

$$D_u \sigma_l(u) = j\, \sigma_{\tilde{l}}(u), \qquad \tilde{l} = l - 1 \ (\text{mod } 3), \tag{29}$$

and therefore

$$D_u^3 \sigma_l(u) + \sigma_l(u) = 0. \tag{30}$$

The usefulness of these functions will appear soon.

We next define the generating functions

$$\Sigma_l(x,t) = \sum_{n\geq 0} d_{3n+l}(jx^{1/3}) \frac{t^{3n+l}}{(3n+l)!}, \qquad l = 0, 1, 2. \tag{31}$$

Routine computations involving the recurrence relations (25) give the following system

$$\begin{aligned} D_t\Sigma_0 &= jx^{1/3}\Sigma_2 + tD_t(t^2\Sigma_0), \\ D_t\Sigma_1 &= jx^{1/3}\Sigma_0, \\ D_t\Sigma_2 &= jx^{1/3}\Sigma_1 + t^2 D_t(t\Sigma_2), \end{aligned} \tag{32}$$

a solution of which is checked to be

$$\begin{aligned} \Sigma_0 &= \frac{\sigma_0(jx^{1/3}\theta_3(t))}{(1-t^3)^{2/3}}, \\ \Sigma_1 &= \sigma_1(x^{1/3}\theta_3(t)), \qquad\qquad \theta_3(t) = \int_0^t \frac{du}{(1-u^3)^{2/3}}, \\ \Sigma_2 &= \frac{\sigma_2(x^{1/3}\theta_3(t))}{(1-t^3)^{1/3}}. \end{aligned} \tag{33}$$

Combining (33) and (25) with (26), we get

$$\sum_{n\geq 0} G_n(x)\frac{t^n}{(3n)!} = \frac{\sigma_1(x^{1/3}\theta_3(t^{1/3}))}{jx^{1/3}t^{1/3}}, \qquad |t|\leq 1, \tag{34}$$

and upon use of (23) we conclude

$$\sum_{n\geq 0} \frac{n!}{(2/3)_n}F_n(x) = \frac{\sigma_1(x^{1/3}\theta_3(t^{1/3}))}{jx^{1/3}t^{1/3}}, \qquad |t|\leq 1, \quad (a)_n \equiv \frac{\Gamma(n+a)}{\Gamma(a)}. \tag{35}$$

This implies

$$\sum_{n\geq 0} F_n(x)t^n = \frac{1}{B(1/3,2/3)}\int_0^1 u^{-1/3}(1-u)^{-2/3}\frac{\sigma_1(x^{1/3}\theta_3((tu)^{1/3}))}{jx^{1/3}(tu)^{1/3}}\,du, \quad |t|\leq 1, \tag{36}$$

(expand σ_1 using relation (28) and integrate term-by-term). Using the variable

$$y = x^{1/3}\theta_3((u)^{1/3})$$

gives

$$\sum_{n\geq 0} F_n(x) = \frac{3}{x^{1/3}B(1/3,2/3)}\int_0^{x^{1/3}\theta_3(1)} \frac{\sigma_1(y)}{jx^{1/3}}\,dy = \frac{3}{B(1/3,2/3)}\frac{\sigma_2(x^{1/3}\theta_3(1))}{(jx^{1/3})^2},$$

the last equality following from (29). Then relation (22) implies

$$D(z) = \frac{3}{B(1/3,2/3)}\, z^{1/3}\, \frac{\sigma_2(z^{1/3}\theta_3(1))}{j^2}. \tag{37}$$

Remarks:

1. One has, more explicitly, $B(1/3,2/3) = 2\pi/\sqrt{3}$.
2. Relation (28) shows that the function $D(z)$ is entire.
3. The generating function (36) was first derived in [10].

One can proceed similarly for the dual polynomials $\tilde{F}_n(x)$. One gets first

$$\sum_{n\geq 0} \frac{(5/3)_n}{n!}\tilde{F}_n(x)t^n = 2\frac{\sigma_2(x^{1/3}\theta_3(t^{1/3}))}{(jx^{1/3}t^{1/3})^2(1-t)^{1/3}}, \qquad |t|<1, \tag{38}$$

then

$$\sum_{n\geq 0} \frac{\tilde{F}_n(x)}{\tilde{\mu}_0}t^n = \frac{1}{3}\int_0^1 (1-u)^{-1/3}\frac{\sigma_2(x^{1/3}\theta_3((tu)^{1/3}))}{(jx^{1/3}(tu)^{1/3})^2(1-tu)^{1/3}}\,du, \quad |t|\leq 1, \tag{39}$$

and upon use of (29),

$$\sum_{n\geq 0} \frac{\tilde{F}_n(x)}{\tilde{\mu}_0} = \frac{1}{x^{1/3}}\int_0^{x^{1/3}\theta_3(1)} \frac{\sigma_2(y)}{(jx^{1/3})^2}\,dy = \frac{1-\sigma_0(x^{1/3}\theta_3(1))}{x}. \tag{40}$$

Therefore, (22) finally gives

$$B(z) - \frac{1}{\alpha} D(z) = -\sigma_0(z^{1/3}\theta_3(1)). \tag{41}$$

We summarize our findings in

Proposition 1. *For the indeterminate Stieltjes moment problem corresponding to*

$$\lambda_n = (3n+1)(3n+2)^2, \qquad \mu_n = (3n)^2(3n+1),$$

the entire functions $D(z)$ *and* $B(z)$ *are given by*

$$D(z) = \frac{3}{B(1/3,2/3)}\, z^{1/3}\, \frac{\sigma_2(\theta_3(1)z^{1/3})}{j^2}, \qquad B(z) - \frac{1}{\alpha} D(z) = -\sigma_0(\theta_3(1)z^{1/3}),$$

with

$$\theta_3(1) = \int_0^1 \frac{du}{(1-u^3)^{2/3}}.$$

The following remarks are now in order:

1. The spectra of the Nevanlinna extremal measures are given by
$$u\frac{\sigma_2(u)}{\sigma_0(u)} = -\frac{3\sqrt{3}}{2\pi}\theta_3(1)\frac{\alpha t}{\alpha - t}, \qquad u = z^{1/3}\theta_3(1), \quad t \in \mathbb{R} \cup \{\infty\}.$$
Contrary to the quartic case, no value of the parameter $t \in [\alpha, 0]$ (positively supported measures) leads to a closed-form spectrum.
2. With the help of $D(z)$ it is an easy task to show that the order and type of the Nevanlinna matrix are given by
$$\rho = \frac{1}{3}, \quad \sigma = \theta_3(1) \equiv \int_0^1 \frac{du}{(1-u^3)^{2/3}}, \tag{42}$$
while its Phragmen-Lindelöf indicator is
$$h(\theta) = \frac{\sigma}{2}[\cos(\theta/3) + \sqrt{3}\sin(\theta/3)] = \sigma \cos\frac{(\theta - \pi)}{3}, \qquad \theta \in [0, 2\pi]. \tag{43}$$

From these results we can now make a complete statement of the conjecture, including the Phragmen-Lindelöf indicator:

Conjecture 1. *Consider the moment problem* $\mathcal{P}$ *with recurrence coefficients*

$$\lambda_n = (pn+b_1)\cdots(pn+b_p), \quad \mu_n = (pn+d_1)\cdots(pn+d_p), \quad p = 3, 4, \dots, \tag{44}$$

satisfying the positivity constraints (12). *If* $\mathcal{P}$ *is* indet S, *the order and type of its Nevanlinna matrix are*

$$\rho = \frac{1}{p}, \qquad \sigma = \theta_p(1) \equiv \int_0^1 \frac{du}{(1-u^p)^{2/p}},$$

and the Phragmen-Lindelöf indicator is

$$h(\theta) = \sigma \cos\frac{(\theta - \pi)}{p}, \qquad \theta \in [0, 2\pi].$$

Remarks:

1. Using the criteria given in [5], one can show that, for $\mu_0 = 0$, the moment problem $\mathcal{P}$ is indet S if and only if
$$p < \sum_{i=1}^{p} (b_i - d_i) < p(p-1). \tag{45}$$
2. We would like to further explain the important role of the integral $\theta_p(1)$, which appears in this context, and which relates orthogonal polynomials to algebraic geometry. Indeed, if we define the generating function
$$F(x,z) = \sum_{n \geq 0} F_n(x) z^n, \qquad z \in \mathbb{C},$$
one gets for the transition rates (44) a differential equation of the form
$$z^{p-1}(1-z)^2 D_z^p F(x,z) + \cdots + (x + \cdots) F(x,z) = 0.$$
It follows that the variable
$$w^p = z^{p-1}(1-z)^2 \tag{46}$$
should play a prominent role, independent of the parameters b_i and d_i appearing in (44).

The algebraic equation (46) defines a Riemann surface, which is characterized by its genus g. In order to compute g, one has to describe the behaviour of w near the ramification points $z_1 = 0,\ z_2 = 1,\ z_3 = \infty$. In a neighbourhood of these points we have
$$w = (z - z_i)^{s_i/r_i} f_i(z),$$
where $f_i(z)$ is holomorphic in some neighbourhood of z_i and s_i/r_i is irreducible with $r_i > 0$. One has to compute numbers α_i such that $\alpha_i r_i = p$; then the genus of the surface (46) is given by Riemann's formula [2, p. 232],
$$g = \frac{1}{2} \sum_i \alpha_i (r_i - 1) - (p-1).$$
For $p = 2l$, easy computations give
$$r_1 = p, \quad s_1 = p-1, \quad \alpha_1 = 1,$$
$$r_2 = p/2, \quad s_2 = 1, \quad \alpha_2 = 2,$$
$$r_3 = p,\ s_3 = -(p+1),\ \alpha_3 = 1,$$
and the genus is $g = l - 1 = p/2 - 1$. For $p = 2l+1$ we have
$$r_1 = p, \quad s_1 = p-1, \quad \alpha_1 = 1,$$
$$r_2 = p, \quad s_2 = 2, \quad \alpha_2 = 1,$$
$$r_3 = p, \quad s_3 = -(p+1), \quad \alpha_3 = 1,$$

which gives $g = l = (p-1)/2$. Therefore, we conclude

$$g = \mathrm{E}\left((p-1)/2\right), \tag{47}$$

where $\mathrm{E}(x)$ denotes the integer part of the real x normalized by $\mathrm{E}(1/2) = 0$. For a given genus g, one can define g abelian differentials of the first kind, and their integrals over the cycles give $2n$ (real) independent periods. The quantity $\theta_p(1)$ is the simplest real period.

Not unexpectedly, the examples worked out so far correspond to $p = 3, 4$, i.e., genus 1. In this case the Jacobi inversion can be solved using elliptic functions, but for higher genus the situation is much more difficult to handle. At any rate, it would be highly interesting to have more examples and more information about the higher genus polynomials.

References

[1] N.I. Akhiezer, *The classical moment problem and some related questions in analysis*, translated by N. Kemmer, Hafner, New York, 1965.

[2] P. Appell and E. Goursat, *Théorie des fonctions algébriques et de leurs intégrales. I: Étude des fonctions analytiques sur une surface de Riemann*, 2d ed., Gauthier-Villars, Paris, 1929.

[3] P. Barrucand and D. Dickinson, *On cubic transformations of orthogonal polynomials*, Proc. Amer. Math. Soc., **17** (1966), 810–814.

[4] C. Berg and H.L. Pedersen, *On the order and type of the entire functions associated with an indeterminate Hamburger moment problem*, Ark. Mat., **32** (1994), 1–11.

[5] C. Berg and G. Valent, *The Nevanlinna parametrization for some indeterminate Stieltjes moment problems associated with birth and death processes*, Methods Appl. Anal., **1** (1994), 169–209.

[6] T.S. Chihara, *Indeterminate symmetric moment problems*, J. Math. Anal. Appl., **85** (1982), 331–346.

[7] M.E.H. Ismail, J. Letessier, D.R. Masson, and G. Valent, *Birth and death processes and orthogonal polynomials*, in: P. Nevai, Ed., *Orthogonal polynomials: theory and practice*, NATO Adv. Sci. Inst. Ser. C Math. Phys. Sci. **294**, Kluwer, Dordrecht, 1990, 229–255.

[8] M.E.H. Ismail and D.R. Masson, *q-Hermite polynomials, biorthogonal rational functions, and q-beta integrals*, Trans. Amer. Math. Soc., **346** (1994), 63–116.

[9] S. Karlin and J.L. McGregor, *The differential equations of birth-and-death processes, and the Stieltjes moment problem*, Trans. Amer. Math. Soc., **85** (1957), 489–546.

[10] J. Letessier and G. Valent, *Some exact solutions of the Kolmogorov boundary value problem*, Approx. Theory Appl., **4** (1988), 97–117.

[11] A.I. Markushevich, *Theory of functions of a complex variable*, Vol. 2, translated and edited by R.A. Silverman, 2d Engl. ed., Chelsea, New York, 1977.

[12] H.L. Pedersen, *Stieltjes moment problems and the Friedrichs extension of a positive definite operator*, J. Approx. Theory, **83** (1995), 289–307.

[13] M. Riesz, *Sur le problème des moments et le théorème de Parseval correspondant*, Acta Sci. Math. (Szeged), **1** (1923), 209–225.

[14] G. Valent, *Exact solutions of some quadratic and quartic birth and death processes and related orthogonal polynomials*, J. Comput. Appl. Math., **67** (1996), 103–127.

[15] G. Valent, *Co-recursivity and Karlin-McGregor duality for indeterminate moment problems*, Constr. Approx., **12** (1996), 531–553.

Galliano Valent
Laboratoire de Physique Théorique et Hautes Energies
Unité associée au CNRS UA 280
Université Paris 7
2 Place Jussieu
F-75251 Paris Cedex 05, France
E-mail address: valent@lpthe.jussieu.fr

International Series of Numerical Mathematics
Vol. 131,

Spectral Methods Based on Nonclassical Orthogonal Polynomials

J.A.C. Weideman

Abstract. Spectral methods for solving differential equations of boundary value type have traditionally been based on classical orthogonal polynomials such as the Chebyshev, Legendre, Laguerre, and Hermite polynomials. In this numerical study we show that methods based on nonclassical orthogonal polynomials may sometimes be more accurate. Examples include the solution of a two-point boundary value problem with a steep boundary layer and two Sturm-Liouville problems.

1. Introduction

The use of orthogonal polynomials for solving boundary value problems analytically is of course a classical idea. However, it was not until the 1977 publication of D. Gottlieb and S.A. Orszag's monograph [9] that the use of orthogonal polynomials gained momentum as a respectable procedure for solving differential equations numerically, and it soon became a serious competitor to more established procedures such as the methods of finite elements or finite differences. The class of solution methods based on orthogonal polynomials (and other global expansions such as Fourier and sinc series) have become known as spectral methods.

Spectral methods are implemented in various ways. For example, the tau, Galerkin, and collocation methods have all been proposed as implementation strategies—for a comparison of these three methods, see [4]. However, the collocation method (also known as the pseudospectral method) has established itself as the one that permits the most convenient computer implementation, at least as far as nonconstant-coefficient or nonlinear differential equations are concerned.

The spectral collocation method is based on weighted interpolants of the form

$$f(x) \approx p_N(x) = \sum_{j=0}^{N} \frac{w(x)}{w(x_j)} \phi_j(x) f_j, \quad a \le x \le b. \tag{1}$$

Here $\{x_j\}_{j=0}^N$ is a set of distinct interpolation nodes in $[a,b]$, $w(x)$ is some positive weight function, $f_j = f(x_j)$, and the set of interpolating functions $\{\phi_j(x)\}_{j=0}^N$ satisfies $\phi_j(x_k) = \delta_{jk}$ (the Kronecker delta). This means that $p_N(x)$, defined by

(1), is an interpolant of the function $f(x)$ in the sense that

$$f(x_k) = p_N(x_k), \quad k = 0, 1, \ldots, N.$$

The functions $\{\phi_j(x)\}$ are often chosen to be sets of interpolating polynomials of degree N, in which case they can be represented explicitly by Lagrange's formula. Approximation theory dictates that the nodes $\{x_j\}$ and interpolating polynomials $\{\phi_j(x)\}$ cannot be arbitrary—good choices are associated with the classical orthogonal polynomials such as:

Chebyshev: The interval is $[-1, 1]$, the weight function is constant, i.e., $w(x)$ is identically 1, and the nodes are the Chebyshev points of the second kind

$$x_j = \cos\left(\frac{j\pi}{N}\right), \quad j = 0, 1, \ldots, N. \tag{2}$$

The interpolant is given by

$$p_N(x) = \sum_{j=0}^{N}{}' \phi_j(x) f_j, \tag{3}$$

where

$$\phi_j(x) = (-1)^{j+1} \frac{(1-x^2)T_N'(x)}{N^2(x-x_j)}.$$

(The prime in the summation indicates that the first and last terms are multiplied by $\frac{1}{2}$; see [2, p. 69].)

Hermite: The interval is $(-\infty, \infty)$, the weight function is $e^{-x^2/2}$, and the nodes are the zeros of the Hermite polynomial of degree $N+1$, i.e., $H_{N+1}(x_j) = 0$, $j = 0, \ldots, N$. The interpolant is given by

$$p_N(x) = \sum_{j=0}^{N} \frac{e^{-x^2/2}}{e^{-x_j^2/2}} \phi_j(x) f_j, \tag{4}$$

where

$$\phi_j(x) = \frac{H_{N+1}(x)}{H_{N+1}'(x_j)(x-x_j)} = \frac{H_{N+1}(x)}{(N+1)H_N(x_j)(x-x_j)}.$$

Spectral methods have also been based on Laguerre, sinc (cardinal), and trigonometric (Fourier) interpolants; see [5], [11], and [4], respectively.

The collocation approach to computing derivatives consists of differentiating interpolants such as (3) and (4) analytically. After taking say m derivatives, and evaluating the result at the nodes, one obtains

$$f^{(m)}(x_k) \approx \sum_{j=0}^{N} \frac{d^m}{dx^m}\left[\frac{w(x)}{w(x_j)}\phi_j(x)\right]_{x=x_k} f_j, \quad k = 0, \ldots, N. \tag{5}$$

The numerical differentiation process represented by (5) may be summarized by the matrix-vector product

$$\mathbf{f}^{(m)} = D^{(m)}\mathbf{f}, \tag{6}$$

where $\mathbf{f}$ (resp. $\mathbf{f}^{(m)}$) is the vector of function values (resp. approximate derivative values) at the nodes $\{x_j\}$. The matrix $D^{(m)}$, with entries

$$D_{kj}^{(m)} = \frac{d^m}{dx^m}\left[\frac{w(x)}{w(x_j)}\phi_j(x)\right]_{x=x_k}, \tag{7}$$

is known as the differentiation matrix. Explicit formulas for first-derivative Chebyshev and Hermite matrices, derived from (3) and (4) respectively, may be found in [2] and [5].

When solving differential equations, the continuous derivatives are approximated by their discrete counterparts (6). In this manner a two-point boundary value problem may be converted to a linear system. A differential eigenvalue problem may similarly be approximated by a matrix eigenvalue problem. Examples of the procedure are given below.

As for the computation of the differentiation matrices defined by (7), an efficient algorithm, implemented in a FORTRAN program, was developed by B.D. Welfert [16]. The algorithm, which is restricted to the case when $\{\phi_j(x)\}$ is a set of polynomials, requires the following input: (a) a set of distinct nodes $\{x_j\}$, (b) the weight function sampled at the nodes, i.e., $w_j = w(x_j)$, and (c) the quantities $d_j^{(\ell)}$ defined by

$$d_j^{(\ell)} = \frac{w^{(\ell)}(x_j)}{w(x_j)}, \quad j = 0, \dots, N, \quad \ell = 1, \dots, m. \tag{8}$$

The present author has converted Welfert's FORTRAN program to a MATLAB function `poldif.m` which will be made available publicly; see [15].

As far as the author is aware, polynomial-based differentiation matrices have been exclusively derived from classical orthogonal polynomials such as the Chebyshev and Hermite expansions (3) and (4), or analogous Legendre and Laguerre expansions (not listed here). The purpose of the present investigation was to explore, through numerical experiments, whether the use of nonclassical orthogonal polynomials might be advantageous. Our examples include:

§2: A spectral collocation method for the bounded interval, based on polynomials orthogonal with respect to a rational weight function. We compare this method with the classical Chebyshev method for solving (a) a certain Sturm-Liouville problem, and (b) a model boundary-layer problem.

§3: A spectral collocation method for the real line, based on polynomials orthogonal with respect to the logistic density weight function. We compare this method with the classical Hermite method for solving Schrödinger's equation with a certain potential.

2. Rational Weight Functions for a Bounded Interval

In this section we assume that the function $f(x)$ is meromorphic in the whole complex plane, i.e., its only singularities are poles. The interval will be taken to be $[a, b]$, with a and b both finite. The weight function is chosen as

$$w(x) = \frac{1}{\pi(x)}, \tag{9}$$

where

$$\pi(x) = \prod_{n=1}^{M} (x - p_n).$$

The parameters $p_1, \ldots, p_M$ are selected so that $w(x)$ is real and of constant sign on $[a, b]$. This means that the $p_1, \ldots, p_M$ are outside $[a, b]$, and if complex, they should occur in conjugate pairs.

With this weight function the interpolant (1) may be expressed as

$$p_N(x) = \sum_{j=0}^{N} \frac{\pi(x_j)}{\pi(x)} \phi_j(x) f_j \iff f(x)\pi(x) \approx \sum_{j=0}^{N} \phi_j(x)\pi(x_j) f_j, \tag{10}$$

where the $\{\phi_j(x)\}$ are the Lagrangian interpolating polynomials associated with the (as yet undetermined) set of nodes $\{x_j\}$.

The left side of expression (10) makes it clear that what we are proposing here is nothing but a rational interpolant with preassigned poles $p_1, \ldots, p_M$. The right side of the expression emphasizes that one should choose the $p_1, \ldots, p_M$ so that they coincide with the poles of $f(x)$ nearest the interval $[a, b]$. For if they do, the function $f(x)\pi(x)$ becomes analytic in a wider region about the interval $[a, b]$ than the original function $f(x)$, and this makes polynomial interpolation (and also differentiation) converge faster; see [4], [12].

Of course, when solving differential equations, the function $f(x)$ is not known and the location of its poles will have to be estimated by some means. In the two examples below we shall demonstrate two such techniques, namely WKB analysis and boundary-layer theory.

It remains to choose the nodes $\{x_j\}$. The natural choice seems to be the roots of the polynomials $\{q_n(x)\}$ orthogonal on $[a, b]$ with respect to the weight function $w(x)$. There are several algorithms available for doing this; see [3], [6], [8], [13].[1] What these algorithms have in common is that they all compute the coefficients (α_n, β_n) in the three-term recurrence

$$q_{n+1}(x) = (x - \alpha_n)\, q_n(x) - \beta_n^2\, q_{n-1}(x), \qquad n = 0, 1, 2, \ldots \tag{11}$$

[1]We have not used one of the existing algorithms for computing the recurrence coefficients (α_n, β_n). Instead, D.P. Laurie and the author developed a new algorithm which we believe has some attractive features. It consists of a recursion to compute the modified Chebyshev moments, which are then used in the Sack-Donovan-Wheeler algorithm for computing (α_n, β_n). Details will be published elsewhere.

(where $q_0(x) = 1$, $q_{-1}(x) = 0$). The roots of $q_{N+1}(x)$, which become the nodes $\{x_k\}_{j=0}^{N}$, are computed as the eigenvalues of the $(N+1) \times (N+1)$ tridiagonal Jacobi matrix associated with (11); see [7].

The nodes $\{x_j\}$ generated in this manner can be shown to be distinct and to lie within the interval of support $[a, b]$; see [13]. It may be necessary to include the points $x = a$ or $x = b$ (or both) as additional nodes, to facilitate the incorporation of boundary conditions.

Once the nodes $\{x_j\}$ have been computed, the computation of the quantities (8) is straightforward, at least for first and second derivatives. At this point, Welfert's algorithm may be used to construct the differentiation matrices defined by (7).

2.1. Example: A Sturm-Liouville problem

Consider the eigenvalue problem

$$y''(x) + \lambda(x+\pi)^4 y(x) = 0, \quad y(0) = y(\pi) = 0, \tag{12}$$

taken from [1, p. 490]. This problem has an infinite number of eigenvalues, all positive, and given asymptotically by

$$\lambda_n \sim \frac{9n^2}{49\pi^4}, \qquad n \longrightarrow \infty.$$

The corresponding (unnormalized) eigenfunctions are

$$y_n(x) \sim \frac{\sin(n(x^3 + 3x^2\pi + 3\pi^2 x)/7\pi^2)}{(x+\pi)}, \qquad n \longrightarrow \infty. \tag{13}$$

Derived by the WKB method, these asymptotic estimates become more accurate with increasing n. But even for $n = 1$ the actual and approximate eigenfunctions are hardly distinguishable; see Fig. 10.2 in [1, p. 491].

The standard Chebyshev approach to solving (12) consists of constructing the $(N+1) \times (N+1)$ second derivative matrix $D^{(2)}$ associated with the nodes (2), but shifted from $[-1, 1]$ to $[0, \pi]$. The incorporation of the boundary conditions $y(0) = y(\pi) = 0$ requires that the first and last rows of the matrix $D^{(2)}$ be deleted, as well as its first and last columns; see [4].

The collocation approximation to the differential eigenvalue problem (12) is now represented by the $(N-1) \times (N-1)$ matrix eigenvalue problem

$$D^{(2)}\mathbf{y} + \lambda \operatorname{diag}\left((x_j + \pi)^4\right)\mathbf{y} = 0, \tag{14}$$

where $\mathbf{y}$ is the vector of approximate eigenfunction values at the interior nodes x_j.

The eigenproblem (14) was solved by MATLAB, and the relative error in the smallest eigenvalue, λ_1, is shown as the curve marked (a) in Fig. 1. On the horizontal axis we have indicated the order of the matrix eigenvalue problem (14), which is $N-1$ in our notation. Note the logarithmic scale on the vertical axis.

The new method proposed here is also based on the matrix eigenproblem (14), but with a different set of nodes and a new differentiation matrix. As suggested

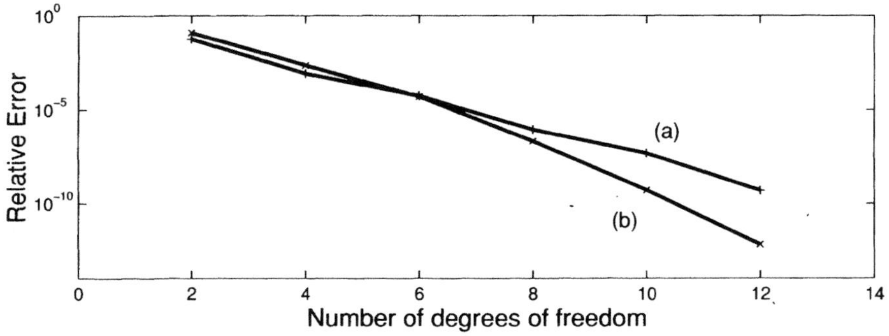

FIGURE 1. Relative errors in the numerical approximations to the smallest eigenvalue of (12). Curve (a) corresponds to the standard Chebyshev method and curve (b) corresponds to the new method proposed here.

by the WKB analysis summarized in (13), we have chosen the weight function to be

$$w(x) = \frac{1}{x+\pi}.$$

The nodes $\{x_j\}$ are the roots of the polynomials orthogonal on $[0, \pi]$ with respect to this weight, plus two additional nodes at $x = 0$ and $x = \pi$, to facilitate the incorporation of the boundary conditions. Compared to the Chebyshev nodes, most of the new nodes are shifted slightly in the direction of the singularity at $x = -\pi$.

The new matrix $D^{(2)}$ is computed by Welfert's algorithm, and the boundary conditions are incorporated as before, by deleting the first and last rows and columns of the matrix. Assuming exact arithmetic, the $(N-1) \times (N-1)$ differentiation matrix $D^{(2)}$ thus generated computes the exact second derivative of a function

$$f(x) = \frac{x(x-\pi)}{x+\pi} p(x),$$

where $p(x)$ is any polynomial of degree less than $N-1$. By contrast, the Chebyshev matrix $D^{(2)}$ of the same order yields exact derivatives for functions of the form

$$f(x) = x(x-\pi)p(x),$$

where $p(x)$ is again a polynomial of degree less than $N-1$.

The new set of nodes and differentiation matrix have been used in the matrix eigenproblem (14). The relative error in the smallest eigenvalue, λ_1, is shown as the curve marked (b) in Fig. 1. One notices that the new method yields more accurate answers by two or more orders of magnitude for large N.

A careful error analysis has not been carried out, but heuristically the two error curves in Fig. 1 may be contrasted as follows. Assuming the asymptotic eigenfunction (13) is an accurate representation of the actual eigenfunction for $n = 1$, the original Chebyshev method is approximating a function with a pole at $x = -\pi$ (and no other singularities). The typical rate of convergence in polynomial interpolation (and also differentiation) for such a function is exponential, where the rate of decay is determined by the location of the singularity with respect to the interval; see [4], [12]. The error curve (a) is consistent with exponential convergence.[2]

By contrast, in the new method the weight function $w(x) = 1/(x+\pi)$ absorbs the pole and effectively the method approximates only the numerator in (13), which is an entire function. The error curve (b) is consistent with the super-exponential convergence rate associated with the error in polynomial approximation of entire functions.

2.2. Example: A boundary-layer problem

This example was inspired by an example in the paper [13]. The authors considered numerical approximations of

$$\int_{-1}^{+1} \omega \mathrm{e}^{-\omega(x+1)}\, dx, \qquad \omega > 0,$$

the integrand of which exhibits a steep gradient near $x = -1$ when ω is large. They showed that a quadrature method based on rational functions—as opposed to polynomials—is markedly superior to the standard Gauss-Legendre rule in this situation. The poles of the rational functions were chosen such as to mimic the "almost singular" behavior of the integrand.

A natural idea is to extend this to two-point boundary value problems, in particular those that possess boundary layers. A simple model problem is [1, p. 431]

$$\epsilon y''(x) + (1+\epsilon)y'(x) + y(x) = 0, \qquad y(0) = 0, \quad y(1) = 1, \tag{15}$$

the exact solution of which is

$$y(x) = \frac{\mathrm{e}^{-x} - \mathrm{e}^{-x/\epsilon}}{\mathrm{e}^{-1} - \mathrm{e}^{-1/\epsilon}}.$$

For small ϵ the solution exhibits a steep gradient, known as a boundary layer, near $x = 0$; see Fig. 2 where the solution corresponding to $\epsilon = 0.001$ is shown. Steep gradients such as these pose severe challenges to numerical methods, and the spectral method is no exception.

The standard spectral collocation method for solving (15) consists of generating the first- and second-derivative Chebyshev matrices $D^{(1)}$ and $D^{(2)}$, both of

---

[2] The convergence rate represented by the error curve (a) in Fig. 1 can in fact be estimated theoretically. Fitting the regularity ellipse (defined in [12]) for Chebyshev interpolation through the pole at $x = -\pi$ indicates a convergence rate of $O(1/(3+\sqrt{8})^N) \approx O(0.17^N)$. Fitting a least squares line to the error curve yields approximately $O(0.16^N)$.

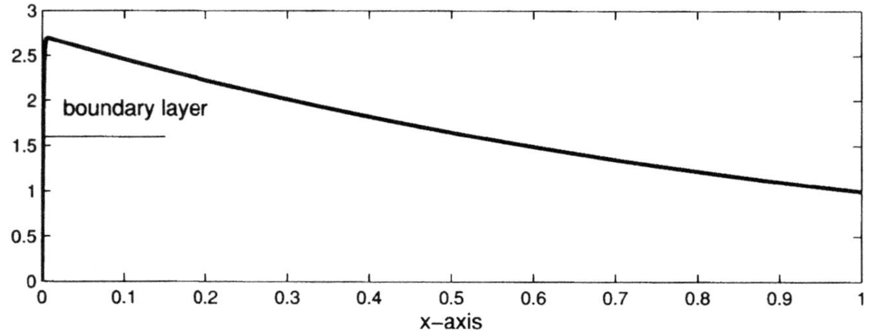

FIGURE 2. Solution of the boundary value problem (15), with $\epsilon = 0.001$

order $(N+1) \times (N+1)$, on the Chebyshev points (2), shifted from $[-1, 1]$ to $[0, 1]$. The operator on the left side of (15) may then be approximated by the matrix

$$A = \epsilon D^{(2)} + (1 + \epsilon) D^{(1)} + I,$$

where I is the identity matrix. Incorporating the boundary conditions is done as before by deleting the first and last rows and columns of A. The boundary condition $y(1) = 1$ means the column of A corresponding to $x = 1$ is transferred to the right side, which will be denoted by the vector $\mathbf{b}$. The equation (15) is thus approximated by the $(N-1) \times (N-1)$ linear system

$$A\mathbf{y} = \mathbf{b}, \tag{16}$$

where $\mathbf{y}$ is the vector of approximations to $y(x)$ at the interior nodes.

The linear system (16) was solved by MATLAB, and the pointwise errors for $N = 26$ and $N = 51$ are shown as the curves marked (a) in Fig. 3. The presence of the thin boundary layer causes the Chebyshev method to converge slowly.

The new method proposed here consists of choosing a rational weight function of the form (9), constructing the corresponding set of nodes $\{x_j\}$ as described above, and supplementing this set with two boundary nodes $x = 0$ and $x = 1$. The differentiation matrices $D^{(1)}$ and $D^{(2)}$ are then generated by Welfert's algorithm, and a linear system similar to (16) can be set up.

The interesting aspect is the weight (9) and in particular the selection of the parameters $p_1, \ldots, p_M$. The poles should reflect the "almost singular" behavior of the solution at $x = 0$, as was pointed out above. In this regard boundary-layer analysis is a useful tool.

Such an analysis consists of finding two approximate solutions: an outer solution that is valid in the region where the solution is slowly varying, and an inner solution that is valid in the boundary layer. For the problem (15) these two

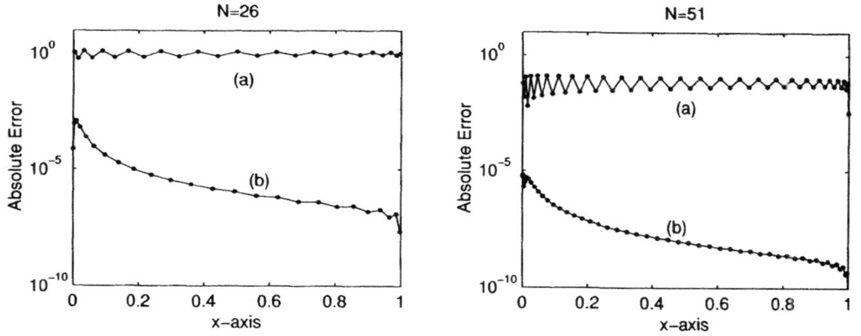

FIGURE 3. Pointwise errors in the numerical solution of (15). The curves (a) correspond to the standard Chebyshev method. The curves (b) correspond to the new method proposed here. The figure on the left (resp. right) corresponds to a linear system (16) of order 25×25 (resp. 50×50).

solutions are given by [1, p. 427][3]

$$y_{\text{in}}(x) = \mathrm{e} - \mathrm{e}^{1-x/\epsilon}, \qquad y_{\text{out}}(x) = \mathrm{e}^{1-x}.$$

The inner solution is the one that poses problems to the polynomial approximation, and therefore we match the weight function to $y_{\text{in}}(x)$. The strategy we propose is the following: construct an (M, M) rational Padé approximation to $y_{\text{in}}(x)$, and use its poles as parameters $p_1, \dots, p_M$. In particular, consider the $(5, 5)$ Padé approximation to the exponential

$$\mathrm{e}^x \approx \frac{30240+15120x+3360x^2+420x^3+30x^4+x^5}{30240-15120x+3360x^2-420x^3+30x^4-x^5}.$$

By making a change of variable $x \longleftrightarrow 1 - x/\epsilon$ the Padé approximation to $y_{\text{in}}(x)$ can be obtained. With $\epsilon = 0.001$ the poles of this rational function are

$$p_1 \approx -0.0073, \quad p_{2,3} \approx -0.0067 \pm 0.0035i, \quad p_{4,5} \approx -0.0046 \pm 0.0071i,$$

which were used to define the weight function (9), and hence the nodes $\{x_j\}$. Compared to the Chebyshev nodes, the new nodes are shifted towards the boundary layer as expected.

After constructing the differentiation matrices via Welfert's algorithm, we solved the linear system (16) and plotted the pointwise errors. The results are shown as the curves marked (b) in Fig. 3. Clearly the new method improves on the Chebyshev method by several orders of magnitude.

A few comments on the choice of M are in order. By increasing M the accuracy improves, but at the same time the condition number of the matrix

[3] Since the problem (15) can be solved exactly, the construction of the inner and outer solutions seems pointless. In realistic situations, however, the exact solution cannot be found explicitly, while it is often possible to construct the inner and outer solutions.

A grows. In particular, choosing M proportional to N leads to very accurate results for small N, but for large N roundoff error corrupts the results. The choice of $M = 5$ (independent of N) that we used here appeared to be a reasonable compromise between accuracy and stability.

Other choices of rational approximations, such as the Thiele or Chebyshev approximations, could lead to different results but this was not investigated. We also tried the strategy of choosing only real parameters $p_1, \ldots, p_M$ suggested in [13]. This improved on the standard Chebyshev method, but not by as much as the Padé approach suggested here.

3. The Logistic Density Weight Function for the Unbounded Interval

We consider the logistic density weight function

$$w(x) = \frac{e^{-x}}{(1+e^{-x})^2} = \tfrac{1}{4}\operatorname{sech}^2 \tfrac{1}{2}x, \qquad -\infty < x < \infty.$$

The polynomials orthogonal on the real line with respect to this weight function are generated by the three-term recurrence

$$Q_{n+1}(x) = x\,Q_n(x) - \tfrac{n^4\pi^2}{4n^2-1} Q_{n-1}(x), \qquad n = 0, 1, \ldots, \tag{17}$$

where $Q_{-1}(x) = 0$, $Q_0(x) = 1$; see [7, p. 45].

We thus propose the following interpolant, which should be seen as alternative to the Hermite interpolant (4) for the real line:

$$p_N(x) = \sum_{j=0}^{N} \frac{\operatorname{sech}^2(x/2)}{\operatorname{sech}^2(x_j/2)} \phi_j(x) f_j. \tag{18}$$

The nodes are the roots of $Q_{N+1}(x)$, and the interpolating functions are

$$\phi_j(x) = \frac{Q_{N+1}(x)}{Q'_{N+1}(x_j)(x-x_j)}.$$

The nodes were computed as the eigenvalues of the Jacobi matrix associated with the recurrence (17), and the differentiation matrices generated by the interpolant (18) were computed by Welfert's algorithm.

As far as we are aware, spectral methods have not been based on these orthogonal polynomials before, and we believe these differentiation matrices to be new. We made no attempt to analyze the properties of these matrices in any systematic manner, but numerical investigations revealed that they have properties similar to those of the Hermite differentiation matrices studied in [14].

For example, the Hermite first-derivative matrix, though not skew-symmetric, has pure imaginary eigenvalues. The second-derivative matrix, though not symmetric, has real eigenvalues, all negative. The new differentiation matrices have the same properties (at least for $1 \leq N \leq 80$, which is how far we have checked it numerically). However, the Hermite method seems to be marginally better conditioned than the new method. For example, the second derivative matrices have condition numbers that grow like $O(N^2)$ for the Hermite method, compared to $O(N^{1+\sqrt{2}})$ for the new method.[4]

3.1. Example: A Schrödinger problem

The Schrödinger equation

$$-y''(x) + q(x)y(x) = \lambda y(x), \qquad \lim_{|x|\to\infty} y(x) = 0, \tag{19}$$

may be approximated by

$$-D^{(2)}\mathbf{y} + \mathrm{diag}(q(x_k))\mathbf{y} = \lambda \mathbf{y}, \tag{20}$$

where $D^{(2)}$ is any second derivative matrix appropriate for the real line, and $\mathbf{y}$ is the vector of approximations to $y(x_k)$. When $D^{(2)}$ is constructed from (4) or (18), no boundary conditions need to be enforced, since both interpolants satisfy these conditions automatically.

There is one technical point that has to be pointed out. The real line can be mapped to itself via the change of variables $x \longleftrightarrow bx$, for any positive number b. This means that the nodes x_j are rescaled to x_j/b, and the second derivative matrix in (20) is multiplied by b^2. The freedom offered by the parameter b may be exploited to maximize the accuracy of the method. No complete theory for choosing an optimal b exists, however, and in practice b is often selected by trial-and-error.

We have solved (20) for various potentials $q(x)$. For the harmonic and quartic harmonic oscillators, i.e., $q(x) = x^2$ and $q(x) = x^4 + x^2$, the Hermite method is superior, as to be expected. We have not found a potential for which the new method comprehensively beats the Hermite method, but for the Morse potential, discussed below, the new method seems competitive.

The Morse potential, which arises in chemical physics, is defined by

$$q(x) = 9\mathrm{e}^{-2x} - 18\mathrm{e}^{-x}. \tag{21}$$

With this potential, equation (19) possesses a continuous spectrum $(0, \infty)$, plus a discrete spectrum consisting of three eigenvalues [10, p. 289]

$$\lambda_n = -\tfrac{1}{4} - (3-n)(2-n), \quad n = 0, 1, 2.$$

We have applied both the Hermite method and the new method to this problem. In Fig. 4 the relative error in the smallest eigenvalue, λ_0, is shown for a range of scaling parameters b. The new method yields a slightly higher accuracy, at least for well-chosen values of b. We should point out that when similar error curves for λ_1 are generated, there is not much to choose between the Hermite and new methods, and for λ_2 the Hermite method is in fact better.

---

[4]The number $1 + \sqrt{2}$ is a conjecture, based on computations.

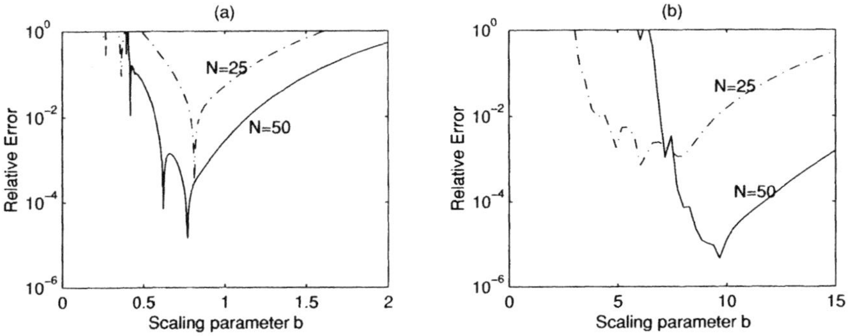

FIGURE 4. The relative error in the computed eigenvalue λ_0 of the Schrödinger equation (19) with Morse potential (21). The figures (a) and (b) correspond to the Hermite and new methods, respectively, and N is the order of the eigenvalue problem (20).

4. Conclusions

We have provided numerical evidence that spectral methods based on nonclassical orthogonal polynomials may be more accurate than their classical counterparts. However, accuracy is not the sole deciding factor when numerical methods are compared—one should also look at stability and computational efficiency.

Aside from the few remarks regarding condition numbers made in §§2.2 and 3.1, we undertook no systematic study to analyze the stability of any of these methods. As for efficiency, the construction of the new orthogonal polynomials and the computation of their differentiation matrices require a certain computational overhead. Whether this additional expense is compensated for by the increase in accuracy will have to be investigated on a case-by-case basis. In at least one example (the boundary-layer problem of §2.2) this appears to be the case, but again no comprehensive comparisons have been made.

Much theoretical work remains to be done to put the methods proposed here on a firm basis. But as the numerical results have shown, these methods have the potential to become an attractive alternative to the classical spectral methods.

Acknowledgments. Part of this work was done while the author was visiting the Department of Computer Science of the University of Utah, where Frank Stenger was very supportive. Financial support from NSF grant DMS–9404599 is gratefully acknowledged.

References

[1] C.M. Bender and S.A. Orszag, *Advanced mathematical methods for scientists and engineers*, McGraw-Hill, New York, 1978.

[2] C. Canuto, M.Y. Hussaini, A. Quarteroni, and T.A. Zang, *Spectral methods in fluid dynamics*, Springer-Verlag, Berlin, 1987.

[3] B. Fischer and G.H. Golub, *How to generate unknown orthogonal polynomials out of known orthogonal polynomials*, J. Comput. Appl. Math., **43** (1992), 99–115.

[4] B. Fornberg, *A practical guide to pseudospectral methods*, Cambridge University Press, Cambridge, 1996.

[5] D. Funaro, *Polynomial approximation of differential equations*, Springer-Verlag, Berlin, 1992.

[6] W. Gautschi, *Gauss-type quadrature rules for rational functions*, in: H. Brass and G. Hämmerlin, Eds., *Numerical integration IV*, Internat. Ser. Numer. Math. **112**, Birkhäuser, Basel, 1993, 111–130.

[7] W. Gautschi, *Algorithm 726: ORTHPOL—a package of routines for generating orthogonal polynomials and Gauss-type quadrature rules*, ACM Trans. Math. Software, **20** (1994), 21–62.

[8] W. Gautschi, *Algorithm xxx: GQRAT—Gauss quadrature for rational functions*, ACM Trans. Math. Software, to appear.

[9] D. Gottlieb and S.A. Orszag, *Numerical analysis of spectral methods: theory and applications*, SIAM, Philadelphia, 1977.

[10] J.D. Pryce, *Numerical solution of Sturm-Liouville problems*, Clarendon Press, Oxford, 1993.

[11] F. Stenger, *Numerical methods based on sinc and analytic functions*, Springer-Verlag, New York, 1993.

[12] E. Tadmor, *The exponential accuracy of Fourier and Chebyshev differencing methods*, SIAM J. Numer. Anal., **23** (1986), 1–10.

[13] W. Van Assche and I. Vanherwegen, *Quadrature formulas based on rational interpolation*, Math. Comp., **61** (1993), 765–783.

[14] J.A.C. Weideman, *The eigenvalues of Hermite and rational spectral differentiation matrices*, Numer. Math., **61** (1992), 409–432.

[15] J.A.C. Weideman and S.C. Reddy, *A* MATLAB *differentiation matrix suite*, submitted for publication. See `http://osu.orst.edu/~weidemaj/differ.html`.

[16] B.D. Welfert, *Generation of pseudospectral differentiation matrices*, SIAM J. Numer. Anal., **34** (1997), 1640–1657.

J. André C. Weideman
Department of Mathematics
Oregon State University
Corvallis, OR 97331-4605, USA
E-mail address: `weideman@na-net.ornl.gov`

present address:
Department of Applied Mathematics, University of Stellenbosch
Private Bag X1, Matieland 7602, South Africa

Author Index

Subject Index

ISNM 117 Bank, R.E. et al. (Ed.), **Math. Modelling and Simulation of Electrical Circuits and Semiconductor Devices.**
1994. 330 pages. Hardcover. ISBN 3-7643-5053-9

ISNM 118 Desch, W. / Kappel, F. / Kunisch, K. (Ed.), **Control and Estimation of Distributed Parameter Systems: Nonlinear Phenomena.** International Conference in Vorau, July 18-24, 1993
1994. 416 pages. Hardcover. ISBN 3-7643-5098-9

ISNM 119 Zahar, R. (Ed.), **Approximation and Computation.** A Festschrift in Honor of Walter Gautschi
1994. 640 pages. Hardcover. ISBN 3-7643-3753-2

ISNM 120 Hackbusch, W., **Integral Equations.** Theory and Numerical Treatment
1995. 376 pages. Hardcover. ISBN 3-7643-2871-1

ISNM 121 Jeltsch, R. / Mansour, M. (Ed.), **Stability Theory.** Hurwitz Centenary Conference, Centro Stefano Franscini, Ascona, 1995
1996. 258 pages. Hardcover. ISBN 3-7643-5474-7

ISNM 122 Khludnev, A.M. / Sokolowski, J., **Modelling and Control in Solid Mechanics.**
1997. 374 pages. Hardcover. ISBN 3-7643-5238-8

ISNM 123 Bandle, C. et al. (Ed.), **General Inequalities 7.** 7th International Conference, Oberwolfach, November 13–18, 1995
1997. 416 pages. Hardcover. ISBN 3-7643-5722-3

ISNM 124 Schmidt, W. et al. (Ed.), **Variational Calculus, Optimal Control and Applications.**
International Conference in Honour of L. Bittner and R. Klötzler, Trassenheide, Germany, September 23-27, 1996
1998. 360 pages. Hardcover. ISBN 3-7643-5906-4

ISNM 125 Nürnberger, G. / Schmidt, J.W. / Walz, G. (Ed.), **Multivariate Approximation and Splines.**
1997. 336 pages. Hardcover. ISBN 3-7643-5654-5

ISNM 126 Desch, W. / Kappel, F. / Kunisch, K. (Ed.), **Control and Estimation of Distributed Parameter Systems.**
International Conference in Vorau (Austria), July 14–20, 1996
1998. 320 pages. Hardcover. ISBN 3-7643-5835-1

ISNM 127 Gerke, H.H. et al., **Optimal Control of Soil Venting: Mathematical Modeling and Applications.**
1999.168 pages. Hardcover. ISBN 3-7643-6041-0

ISNM 128 Meyer-Spasche, R., **Pattern Formation in Viscous Flows.** The Taylor-Couette Problem and Rayleigh-Bénard Convection
1999. 220 pages. Hardcover. ISBN 3-7643-6047-X

ISNM 129/130 Fey, M. / Jeltsch, R. (Ed.), **Hyperbolic Problems: Theory, Numerics, Applications.**
Seventh International Conference in Zürich, February 1998. 2 Vols. Set
1999. 1034 pages. Hardcover. ISBN 3-7643-6123-9

ISNM 132 Müller, M.W. et al. (Ed.), **New Developments in Approximation Theory.** Second International Dortmund Meeting (IDoMAT) '98, February 23–27, 1998. 1999. Approx. 240 pages. Hardcover. ISBN 3-7643-6143-3

Numerical Analysis
An Introduction
TEXTBOOK!
Walter Gautschi

GPSR Compliance

The European Union's (EU) General Product Safety Regulation (GPSR) is a set of rules that requires consumer products to be safe and our obligations to ensure this.

If you have any concerns about our products, you can contact us on ProductSafety@springernature.com

In case Publisher is established outside the EU, the EU authorized representative is:

Springer Nature Customer Service Center GmbH
Europaplatz 3
69115 Heidelberg, Germany

Batch number: 10732314

Printed by Printforce, the Netherlands